Numerical Solutions for Partial Differential Equations

Problem Solving Using *Mathematica*

Numerical Solutions for Partial Differential Equations

Problem Solving Using *Mathematica*

Victor G. Ganzha
Evgenii V. Vorozhtsov

CRC PRESS

Boca Raton   London   New York   Washington, D.C.

Published in 1996 by
CRC Press
Taylor & Francis Group
6000 Broken Sound Parkway NW, Suite 300
Boca Raton, FL 33487-2742

International Standard Book Number-10: 0-8493-7379-4 (Hardcover)
International Standard Book Number-13: 978-0-8493-7379-4 (Hardcover)
Library of Congress Card Number 95-48195
The software mentioned in this book is now available for download on the CRC Web site at:
http://www.crcpress.com/e_products/downloads/default.asp

This book contains information obtained from authentic and highly regarded sources. Reprinted material is quoted with permission, and sources are indicated. A wide variety of references are listed. Reasonable efforts have been made to publish reliable data and information, but the author and the publisher cannot assume responsibility for the validity of all materials or for the consequences of their use.

Library of Congress Cataloging-in-Publication Data

Ganzha, V. G. (Victor Grigorevich), 1956—
    Number solutions for partial differential equations : problem solving using Mathematica / Victor G. Ganzha and Evgenii V. Vorozhtsov.
        p. cm. — (Symbolic and numeric computation series.)
    Includes bibliographical references (p.   -) and index.
    ISBN 0-8493-7379-4 (alk. paper)
    1. Differential equations, Partial—Numerical solutions—Data processing. 2. Mathematica (computer file). I. Vorozhtsov, E. V. (Evgenii Vasilevich) 1946— . II. Title. III. Series.
QA377.G2345   1996
515'.353—dc20

                                                                            95-48195

Taylor & Francis Group
is the Academic Division of Informa plc.

The software mentioned in this book is now available
for download on the CRC Web site at:
http://www.crcpress.com/e_products/downoads/
default.asp

Visit the Taylor & Francis Web site at
http://www.taylorandfrancis.com

and the CRC Press Web site at
http://www.crcpress.com

Symbolic and Numeric Computation Series

Edited by Richard Fateman

Published Titles

Computational Mathematics in Engineering and Applied Science, William E. Schiesser
Introduction to Boundary Element Methods, Prem K. Kythe
A Numerical Library in C for Scientists and Engineers, H. T. Lau
Transform Methods for Solving Partial Differential Equations, Dean Duffy
Elliptic Marching Methods and Domain Decomposition, Patrick J. Roache
Numerical Solutions for Partial Differential Equations: Problem Solving Using Mathematica,
  Victor G. Ganzha and Evgenii V. Vorozhtsov
Conservative Finite-Difference Methods on General Grids, Mikhail Shashkov

Preface

The numerical simulation of various natural and technological processes or phenomena with the aid of computers has now become a powerful tool for studying these processes or phenomena. This numerical simulation is based on using methods for solving the ordinary differential equations (ODEs) or partial differential equations (PDEs) from mathematical physics.

The advanced numerical methods can be subdivided into several classes, depending on the way in which these methods were developed, in particular:

1. The finite difference methods.
2. The finite element methods (FEM).
3. The finite volume methods.

The majority of the books on the numerical methods for the solution of PDEs are oriented towards the computer implementation of the presented methods with the aid of such conventional programming languages as FORTRAN, C, etc. However, in recent years the development of new powerful computer algebra systems (CAS) like *Mathematica*, MAPLE and others has changed drastically the traditional concepts of the use of computers in engineering applications.

The availability of the means for symbolic and numerical computations as well as the powerful computer graphics in these systems makes unnecessary the use of different software packages for the execution of symbolic or numeric computations. Although the numerical computations are performed in, e.g., the *Mathematica* system, somewhat slower than in the conventional programming systems, the availability of symbolic and numerical functions in the same computer algebra system make the new CASs comparable with the conventional programming systems in terms of the total computer time and the man-hours of work. We show in this book the capabilities of the *Mathematica* system for solving the PDEs, investigating the corresponding numerical methods, and in the graphical representation of the results. We show by simple means how the various problems can be solved that arise in the use of the numerical methods for the solution of the partial differential equations. This appears to be important both for the educational process and for the scientific and engineering applications.

Another feature of our book is the fact that we have presented unified approach to the application of a computer algebra system for the numerical solution of all classical types (hyperbolic, parabolic, and elliptic) of the PDEs. We have presented not only the well-known standard algorithms, but also a number of new approaches for the numerical solution of the partial differential equations of mathematical physics. Our book also has the following features:

1. It is self-contained, that is the reader will have no need to read other textbooks on the numerical methods to understand the material.

2. The derivation of all required formulas is shown in detail by means of a personal computer. In addition, the detailed steps of the algorithms for studying the most important properties of the numerical methods, such as approximation and stability, are presented. That is, we follow in our book the "open box" philosophy. Because the detailed knowledge of the algorithms will help the students and researchers to develop better algorithms.

3. We intensively use symbolic computation and computer graphics for the purpose of mathematics education. The needed symbolic computations are performed in our lecture course with the aid of the advanced CAS *Mathematica*. We show in our book how this system can be used

–for the derivation of the difference operators of a given order of accuracy by the method of indeterminate coefficients;

–for the local approximation study of difference schemes;

–for the stability investigation of difference schemes with the aid of the Fourier method;

– for the detection of boundaries of arbitrary multiply connected stability regions with the aid of the Fourier method and digital pattern recognition;

–for the numerical solution of initial- and boundary-value problems for scalar PDEs by the specific difference scheme;

–for the reduction of multi-level difference schemes to two-level difference schemes.

We can see the following benefits from using a computer algebra system in our introductory lecture course:

1. User friendliness, very little programming is required; many CAS operations have the same names as the standard mathematical operations that they implement. Our experience has been that most students quickly learn the necessary procedures, and that the laboratories help them to become familiar with the standard terminology of the subject.

2. Symbolic computation helps the students to examine without drudgery numerous instances of basic numerical schemes for PDEs, and also avoid the frustration caused by algebraic errors. Students can concentrate on the concepts, while the machine looks after the arithmetic.

3. A comfortable environment, both for numerical computations (approximations) and for algebraic manipulations (especially simplifications).

4. Powerful computer graphics software available in the *Mathematica* system facilitates greatly plotting of functions and surfaces, part of the visualization of numerical mathematics.

Thus the use of a CAS in our lecture course can be viewed as a paradigm beyond paper, pencil, and calculator. We hope that our lectures will not only help the student to learn the basic concepts of the numerical methods for PDEs, but also encourage the students to generate new ideas on improving the current numerical algorithms.

The materials of our textbook are based on a number of lecture courses devoted to Computer Algebra Systems and their applications presented by one of us (V.G.G) since the year 1991 at the University of Kassel, Germany, and the lecture course "Advanced Numerical Methods for the Solution of Fluid Mechanics Problems", which was presented by one of us (E.V.V.) during the academic year 1994/95 at the Novosibirsk State Technical University, Russia.

We are grateful to W. Blum (The University of Kassel), A. Bocharov (Wolfram Research, Inc.), M. Shashkov (Los Alamos National Laboratory), W. Strampp (The University of Kassel), and Chr. Zenger (The University of Munich) for fruitful discussions. The financial support of the DFG (Germany) was invaluable for Victor Ganzha.

V.G.Ganzha
E.V.Vorozhtsov

September 1995

Contents

Chapter 1

INTRODUCTION TO *MATHEMATICA*

1.1 GENERAL INFORMATION ABOUT *MATHEMATICA*

The development of the program package *Mathematica* was initiated by Stephen Wolfram in the 1980s. Now versions of *Mathematica* are available for a wide variety of computer systems, including: Apple Macintosh, IBM PC compatibles (MS-DOS and Microsoft Windows), Hewlett-Packard/Apollo, NeXT, Sony, Sun, and a number of other computer systems. *Mathematica* Version 2, as described in Wolfram's book[1], was first released in January 1991.

The *Mathematica* software package unites a number of functions in the same interactive environment:

1. Numerical computations,
2. Symbolic computations,
3. Computer graphics.

The *Mathematica* program itself, or the kernel, now includes over 300,000 lines of the source text. It is written in the object oriented extension of the programming language C developed for *Mathematica*[2]. By using a pre-compiler program a C-program is generated, which can be compiled into an executable computer code on different computer systems. In this way it became possible to extend the same huge program during a short time for various types of computers (from Macintosh and PC to supercomputers).

There are now a number of handbooks on *Mathematica*, see, in particular, References 1-21. The handbooks[2,19,20,21] show how the *Mathematica* graphics functions can be used in various situations. There have also appeared a number of books, in which *Mathematica* is applied for the solution of a number of applied problems in the natural sciences, see, in particular, References 22-38. It should be noted that none of the

1

existing books addresses systematically of the application of the *Mathematica* for the development, investigation, and computer implementation of the numerical methods for solving elementary PDEs. Only Chapter 5 of the book[25] is devoted to the numerical solution of the Korteweg-de Vries equation.

We have described in our book the applications of *Mathematica* for the numerical solution of the partial differential equations. The elements of computer graphics with the *Mathematica* system are also included to help the reader to better understand the physics of the phenomena, which are modeled on a computer with the aid of finite difference schemes. The necessary explanations on the practical use of the *Mathematica* functions will be given whenever *Mathematica* is used. The outstanding feature of *Mathematica* is that it uses symbolic expressions to provide a very general representation of mathematical and other structures. The generality of symbolic expressions allows *Mathematica* to cover a wide variety of applications with a fairly small number of methods from mathematics and computer science.

It would be useless to describe here all the possibilities of symbolic manipulations, which are offered by the *Mathematica* system. Since the symbolic computations are used throughout our book, we describe in the next section how the elementary symbolic operations can be implemented with the aid of *Mathematica*.

1.2   SYMBOLIC COMPUTATIONS WITH *MATHEMATICA*

Although the details of running *Mathematica* differ from one computer system to another, the structure of *Mathematica* calculations is the same in all cases. You enter input, then *Mathematica* processes it, and returns the result. Each such dialog between the user and the *Mathematica* system is called the *session*[1].

With a text-based interface, you interact with the computer primarily by typing text on the keyboard. This kind of interface is available for *Mathematica* on many computer systems.

On such systems, *Mathematica* is typically started by typing the operating system command math. After you type the command math, press the key ENTER.

When *Mathematica* has started, it will print the text In[1]:=, signifying that it is ready for your input. You can then type your input. Ending a line (usually by pressing the key RETURN or ENTER tells *Mathematica* that you have finished giving your input.

Mathematica will then process the input, and generate a result. If it prints the result out, it will label it with Out[1]. The labels In[n]

and Out[n] are called the *prompts*. If the space of the monitor screen is not enough to see all the computed results, you can look through the file of the results by pressing the keys PGUP or PGDN. To finish a *Mathematica* session, either type EXIT or QUIT at an input prompt.

Let us now show how *Mathematica* can handle algebraic formulas. Type the following algebraic expression:

```
In[1]:= 4x - x + 3
Out[1]= 3 + 3 x
```

You can type any algebraic expression into *Mathematica*. The expression is printed out in an approximation to standard mathematical notation.

Mathematica automatically carries out standard algebraic simplifications. In the above output Out[1] it has combined $4x$ and $-x$ to get $3x$. The basic arithmetic operations in *Mathematica* are:

x ^ y	power
-x	minus
x/y	divide
x y z or x*y*z	multiply
x+y+z	add

Be careful not to forget the space in **x y**. If you type in **xy** with no space, *Mathematica* will interpret this as a single symbol, with the name **xy**, not as a product of the two symbols **x** and **y**.

```
In[2]:= x y + 2 x^2 + y^2 x^2 - 2 y x
                  2       2 2
Out[2]= -(x y) + 2 x  y + x  y
```

Now take the following algebraic expression:

```
In[3]:= (-2 + x)^2 (1 + x + 2 y)
```

The function **Expand** multiplies out products and powers:

```
In[4]:= Expand[%]
```

The symbol % is used in *Mathematica* to refer to the last result which was typed in or generated by *Mathematica*. In our case this is the above input expression $(-2 + x)^2(1 + x + 2y)$. The arguments of functions are always contained in square brackets [].

```
            2    3              2
Out[4]= 4 - 3 x  + x  + 8 y - 8 x y + 2 x  y
```

Another built-in *Mathematica* function `Factor[%]` does essentially the inverse of **Expand**:

```
In[5]:= Factor[%]
                2
Out[5]= (-2 + x)  (1 + x + 2 y)
```

It is important to remember that every built-in *Mathematica* function begins with a capital letter. The possibility of conflicting with a built-in *Mathematica* command or function is completely eliminated, if every user-defined function or expression is defined using lower case letters. When you **first** define a function, you must always enclose the argument in square brackets and place an underline **after** the argument on the left-hand side of the equals sign in the definition of the function.

Example: $f[x_-] = x^2$.

1.3   NUMERICAL COMPUTATIONS WITH *MATHEMATICA*

The *Mathematica* system can be used also for numerical computations. There are four kinds of numbers represented in *Mathematica* – integers, rationals, reals, and complex. Integers are considered to be exact and are represented without a decimal point; rational numbers are quotients of integers and are also considered to be exact.

```
In[6]:= 3/27

         1
Out[6]= -
         9


In[7]:= 2/3 + 4/7

         26
Out[7]= --
         21

In[8]:= 1111111^2
Out[8]= 1234567654321
```

Real numbers (often referred to as "floating point numbers") contain decimal points. Mathematical constants such as e, π, i, and γ are all built-in:

```
In[9]:= N[Pi]
Out[9]= 3.14159
```

```
In[10]:= N[E]
Out[10]= 2.71828
```

```
In[11]:= N[EulerGamma]
Out[11]= 0.577216
```

The built-in *Mathematica* function N[*expr*] computes the numerical value of *expr*. Notice that *Mathematica* displays 6 significant digits when N is used (even though it stores more than 6 digits internally). We can use N to get higher precision than the above results. To calculate π to 35-digit precision, we enter:

```
In[12]:= N[Pi, 35]
Out[12]= 3.1415926535897932384626433832795028
```

All the standard mathematical functions are available. For example, the natural logarithm is given by Log:

```
In[13]:= Log[E]
Out[13]= 1
```

The trigonometric functions (Sin, Cos, Tan, Sec, etc.) and their inverses (ArcSin, ArcCos, etc.) can operate on any type of number or expression:

```
In[14]:= Cos[Pi/3]

             1
Out[14]=  -
             2
```

```
In[15]:= ArcTan[1]

            Pi
Out[15]=  --
            4
```

```
In[16]:= Expand[Sin[x]^2, Trig -> True]

            1     Cos[2 x]
Out[16]=  - - ---------
            2        2
```

The Trig -> True option tells *Mathematica* to use built-in rules for simplifying terms containing trigonometric functions.

REFERENCES

1. **Wolfram, S.,** *Mathematica, a System for Doing Mathematics by Computer.* Addison-Wesley, Redwood City, 1991.
2. **Cameron, S.,** *The Mathematica Graphics Guidebook.* Addison-Wesley, Redwood City, 1990.
3. **Blachman, N.,** *Mathematica: A Practical Approach.* Prentice-Hall, Englewood Cliffs, 1991.
4. **Blachman, N.,** *Mathematica Quick Reference, Version 2.* Prentice-Hall, Englewood Cliffs, 1991.
5. **Crooke, P., and Ratcliffe, J.,** *Guidebook to Calculus with Mathematica.* Wadsworth, London, 1991.
6. **Ellis, W.Jr., and Lodi, E.,** *A Tutorial Introduction to Mathematica.* Brooks/Cole, Pacific Grove, 1991.
7. **Maeder, R.,** *Programming in Mathematica*, Second Edition. Addison-Wesley, Redwood City, 1991.
8. **Wagon, S.,** *Mathematica in Action.* Freeman, New York, 1991.
9. **Abell, M., and Braselton, J.,** *Mathematica by Example.* Academic Press, London, 1992.
10. **Abell, M., and Braselton, J.,** *The Mathematica Handbook.* Academic Press, London, 1992.
11. **Finch, J., and Lehman, M.,** *Exploring Calculus with Mathematica.* Addison-Wesley, Redwood City, 1992.
12. **Gray, T., and Glynn, J.,** *The Beginners Guide to Mathematica 2.* Addison-Wesley, Redwood City, 1992.
13. **Heimrich, E., and Janetzko, H.D.,** *Das Mathematica-Arbeitsbuch.* Vieweg, Braunschweig, Wiesbaden, 1992.
14. **Kaufmann, S.,** *Mathematica als Werkzeug: eine Einführung mit Anwendungsbeispielen.* Birkhäuser Verlag, Basel, Boston, 1992.
15. **Kofler, M.,** *Mathematica.* Addison-Wesley, Bonn, Reading, 1992.
16. **Wolfram, S.,** *Mathematica Reference Guide.* Addison-Wesley, Redwood City, 1992.
17. **Burkhardt, W.,** *Erste Schritte mit Mathematica.* Springer-Verlag, Berlin, Heidelberg, New York, 1993.
18. **Gaylord, R.J., Kamin, S. and Wellin, P.R.,** *Introduction to Programming with Mathematica.* Springer-Telos, New York, 1993.

19. Stelzer, E.H.K., *Mathematica. Ein systematisches Lehrbuch mit Anwendungsbeispielen.* Addison-Wesley, Bonn, Reading, Massachusetts, 1993.

20. Schaper, R., *Grafik mit Mathematica. Von den Formeln zu den Formen.* Addison-Wesley Publishing Company, Bonn, Reading, Massachusetts, 1994.

21. Wickham-Jones, T., *Mathematica Graphics: Techniques & Applications.* Telos-Springer, New York, Santa Clara, 1994.

22. Smith, C., and Blachman, N., *The Mathematica Graphics Guidebook.* Addison-Wesley, Reading, 1995.

23. Crandall, R.E., *Mathematica for the Sciences.* Addison-Wesley, Redwood City, 1991.

24. Stroyan, K.D., *Calculus Using Mathematica.* Academic Press, Boston, 1992.

25. Vvedensky, D., *Partial Differential Equations with Mathematica.* Addison-Wesley, Redwood City, 1992.

26. Baumann, G., *Mathematica in der theoretischen Physik.* Springer-Verlag, Berlin, Heidelberg, New York, 1993.

27. Maeder, R.E., *Informatik für Mathematiker und Naturwissenschaftler.* Addison-Wesley, Reading, Bonn, Paris, 1993.

28. Skeel, R.D., and Keiper, J.B., *Elementary Numerical Computing with MATHEMATICA.* McGraw-Hill, Inc. New York, San Francisco, 1993.

29. Wolff, R.S., and Yaeger, L., *Visualization of Natural Phenomena.* Telos-Springer, New York, Santa Clara, 1993.

30. Davis, B., Porta, H., and Uhl, J., *Calculus & Mathematica,* Addison-Wesley Publishing Company, Reading, 1994.

31. Feagin, J.M., *Quantum Methods with Mathematica,* Telos-Springer, New York, Santa Clara, 1994.

32. Freeman, J.A., *Simulating Neural Networks with Mathematica.* Addison-Wesley Publishing Company, Reading, 1994.

33. Gaylord, R.J., and Wellin, P.R., *Computer Simulations with Mathematica: Explorations in Complex Physical and Biological Systems.* Addison-Wesley Publishing Company, Reading, 1994.

34. Kreyszig, E., and Normington, E., *Mathematica Manual to Accompany Advanced Engineering Mathematics.* John Wiley & Sons, Inc., New York, 1994.

35. Shaw, W.T., and Tigg, J., *Applied Mathematica: Getting Started, Getting It Done.* Addison-Wesley Publishing Company, Reading, 1994.

36. Bahder, T.B., *Mathematica for Scientists and Engineers.* Addison-Wesley, Reading, 1995.

37. **Coombs, K.R., Hunt, B.R., Lipsman, R.L., Osborn, J.E., and Stuck, G.J.,** *Differential Equations with Mathematica.* John Wiley & Sons, New York, 1995.
38. **Strampp, W., and Ganzha, V.,** *Differentialgleichungen mit Mathematica.* Vieweg, Braunschweig, Wiesbaden, 1995.
39. **Zimmerman, R.L., and Olness, F.I.,** *Mathematica for Physicists.* Addison-Wesley Publishing Company, Reading, 1995.

Chapter 2

FINITE DIFFERENCE METHODS FOR HYPERBOLIC PDEs

2.1 CONSTRUCTION OF DIFFERENCE SCHEMES FOR THE ADVECTION EQUATION

2.1.1 Divided Differences

Let us consider the following linear hyperbolic partial differential equation (PDE):

$$\partial u/\partial t + a\partial u/\partial x = 0, \quad -\infty < x < \infty \qquad (2.1.1)$$

where x is the spatial coordinate, t is the time, $a = const$. Eq. (2.1.1) was given a number of names in the literature: the *advection* equation[1], the *convection* equation[2], the *one-way wave* equation[3]. To ensure the uniqueness of the solution of (2.1.1), we must specify the function $u(x,t)$ at some moment of time $t = t_0 \geq 0$. Typically the value $t = 0$ is chosen. Thus we assume that at $t = 0$ the function

$$u(x,0) = u_0(x), \quad -\infty < x < \infty \qquad (2.1.2)$$

is given. The equation (2.1.2) is called the *initial condition* for the equation (2.1.1). The mathematical problem (2.1.1)-(2.1.2) is called the *initial-value problem*, or the *Cauchy problem*. We now consider the construction of a finite difference scheme for the numerical solution of the initial-value problem (2.1.1)-(2.1.2). In the case of an analytical solution of a PDE, its solution is found in a region of continuous change of the variables x and t. In contrast to the analytic case, the approximate solution of the PDE by the method of finite differences is determined only at some discrete points of the (x,t) plane. If the spatial interval $[a_1, b_1]$, in which the PDE solution is to be found, is finite, then the number of points in this interval, which is used by the finite difference method, is also finite. Consider a finite set of points in the spatial interval $[a_1, b_1]$,

9

which satisfies the conditions

$$a_1 = x_0 < x_1 < x_2 < \ldots < x_{M-1} < x_M = b_1 \qquad (2.1.3)$$

This set of points is called the *computing mesh* in the interval $[a_1, b_1]$, or the computational grid, or the spatial grid (because x is the spatial variable). The points x_i from the set (2.1.3) are called the *nodes* of the spatial grid.

Let us assume that the numerical solution of equation (2.1.1) should be determined in a finite time interval $0 \le t \le T$. Similarly to the set (2.1.3) we can define a mesh on the t-axis:

$$0 = t_0 < t_1 < t_2 < \ldots < t_{N-1} < t_N = T \qquad (2.1.4)$$

The meshes (2.1.3) and (2.1.4) are called *uniform*, if the following conditions are satisfied:

$$x_j = a_1 + jh, \quad j = 0, 1, \ldots, M; \quad t_n = n\tau, \quad n = 0, \ldots, N \qquad (2.1.5)$$

The quantity $h = (b_1 - a_1)/M$ is called the *step* of a spatial mesh. The quantity $\tau = T/N$ is called the *time step*.

Let us now introduce the uniform mesh G_h in the (x, t) plane as a finite set of the points of intersection of the straight lines (see Fig. 2.1)

$$x = a_1 + jh, \quad j = 0, 1, \ldots, M; \quad t = n\tau, \quad n = 0, 1, \ldots, N$$

The function $u_j^n = u(x_j, t_n)$ which is determined at the nodes of the grid G_h is called a *grid function*. When we solve equation (2.1.1) by a finite difference method, we must find the table

$$\{u_j^n \equiv u(x_j, t_n), \quad j = 0, 1, \ldots, M, \quad n = 0, 1, \ldots, N\}$$

of the values of the solution $u(x, t)$ of the problem (2.1.1) - (2.1.2) at points of the G_h grid.

Let us now go over to the construction of a finite difference scheme which approximates the initial-value problem (2.1.1) - (2.1.2). The basic technique here is the use of the Taylor series expansions. We at first recall the Taylor formula for the function $u(x)$:

$$u(x + h) = u(x) + hu'(x) + \frac{h^2}{2!}u''(x) + \frac{h^3}{6}u'''(\xi_1)$$

$$u(x - h) = u(x) - hu'(x) + \frac{h^2}{2!}u''(x) - \frac{h^3}{6}u'''(\xi_2) \qquad (2.1.6)$$

$$u(x + h) = u(x) + hu'(x) + \frac{h^2}{2}u''(\xi_3)$$

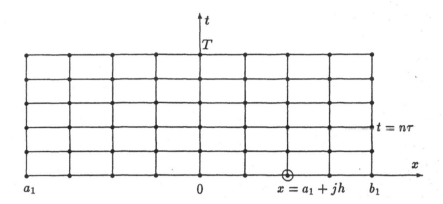

Figure 2.1: The uniform mesh G_h in the (x, t) plane

where $\xi_1 \in [x, x+h]$, $\xi_2 \in [x-h, x]$, $\xi_3 \in [x, x+h]$. We have from the formulas (2.1.6) that

$$\frac{u(x+h) - u(x-h)}{2h} = u'(x) + \frac{h^2}{12}[u'''(\xi_1) + u'''(\xi_2)]$$

$$\frac{u(x+h) - u(x)}{h} = u'(x) + \frac{h}{2}u''(\xi_3) \qquad (2.1.7)$$

$$\frac{u(x) - u(x-h)}{h} = u'(x) - \frac{h}{2}u''(\xi_4)$$

where $\xi_4 \in [x-h, h]$. The expressions in the left hand sides of equalities (2.1.7) are called *divided differences*. The divided difference

$$\frac{u(x+h) - u(x)}{h}$$

is also called the *forward difference*. The divided difference

$$\frac{u(x) - u(x-h)}{h}$$

is also called the *backward difference*. And finally the difference

$$\frac{u(x+h) - u(x-h)}{2h}$$

is often called the *central difference*.

The forward difference can be conveniently implemented in *Mathematica* by introducing the function `discretize`:

```
In[1]:= discretize[Derivative[1][u_],h_,
            schemetype->forward] := ((u[#+h]-u[#])/h)&
```

The above function `discretize` can be regarded as a *pure function*. Pure functions are powerful tools which are absent in many conventional programming languages. The basic idea is here to represent the functions in a compact form. In the body of a pure function, we only specify the operations that must be executed, and where the arguments are to be placed. The *Mathematica* system makes use of # for the argument or for the location of a pure function (or #1, #2, etc., for two or more arguments). The symbol & marks the end of a pure function. Example: `cube:= #^3&`.

We have used above the function `Derivative[n][f][x]` which defines the nth derivative of the function $f(x)$. We can define in a similar way the functions for backward and central differences:

```
In[2]:= discretize[Derivative[1][u_],h_,
            schemetype->backward] := ((u[#]-u[#-h])/h)&

In[3]:= discretize[Derivative[1][u_],h_,
            schemetype->central] := ((u[#+h]-u[#-h])/(2h))&
```

The transformation rule `schemetype->forward` enables us to specify the type of the divided difference.

For example, to obtain the backward and central differences we type in the commands

```
In[4]:= ux1 = discretize[u',h,schemetype->backward][x]
             -u[x] + u[h + x]
Out[4] = ------------------
                 h
```

```
In[5]:= ux2 = discretize[u',h, schemetype -> central][x]
           -u[-h+x] + u[h + x]
Out[5] = --------------------
                 2 h
```

Here the expression u[x] denotes the function $u(x)$. The use of the transformation rule **schemetype** enables us to use the same function discretize for all divided differences.

For the study of the approximation errors caused by using the above introduced divided differences we can use the *Mathematica* function Series[*expr*, $\{x, x_0, n\}$]. This function determines the Taylor expansion of *expr* at point $x = x_0$ up to the order $(x - x_0)^n$. For example, for the forward difference we obtain the expansion

```
In[5]:= Series[ux1, {h,0,3}]

                         (3)    2   (4)    3
              u''[x] h   u   [x] h   u   [x] h            4
Out[5]= u'[x] + -------- + ----------- + ----------- + O[h]
                  2            6             24
```

The prime is used to denote the differentiation with respect to x. For example, u''[x] denotes the second derivative $u''(x) = d^2u/dx^2$. Now we can estimate the approximation error of the above divided difference. We can easily see that the approximation of the derivative $u'(x)$ by the forward difference introduces an error of the order $O(h)$, which coincides with the corresponding formula in (2.1.7).

Similarly, to get information on the approximation error of the central difference we type in the command

```
In[6]:= Series[ux2, {h,0,3}]

                    (3)    2
              u   [x] h            4
Out[6]= u'[x] + ----------- + O[h]
                  6
```

2.1.2   The Method of Indeterminate Coefficients

We have presented above a number of simple divided differences for the approximation of the derivative $\partial u/\partial x$. For the derivation of the divided differences, which have a higher accuracy or which approximate the higher-order derivatives (for example, the derivative $\partial^2 u/\partial x^2$), one can use a number of methods. One of these methods, which has widespread use, is the *method of indeterminate coefficients*.

Let us now show how the *Mathematica* system can be used for obtaining the divided difference which approximates the second derivative $u''(x)$ at point $x = x_j$ and involves only the values u_{j-1}, u_j and u_{j+1}. For this purpose we use the formulas (2.1.6). Indeed, let us assume that the desired approximation to $u''(x_j)$ has the form

$$u_{xx}(x_j) = au_{j+1} + bu_j + cu_{j-1}$$

Since $u_{j+1} = u(x_{j+1}) = u(a_1 + (j+1)h) = u(a_1 + jh + h) = u(x_j + h) = u(x + h)$, we can rewrite this expression in the form

$$u_{xx}(x_j) = au(x + h) + bu(x) + cu(x - h)$$

Now type in the commands

```
In[7]:= uxx = a*u[x+h] + b*u[x] +c*u[x-h]
Out[7]= b u[x] + c u[-h + x] + a u[h + x]
In[8]:= s = Normal[Series[uxx, {h,0,2}]]

Out[8]=a u[x] + b u[x] + c u[x] +

                          2  a u''[x]     c u''[x]
    h (a u'[x] - c u'[x]) + h  (--------  + --------)
                              2            2
```

We have used above the function **Normal[*expr*]** to revert a power series into a normal expression without the term $O(h^k)$.

We can see from the obtained Taylor expansion of the expression **uxx**, the expression **s**, that **s** is a polynomial in h. Since we want that **s**$= u_{xx}$, we need the equation **s** = **u''[x]**. This equation should be valid at any h. To ensure this we require that each of the coefficients of the various powers of h in the expression for **s-u''[x]** be equal to zero. In this way we will obtain an algebraic system for determining the coefficients **a,b,c** in the expression **uxx**. Each equation of this system involves **u,** u', u'',.., as coefficients and is valid for any u(x). Therefore, we further equal to zero the coefficients of u, u', u'',.... As a result we obtain a much simpler algebraic system involving only the indeterminate coefficients and the step h. Thus we have presented the basic idea of the *method of indeterminate coefficients* for constructing the finite difference approximations.

In accordance with the above ideas, we now should determine the form of the algebraic equations for the coefficients a, b, and c in uxx. For this purpose we can use the following operators:

```
In[9]:= s1 = s - u''[x]
Out[9]= a u[x] + b u[x] + c u[x] +

                                   2    au''[x]   c u''[x]
    h (a u'[x] - c u'[x]) - u''[x] + h  (------- - --------)
                                         2          2
In[10]:= derlist = Prepend[
Union[Cases[{s1}, Derivative[_][u][x], -1]], u[x]]
Out[10]= {u[x], u'[x], u''[x]}
In[11]:= coefh = DeleteCases[Flatten[CoefficientList[s1,
                                      derlist]], 0]

            2       2
          a h     c h
Out[11]= {-1 + ---- + ----, a h - c h, a + b + c}
            2       2
```

Mathematica intensely uses *lists*. These are very general objects representing sets of objects. The objects in the list are separated by commas. An example of a list is: {u[x], u'[x], u''[x]}.

The *Mathematica* function Cases[{$e_1, e_2, \dots$}, *pattern*] gives a list of the e_i that match the pattern.

The built-in function Prepend[*expr, elem*] produces the expression *expr* with *elem* as the first element of the expression. Example: let expr ={a, b}. The command expr1 = Prepend[expr, x] creates the list expr1 = {x, a, b}.

The *Mathematica* function

```
CoefficientList[poly, var1, var2,...]
```

creates a matrix with the coefficients of the var_i. Example:

```
CoefficientList[x^2 + 2 x y - y, {x, y}]
```

produces the matrix $\{\{0, -1\}, \{0, 2\}, \{1, 0\}\}$.

The built-in function `DeleteCases[`*expr, pattern*`]` removes all elements of *expr* which match *pattern*. Example: the command

```
DeleteCases[{1.0, a^2, 0, b}, 0]
```

creates the list

```
{1., a^2, b}
```

The built-in function `Flatten[`*list*`]` flattens out the nested lists by omitting the inner braces $\{\}$. Example: the command `Flatten[{a, {b, c}, {d}}]` gives the list $\{a, b, c, d\}$. From the list in `Out[11]` we can see that the algebraic system for indeterminate coefficients a, b, and c has the form

$$a + b + c = 0$$

$$a - c = 0$$

$$\frac{h^2}{2}(a + c) - 1 = 0$$

The built-in *Mathematica* function `Solve` can find explicit solutions for a large class of simultaneous polynomial equations. There are three ways to present simultaneous equations to `Solve`:

$$\text{Solve}[\{lhs_1 == rhs_1, lhs_2 == rhs_2, \ldots\}, vars]$$
$$\text{Solve}[lhs_1 == rhs_1 \;\&\&\; lhs_2 == rhs_2 \;\&\&\; \ldots, vars]$$
$$\text{Solve}[\{lhs_1, lhs_2, \ldots\} == \{rhs_1, rhs_2, \ldots\}, vars]$$

Here *vars* denotes the list of unknown variables; in our case *vars* is typed in as $\{a, b, c\}$, see the next program fragment:

```
In[12]:= ss = Solve[coefh==0, {a,b,c}]
                -2           -2          -2
Out[12]= {{b -> --, a -> h  , c -> h  }}
                2
                h
In[13]:= uxx = First[uxx/.ss]
            -2 u[x]    u[-h + x]    u[h + x]
Out[13]= ------- + --------- + --------
              2          2          2
             h          h          h
```

The expressions in `Out[11]` represent a sequence of the *transformation rules* for the variables `a`, `b`, `c`. The replacement operator `/.` (pronounced "slash-dot") applies rules to expressions. In the result of the application of the replacement operator to the expression `uxx` the identifiers `a`, `b`, `c` are replaced by the corresponding expressions from the list `ss`. The built-in *Mathematica* function `First[expr]` gives the first element of *expr*. We can rewrite the difference approximation for $u''(x_j)$ obtained above (see `Out[12]`) with the aid of the method of indeterminate coefficients and the *Mathematica* package in the usual mathematical notation as follows:

$$u''(x_j) = \frac{u_{j+1} - 2u_j + u_{j-1}}{h^2}$$

We now present the complete listing of the file `tayl.ma` for the symbolic computation of the difference expression for the second derivative with the aid of the *Mathematica* system.

─────────────── File tayl.ma ───────────────

```
uxx = a*u[x+h] + b*u[x] +c*u[x-h]
s = Normal[Series[uxx, {h,0,2}]]
s1 = s - u''[x]

derlist = Prepend[
Union[Cases[{s1}, Derivative[_][u][x], -1]], u[x]]
coefh = DeleteCases[Flatten[CoefficientList[s1,
                                derlist]], 0]

ss = Solve[coefh==0, {a,b,c}]
OutputForm[ss] >> taylr.m
uxx = First[uxx/.ss]
OutputForm[uxx] >>> taylr.m
```

We have used above the *Mathematica* command

$$expr >> \texttt{taylr.m}$$

This writes an expression to file `taylr.m`. When you use *expr* $>>>$ *file*, *Mathematica* appends each new expression you give to the end of your file. In this way the results of intermediate computations can be stored in the file `taylr.m`. If you use *expr* $>>$ *file*, however, then *Mathematica* instead wipes out anything that was in the file before, and then puts *expr* into the file.

When you have finished a computation, you typically need to output the result. The built-in command OutputForm[*expr*] enables the user to get a standard two-dimensional mathematical form of the expression *expr*, with exponents raised, fractions built up, and so on.

The simplest technique for the construction of difference initial- and boundary-value problems which approximate the initial- and boundary-value problems for PDEs consists in a replacement of the partial derivatives by the corresponding divided differences.

Let us, for example, replace the time derivative $\partial u/\partial t$ in (2.1.1) by the forward difference

$$\partial u/\partial t \approx \frac{u(x,t+\tau) - u(x,t)}{\tau}$$

and replace the spatial derivative $\partial u/\partial x$ by the backward difference

$$\partial u/\partial x \approx \frac{u(x,t) - u(x-h,t)}{h}$$

In the grid node (x_j, t_n) we can write, in particular, that

$$\left(\frac{\partial u}{\partial t}\right)_{x=x_j, t=t_n} \cong \frac{u_j^{n+1} - u_j^n}{\tau}, \quad \left(\frac{\partial u}{\partial x}\right)_{x=x_j, t=t_n} \cong \frac{u_j^n - u_{j-1}^n}{h} \quad (2.1.8)$$

Substituting the expressions (2.1.8) into the PDE (2.1.1) for of the partial derivatives, we obtain the equation

$$\frac{u_j^{n+1} - u_j^n}{\tau} + a\frac{u_j^n - u_{j-1}^n}{h} = 0 \qquad (2.1.9)$$

$$j = 0, \pm 1, \pm 2, \ldots; \quad n = 0, 1, \ldots, N$$

$$u_j^0 = u_0(jh), \quad j = 0, \pm 1, \ldots \qquad (2.1.10)$$

The equation (2.1.9) is called *the finite difference equation*, or the *finite difference scheme*, or the *difference scheme*. The difference problem (2.1.9)-(2.1.10) is called the *difference initial-value problem*.

We recall that our purpose is to obtain the table of the values u_j^n for $j = 0, \pm 1, \ldots$, $n = 0, 1, \ldots, N$, from equation (2.1.9). The initial condition (2.1.2) enables us to find the discrete values u_j^0 in accordance with (2.1.10). To compute the values u_j^n for $n = 1, 2, \ldots, N$ let us rewrite the difference equation (2.1.9) as follows:

$$u_j^{n+1} = u_j^n - \kappa(u_j^n - u_{j-1}^n) \qquad (2.1.11)$$

$$j = 0, \pm 1, \pm 2, \ldots; \quad n = 0, 1, \ldots, N$$

where $\kappa = a\tau/h$. In particular, we have from (2.1.11) at $n = 0$ the difference equation

$$u_j^1 = u_j^0 - \kappa(u_j^0 - u_{j-1}^0) \qquad (2.1.12)$$

Since the grid values u_j^0, u_{j-1}^0 in the right hand side of (2.1.12) are known, we can find u_j^1 for any j. We can calculate in this way sequentially the grid values $u_j^2, u_j^3, \ldots, u_j^N$ with the aid of the explicit formula (2.1.11). As a result of the execution of this computational process we obtain an approximate solution to the initial-value problem (2.1.1), (2.1.2).

It is easy to obtain the difference equation (2.1.9) by using the *Mathematica* system:

```
In[14]:= eqn = Derivative[1][u][t] +
              a Derivative[1][u][x] == 0
Out[14] = u'[t] + a u'[x] == 0

In[15]:= eqn1 = eqn/. {u'[t] -> (u[j,n+1] - u[j,n])/tau,
               u'[x] -> (u[j,n] - u[j-1,n])/h}

          a (-u[-1 + j, n] + u[j, n])
Out[15]= --------------------------- +
                    h

          -u[j, n] + u[j, 1 + n]
          ---------------------- == 0
                   tau
```

We can solve equation (2.1.9) with respect to u_j^{n+1} also by using the *Mathematica* system:

```
In[16]:= ss = Solve[eqn1, u[j,n+1]]
Out[16]= {{u[j, 1 + n] ->
 -((-a tau u[-1 + j, n]) - h u[j, n] +
     a tau u[j, n])/h)}}

In[17]:= u[j, n+1] == Simplify[u[j,n+1]/. ss/.
                  h-> (a tau)/kappa] [[1]]
```

```
Out[17]= u[j, 1 + n] == kappa u[-1 + j, n] + u[j, n] -
                        kappa u[j, n]
```

There arises the following question: Is it always possible to find the solution of a difference equation like equation (2.1.11)? At first sight it appears that this is always possible. But in practice the formula (2.1.11) is applicable only in the case when the finite difference scheme (2.1.9) is *stable*. We will discuss this important notion later in Section 2.3.

Exercise 2.1. Show with the aid of the Taylor series expansions that if the half-step $h/2$ is used instead of h, then the central difference for the approximation of $u'(x_j)$ may be obtained in the form

$$u'(x_j) = \frac{u_{j+1/2} - u_{j-1/2}}{h} \qquad (2.1.13)$$

where $u_{j\pm 1/2} = u((j \pm 1/2)h)$.

Exercise 2.2. Derive a three-point central difference approximation for $u''(x_j)$ by applying equation (2.1.13) to $g(x) = u'(x)$ and using the formula [1]

$$u'(x_{j+1/2}) = \frac{u_{j+1} - u_j}{h}$$

Exercise 2.3. Derive the one-sided difference approximation to u_{xx} involving the values u_{j-2}, u_{j-1}, u_j with the aid of the above program tayl.ma.

---

[1] The answers to the exercises are given at the end of the chapter.

2.2 THE NOTION OF APPROXIMATION

2.2.1 Approximation Order of Scheme

We have already mentioned that the finite difference methods enable one to obtain only an approximate solution of the original initial-value problem for the PDE. The accuracy of the numerical solution can be checked with the aid of the comparison with some exact analytic solution. However, in cases of complex mathematical physics problems it is often impossible to find the closed form analytic solutions even under simplifying assumptions. How to answer, in these cases, the question of whether the constructed difference scheme would indeed deliver an approximate solution of the initial-value problem for PDE?

For this purpose the notion of *approximation* was introduced in the theory of difference schemes. The mathematical formulation of this property enables one to study the approximation property of any difference method prior to any numerical computation using this method.

Furthermore, this mathematical study of approximation helps to determine the *accuracy* of the numerical solution by a specific difference scheme, which can in principle be attained on a given spatial/temporal computing mesh.

We now want to introduce the approximation notion for the example of the difference initial-value problem (2.1.9), (2.1.10). The approximation studies are always performed using some *grid norm*. In the case of the difference scheme (2.1.9), (2.1.10) it is convenient to introduce the grid norm as

$$\| u \|_h = \max_{j, n \geq 0} |u_j^n| + \max_j |u_0(x_j)|$$

Let us now assume that the solution of the problem (2.1.1), (2.1.2) has the bounded second derivatives. Then in accordance with the Taylor formula we have that

$$\frac{u(x_j, t_n) - u(x_j - h, t_n)}{h} = \frac{\partial u(x_j, t_n)}{\partial x} - \frac{h}{2} \frac{\partial^2 u(x_j - \xi, t_n)}{\partial x^2}$$
$$\frac{u(x_j, t_n + \tau) - u(x_j, t_n)}{\tau} = \frac{\partial u(x_j, t_n)}{\partial x} + \frac{\tau}{2} \frac{\partial^2 u(x_j, t_n + \eta)}{\partial t^2} \quad (2.2.1)$$

where ξ and η are certain numbers depending on j, n, and h and satisfying the inequalities $0 < \xi < h, 0 < \eta < \tau$. Let

$$L_h u = \frac{u_j^{n+1} - u_j^n}{\tau} + a \frac{u_j^n - u_{j-1}^n}{h}$$

With the aid of formulas (2.2.1) the left hand side of the difference equation (2.1.9) may be rewritten in the form

$$L_h u = \left(\frac{\partial u}{\partial t} + a \frac{\partial u}{\partial x} \right)_{x_j, t_n} + \frac{\tau}{2} \frac{\partial^2 u(x_j, t_n + \eta)}{\partial t^2} -$$
$$a \frac{h}{2} \frac{\partial^2 u(x_j - \xi, t_n)}{\partial x^2} = 0 \qquad (2.2.2)$$

Let us introduce the operator L by the formula

$$L u = \frac{\partial u}{\partial t} + a \frac{\partial u}{\partial x}$$

Then we have from the formula (2.2.2) that

$$\| L_h u - L u \|_h \leq \sup \left| \frac{\partial^2 u}{\partial t^2} \right| \cdot \frac{\tau}{2} + \sup \left| \frac{\partial^2 u}{\partial x^2} \right| \cdot \frac{h}{2} \qquad (2.2.3)$$

where $u(x, t)$ is the solution of the initial-value problem (2.1.1)-(2.1.2). The steps τ and h enter the formula (2.2.3) as the powers τ^1, h^1. In this case one says that the difference scheme (2.1.9) has the first order of approximation in τ and h for a solution $u(x, t)$ possessing bounded second derivatives.

We can see from (2.2.3) that

$$\lim_{h \to 0, \tau \to 0} \| L_h u - L u \| = 0 \qquad (2.2.4)$$

If property (2.2.4) holds, then, by definition, difference problem approximates the original "differential" (initial-value) problem.

If, in addition,

$$\| L_h u - L u \|_h \leq C_1 h^{k_1} + C_2 \tau^{k_2} \qquad (2.2.5)$$

where $k_1 > 0, k_2 > 0$ and the constants C_1 and C_2 do not depend on τ and h, then the approximation has order k_1 with respect to h and the order k_2 with respect to τ. In this case one says that the difference scheme *has the approximation order* $O(h^{k_1}) + O(\tau^{k_2})$.

It can be seen from (2.2.5) that the right hand side of this inequality of tends to zero with $h \to 0, \tau \to 0$ for any positive values of k_1 and k_2, including fractional values, say, $k_1 = 1/2, k_2 = 1/3$. However, the construction of difference schemes for PDEs by using Taylor series expansions usually yields difference schemes with positive integer approximation orders. From this point of view, the difference scheme (2.1.9) has the least possible order of approximation. The determination of the values k_1 and k_2 in (2.2.5) for the chosen difference scheme is important for practice, because a scheme with the approximation order $O(\tau) + O(h^2)$

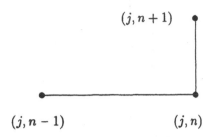

$(j, n+1)$

$(j, n-1)$ (j, n)

Figure 2.2: The stencil of scheme (2.1.9)

proves to be more accurate than a scheme of order $O(\tau) + O(h)$. A scheme of order $O(\tau^2) + O(h^2)$ is usually more accurate, in practice, than the scheme having the approximation order $O(\tau) + O(h^2)$.

The set of grid nodes, which are used in a difference equation to compute the value u_j^{n+1}, is called the *stencil* of a difference scheme. For example, the stencil of scheme (2.1.9) consists of three points (x_{j-1}, t_n), (x_j, t_n) and (x_j, t_{n+1}). Since the formulas (2.1.5) enable us to compute the grid coordinates x_j, t_n at the integer values (j, n), one often writes that the stencil of the difference scheme (2.1.9) consists of three points $(j-1, n), (j, n)$ and $(j, n+1)$, see Fig. 2.2.

2.2.2 Square Wave Test

Now let us try the scheme (2.1.9) on a number of test problems. Let us take the model problem on the propagation of a step (see Fig. 2.3). This test is also called the square wave test. We now give the exact solution for the square wave problem. Let

$$u_0(x) = \begin{cases} u_l, & x < x_{f0} \\ u_r, & x > x_{f0} \end{cases} \qquad (2.2.6)$$

where $0 < x_{f0} < b$ is the given abscissa of the discontinuity front at $t = 0$; $x = b$ is the right end of the spatial integration interval. u_l and u_r are given constants, such that $u_l \neq u_r$.

Now consider at $t > 0$ any point (x, t) in the (x, t) plane. Let us take the straight line $x = x(t)$ given by the ordinary differential equation

$$\frac{dx(t)}{dt} = a \qquad (2.2.7)$$

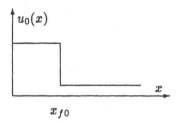

Figure 2.3: The graph of the function $u_0(x)$ in (2.1.2) for the square wave test

Let us now consider the solution $u(x,t)$ of the problem (2.1.1), (2.1.2) along the line (2.2.7). For this purpose let us introduce the notation $\tilde{u}(t) = u(x(t),t)$. Then we can see from (2.1.1) that

$$\frac{d\tilde{u}(t)}{dt} = \frac{\partial u}{\partial t} + \frac{\partial u}{\partial x} \cdot \frac{dx(t)}{dt} = 0 \qquad (2.2.8)$$

It follows from (2.2.8) in the result of integration that $\tilde{u}(t) = const$ along the line (2.2.7). By integrating the ODE (2.2.7) we get the solution

$$x = at + C \qquad (2.2.9)$$

where C is the integration constant. Also the solution $u(x,t)$ is constant along any straight line (2.2.9). Now take a specific point (x,t) and introduce the notation $x_0 = x - at$. Then $C = x_0$ in (2.2.9), and the solution is equal to the constant $\tilde{u}(x_0)$ along the line $x_0 = x - at$. But this constant is known at $t = 0$ from the initial condition (2.1.2), so that we may write that

$$\tilde{u}(x_0) = u_0(x_0)$$

Since $x_0 = x - at$, $\tilde{u}(t) = u(x,t)$, we obtain the exact solution of the problem (2.1.1)-(2.1.2) as

$$u(x,t) = u_0(x - at) \qquad (2.2.10)$$

The formula (2.2.10) says that the initial data $u_0(x)$ are translated without change along the lines (2.2.9), see Fig. 2.4. Therefore, Eq. (2.1.1) can be used for modeling the transport or the advection of some substance or property. Eq. (2.1.1) governs, for example, the advection of

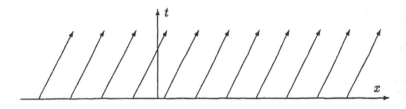

Figure 2.4: The characteristics of equation (2.1.1) in the case $a > 0$

one colored liquid in another (say ink in water; in this case the function $u(x,t)$ is the concentration of ink). In problems on fluid dynamics, the function $u(x,t)$ has the meaning of the fluid density [1,2]. In problems of the filtration theory, $u(x,t)$ has the meaning of the saturation (fraction of water in total fluid)[4]. The one-parameter family of lines (2.2.9), where C is the parameter, are called the *characteristics* of the hyperbolic PDE (2.1.1). Note that the slope of the characteristics, a, is a real number, because the advection speed a in (2.1.1) is assumed to be real.

Table 2.1. Parameters used in program `advbac.ma`

Parameter	Description
a	the advection speed in equation (2.1.1), $a > 0$
a1	the abscissa of the left end of the interval on the x-axis
b1	the abscissa of the right end of the above interval, $a_1 < b_1$
jf0	the number of the grid node, which coincides with the discontinuity position at $t = 0$; jf0 should satisfy the inequalities 3 < jf0 < M
ul,ur	the constant solution values at $t = 0$ to the left and to the right of the discontinuity
M	the number of the grid node coinciding which with the right boundary $x = b_1$ of the spatial integration interval (see also (2.1.5))
cap	the Courant number $\kappa = a\tau/h$, where τ is the time step of the difference scheme, and h is the step of the uniform grid in the interval $a_1 \leq x \leq b_1$
Npic	the number of the pictures of the difference solution, Npic > 1

It follows from the exact solution (2.2.10) that in the case of the square wave problem (2.1.1), (2.2.6) the discontinuous profile (2.2.6) will be transported in the (x,t) plane with the speed $dx/dt = a$. The discontinuity front will thus propagate along the x-axis at the speed $dx/dt = a$. In the case $a > 0$ it moves from the left to the right. We have written the *Mathematica* package advbac.ma using the difference scheme (2.1.9) with the initial data (2.2.6). The complete listing of the package is presented in Appendix 2. The various parameters used in the program advbac.ma are described in Table 2.1. The main function of the program advbac.ma is

$$\text{advbc} [a, a_1, b_1, j_{f0}, u_l, u_r, M, \kappa, Npic].$$

In order to begin a computation using scheme (2.1.9), we must at first specify the initial data in accordance with (2.2.6). This specification is given by the following fragment of the program advbac.ma:

```
(* Computation of the initial values u(xj,0)
                                      at grid points *)
   uj = Table[ul, {j,0,M}];  up = uj;
           Do[ uj[[j]] = ul, {j, 0, jf0}];
           Do[ uj[[j]] = ur, {j, jf0+1, M, 1}];
```

The *Mathematica* function

```
        uj = Table[ul, {j, 0, M}];
```

makes a list of the values of ul with j running from 0 to M. As a result of the execution of the command Table, we obtain a table of values. To take the jth element of the list uj, we can simply write uj[[j]].

The command

```
    ut = Table[ur, {k, 1, Npic}, {j, 1, M1}];
```

enables us to generate a two-dimensional table, or $Npic \times M1$ matrix ut. To take the i,jth element in matrix ut, we can write ut[[i, j]].

The command

```
        mt = Table[j, {j, Npic}];
```

enables us to make a list mt of values of j with j running from 1 to Npic.

It is reasonable to make only a finite number of the time steps while solving problem (2.1.1), (2.2.6), because after the travelling discontinuity reaches the right end $x = a_1$ of the spatial integration region $a_1 \leq x \leq b_1$, the exact solution $u(x,t)$ becomes constant throughout this region in accordance with Eq. (2.2.10). In our program advbac.ma, the number of the time steps to be computed is denoted by N and is determined by the following program fragment:

```
     dt = cap*h/a;    xf0 = a1 + jf0*h;
  Tmax = (b1 - 5.0*h - xf0)/a;
        N = Floor[Tmax/dt];
```

We have used here the built-in *Mathematica* function Floor[x] that determines greatest integer not larger than x. For example,

$$\text{Floor}[2.4] = 2, \ \text{Floor}[-2.4] = -3$$

The numerical computation by scheme (2.1.9) is implemented by the following fragment of our program advbac.ma:

```
(* -- Do-loop over grid nodes on the x-axis --------- *)
          Do[
(* -- The first-order scheme (2.1.9)
                              for equation (2.1.1) *)
     up[[j]] = uj[[j]] - cap*(uj[[j]] - uj[[j-1]]),
     {j, 1, M }];
(*  The boundary condition on the left boundary x = a1 *)
              up[[0]] = ul;
(* -- Exchange of the lists -------------------------- *)
              uj = up;
```

In our programs, we use many Do-loops. The Do-loop of the form

$$\text{Do}[expr, \{i, imax\}]$$

evaluates *expr* repetitively, with i varying from 1 to *imax* in steps of 1. The Do-loop of the form

$$\text{Do}[expr, \{n, nmin, nmax, dn\}]$$

evaluates *expr* with n varying from *nmin* to *nmax* in steps of *dn*. The expression *expr* may indeed be a group of expressions separated by a semicolons. It is to be noted that the last expression in the group *expr* should be followed by a comma rather than by a semicolon, otherwise the Do-loop will not be executed at all.

The line

```
up[[j]] = uj[[j]] - cap*(uj[[j]] - uj[[j-1]]),
```

enables us to compute the value u_j^{n+1} in accordance with the difference equation (2.1.9) using the backward difference to approximate the partial derivative $\partial u/\partial x$. The variable `cap` denotes the quantity $\kappa = a\tau/h$.

Our program `advbac.ma` enables us to perform a computation also by an unstable scheme. For this purpose we check at each time step n the condition

$$\max_j |u_j^n - \max(|u_l|, |u_r|)| \leq 2 \cdot \max(|u_l|, |u_r|)$$

On the right-hand side of this inequality we use a finite value, so that the machine overflow does not occur. As soon as this condition is violated, we exit the `Do`-loop in n with the aid of the *Mathematica* command `Break[ ]`. To check the above condition we use the `If` conditional.

Mathematica provides various ways to set up *conditionals*, which specify that particular expressions should be evaluated only if certain conditions hold. Within the body of our function `advbc` we have used the `If`-condition 18 times. The general form of `If`-condition is as follows:

$$\text{If}\,[test,\ then,\ else]$$

It evaluates *then* if *test* is `True`, and *else* if it is `False`. The expressions *then* as well as *else* may indeed be the groups of expressions; in this case the individual expressions in the groups should be separated by a semicolon from each other.

For the computational process control we have also used the *Mathematica* functions `Goto` and `Max, Abs`. The function `Goto[name]` makes the computer go to the element `Label[name]` in the current `Do`-loop or in the current procedure. As examples of the application of the command `Goto` see in our program the commands `Goto[store]`, `Goto[endt]`, `Goto[endn]`.

The built-in *Mathematica* function `Max[x_1, x_2, ...]` finds the maximum of $x_1, x_2, \ldots$.

The built-in *Mathematica* function `Abs[x]` determines the absolute value $|x|$ of x.

The built-in *Mathematica* function `Print` enables the user to print out expressions:

$$\text{Print}\,[expr_1,\ expr_2, ...]$$

prints the $expr_i$, with no spaces in between, but with a new line (line feed) at the end.

To show the results of the computation using the difference scheme (2.1.9), we have used the built-in *Mathematica* graphics functions

ListPlot, GraphicsArray and Show,

see the program fragment

```
    Print["The solution graph at t = ", t];
mt[[j]] = ListPlot[ug,
                AxesLabel -> {"x", "u"},
                PlotJoined -> True,
                PlotRange -> {{0, M1}, {umin, umax}}],
        {j, nt}];
If[nt == 2,
    Show[GraphicsArray[{mt[[1]], mt[[2]]}]] ];
If[nt == 3,
  Show[GraphicsArray[{{mt[[1]], mt[[2]]}, {mt[[3]]}}]] ];
If[nt == 4,
Show[GraphicsArray[{{mt[[1]], mt[[2]]},
                                {mt[[3]], mt[[4]]}}]]];
```

The function ListPlot plots a list of values. It has a number of options. The option AxesLabel -> {"x", "u"} enables us to plot the label x for the horizontal axis and the label u for the vertical axis.

The option PlotJoined -> True enables us to join the points x_j, u_j^n with lines. The option PlotRange -> {{0, M1}, {umin, umax}} enables us to specify the range of the variation of x and u.

The built-in *Mathematica* function

$$\text{GraphicsArray}[\{g_1, g_2, \ldots\}]$$

represents a row of graphics objects. The command

$$\text{GraphicsArray}[\{\{g_{11}, g_{12}, \ldots\}, \ldots \}]$$

represents a two-dimensional array of graphics objects. The form of this array (the matrix of graphics objects) should be specified explicitly by the user. For example, six graphics objects can be represented by a single row of six pictures, or by a 2×3 matrix having two rows of pictures, with three pictures in each row. In our program this specification of the matrices for the graphics objects is made for the number of pictures nt$\leq$ 8. The interested reader can continue this specification in a similar way for nt= 9, nt= 10, etc.

The function

$$\text{Show}[\text{GraphicsArray}[\{\{plot_1, plot_2, \ldots\}, \ldots \}]]$$

draws an array of plots.

But at first the function **ListPlot** shows on the screen of the computer monitor the graphics objects obtained for several moments of time; you will see a single solution profile for one of the moments of time $n\tau$. To get the next picture on your screen, simply press the key ENTER.

The last picture, which is generated by **Show[GraphicsArray[ ]]**, will show you all the solution graphs on the screen.

The file **advbac.ma** is available on a disc that comes with this book. The internal instruction in the corresponding *Mathematica* Notebook explains in detail how to load the program **advbc[...]** contained in the Notebook **advbac.ma** (see also Appendices 1 and 2). The line

```
advbc[1.0, 0.0, 2.0, 4, 1.0, 0.2, 80, 0.8, 4]
```

specifies the values of $a, a_1, \ldots, Npic$ which are needed to start a numerical computation by the program **advbc**. The result of the computation is shown in Fig. 2.5. In Fig.2.6 we show the solution graphs obtained

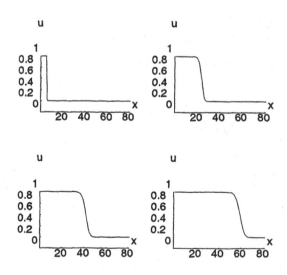

Figure 2.5: Numerical solution graphs in the case when $\kappa = 0.8$ in (2.1.9)

by scheme (2.1.9) in the case $\kappa = 1.03$. To perform this computation we have used the line of input data

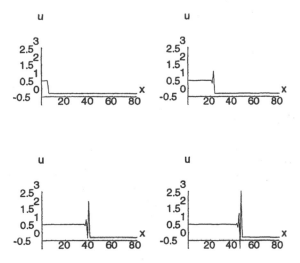

Figure 2.6: $\kappa = 1.03$ in (2.1.9)

```
advbc[1.0, 0.0, 2.0, 4, 1.0, 0.2, 80, 1.03, 4]
```

At $t = 1.10725$ ($n = 42$) we have obtained that $\max|u(x,t)| \approx 3.0$, so that the relative error of the difference solution reaches 200%, see Fig. 2.6. This makes the difference scheme impractical. The reason for this is that the difference scheme (2.1.9) becomes *unstable* for $\kappa > 1$. Therefore, the stability investigation of the difference schemes is an important practical problem of the theory of difference schemes for PDEs.

We now describe the most useful method for the stability investigation of difference schemes, namely the Fourier method.

Exercise 2.4. Determine the order of approximation of the forward-time central-space difference scheme

$$\frac{u_j^{n+1} - u_j^n}{\tau} + a\frac{u_{j+1}^n - u_{j-1}^n}{2h} = 0 \qquad (2.2.11)$$

Exercise 2.5. Implement scheme (2.2.11) by modifying the program advbac.ma. Try to solve numerically the square wave problem with the aid of scheme (2.2.11).

2.3   FOURIER STABILITY ANALYSIS

2.3.1   An Outline of the Fourier Method

Let us consider a slightly more complicated PDE than the equation (2.1.1),

$$\partial u/\partial t + a\partial u/\partial x = \varphi(x,t), \quad -\infty < x < \infty \qquad (2.3.1)$$

where $\varphi(x,t)$ is a given function, it is sometimes called the *forcing function*[3]; generally $|\varphi(x,t)| < \infty$ for $-\infty < x < \infty$. Similarly to the difference scheme (2.1.9), we now approximate the equation (2.3.1) by the forward-time upwind finite difference equation

$$\frac{u_j^{n+1} - u_j^n}{\tau} + a\frac{u_j^n - u_{j-1}^n}{h} = \varphi(jh, n\tau) \qquad (2.3.2)$$

$$j = 0, \pm 1, \pm 2, \ldots; \quad n = 0, 1, \ldots, N$$

$$u_j^0 = u_0(jh), \quad j = 0, \pm 1, \ldots \qquad (2.3.3)$$

We now define the *stability* of the difference problem (2.3.2), (2.3.3) with respect to the perturbation δu_j^0 of the initial data u_j^0. For this purpose we first find an equation for the difference solution perturbations. Let us assume that we take the initial data

$$\bar{u}_j^0 = u_0(jh) + \delta u_j^0, \quad j = 0, \pm 1, \ldots \qquad (2.3.4)$$

instead of (2.3.3). We will obtain with the aid of (2.3.2) the difference solution values $\bar{u}_j^n$ for $n = 1, 2, \ldots, N$. These solution values satisfy the difference equation

$$\frac{\bar{u}_j^{n+1} + \bar{u}_j^n}{\tau} + a\frac{\bar{u}_j^n - \bar{u}_{j-1}^n}{h} = \varphi(jh, n\tau) \qquad (2.3.5)$$

where

$$\bar{u}_j^n = u_j^n + \delta u_j^n, \quad j = 0, \pm 1, \pm 2, \ldots \qquad (2.3.6)$$

Let us now substitute (2.3.6) into (2.3.5) and substract (2.3.2) to obtain the difference equation

$$\frac{\delta u_j^{n+1} - \delta u_j^n}{\tau} + a\frac{\delta u_j^n - \delta u_{j-1}^n}{h} = 0, \quad n = 0, 1, \ldots, N \qquad (2.3.7)$$

with the given initial perturbation δu_j^0. Now define the grid norm for the perturbations δu_j^n as a discrete analog of the C norm:

$$\delta u^n = \max_j |\delta u_j^n|, n = 0, 1, \ldots, N \qquad (2.3.8)$$

Definition 2.1. The difference initial-value problem (2.3.2), (2.3.3) is called *stable* with respect to the perturbations of the initial data, if the condition

$$\delta u^n \leq c\, \delta u^0 \qquad (2.3.9)$$

is satisfied for arbitrary bounded function $u_j^0 = u_0(jh)$, where c is a constant, which does not depend on h and τ.

This property means that the perturbation $\{\delta u_j^0\}$ introduced into the initial data of problem (2.3.2), (2.3.3) will cause the perturbation $\{\delta u_j^n\}$ of the solution to the problem (2.3.2), (2.3.3), which by virtue of (2.3.9) exceeds the initial data perturbation by a factor of no more than c.

The condition (2.3.9) should be satisfied also for the initial perturbation of the form

$$\delta u_j^0 = e^{i\alpha j h}, j = 0, \pm 1, \ldots \qquad (2.3.10)$$

$$i = \sqrt{-1}$$

where α is a real parameter. The solution of the difference problem (2.3.7), (2.3.10) has the form

$$\delta u_j^n = \lambda^n e^{i\alpha j h} \qquad (2.3.11)$$

where $\lambda = \lambda(\alpha)$ is determined by the substitution of the expression (2.3.11) into the homogeneous difference equation (2.3.7):

$$\lambda(\kappa, \xi) = 1 - \kappa(1 - e^{-i\xi}) \qquad (2.3.12)$$

where $\kappa = a\tau/h$ and $\xi = \alpha h$. The expression (2.3.12) for $\lambda(\xi)$ is called the *amplification factor* of the difference scheme (2.3.7). For the solution (2.3.11) the equality

$$\max_j |\delta u_j^n| = |\lambda(\kappa, \xi)|^n \max_j |\delta u_j^0|$$

is valid. From this it follows that the inequality

$$|\lambda(\kappa, \xi)|^n \leq c, n = 0, 1, \ldots, N \qquad (2.3.13)$$

is necessary to ensure the satisfaction of the condition (2.3.9). In its turn, the inequality (2.3.13) is possible only if

$$|\lambda(\kappa, \xi)| \leq 1 + c_1 \tau \qquad (2.3.14)$$

where c_1 is a constant which does not depend on τ and h. In order to prove that the inequality (2.3.14) should hold for the stability let us assume that the opposite assertion also ensures the satisfaction of (2.3.13). Since in accordance with the Taylor formula

$$\ln(1 + z) = z - \frac{1}{2}z^2 + \frac{z^3}{3(1 + \theta)^3}$$

with $z = c_1\tau$ and $0 \le \theta \le z$, we obtain in this case, for $t = n\tau$,

$$|\lambda|^n > (1 + c_1\tau)^{\frac{t}{\tau}} = \exp\left(\frac{t}{\tau}\ln(1 + c_1\tau)\right) \ge$$
$$\exp\left(\frac{t}{\tau}(c_1\tau - \frac{1}{2}c_1^2\tau^2)\right) > c$$

if, for example, $0 < c_1\tau < 2$ and

$$t > \ln|c|/[c_1(1 - \frac{1}{2}c_1\tau)]$$

Note that the condition (2.3.14) should be satisfied for all ξ. The condition (2.3.14) is called the *von Neumann necessary stability condition*.[5] Since

$$e^{-i\xi} = \cos\xi - i\sin\xi$$

we may rewrite the formula (2.3.12) obtained in the case of the difference scheme (2.3.7) as

$$\lambda(\kappa, \xi) = 1 - \kappa(1 - \cos\xi) - \kappa i\sin\xi \qquad (2.3.15)$$

In the practical stability analyses one usually assumes $c_1 = 0$ in (2.3.14). It follows from (2.3.11) that the function $\lambda(\xi)$ obtained from the difference scheme is a periodic function of ξ, and the period is equal to 2π. Therefore, it is sufficient to study the behaviour of $|\lambda(\xi)|$ in the interval $0 \le \xi \le 2\pi$. In particular, in the case of (2.3.15) we have that

$$|\lambda(\kappa, \xi)|^2 = (1 - \kappa\sin^2\frac{\xi}{2})^2 + \kappa^2\sin^2\xi$$

We find, after some simple trigonometric manipulations that

$$|\lambda(\kappa, \xi)|^2 = 1 - 4\kappa(1 - \kappa)\sin^2\frac{\xi}{2} \qquad (2.3.16)$$

The von Neumann condition (2.3.14) should be valid also for small positive values c_1 and τ. Therefore, we can require that $|\lambda|^2 \le 1$. The maximum of the right-hand side of (2.3.16) considered as a function of

ξ exceeds 1 at $\kappa < 0$ and $\kappa > 1$. Therefore, the von Neumann condition (2.3.14) is satisfied, if the number κ satisfies the inequalities

$$0 \leq \kappa \leq 1 \qquad (2.3.17)$$

This stability condition was obtained for the first time by R. Courant, K. Friedrichs, and H. Lewy.[6] The number $\kappa = a\tau/h$ is called the Courant number. The condition (2.3.17) ensures that the closed curve formed by the points $(Re\lambda(\kappa,\xi), Im\lambda(\kappa,\xi))$ remains inside the unit circle $|\lambda(\kappa,\xi)| \leq 1$ of the complex plane.

It follows that the difference scheme (2.3.7) is unstable for any negative coefficient a. The difference equation for the perturbations δu_j^n, which can be obtained from the difference scheme (2.1.9), coincides with the equation (2.3.7). Thus the difference scheme (2.1.9) is unstable for $a < 0$.

The above method of the stability analysis, which is based on the substitution of the Fourier harmonics (2.3.11) into the difference scheme, is called in the literature the *Fourier method*, or the *spectral method*.[5,7]

2.3.2 Computer Implementation of the Fourier Method

In what follows we describe our program st1.ma, which enables us

-to obtain the symbolic expression for the amplification factor $\lambda(\kappa,\xi)$;

-to plot the curves

$$(Re\,\lambda(\kappa,\xi), Im\,\lambda(\kappa,\xi))$$

for a number of fixed values of κ;

-to determine the stability interval on the κ-axis to a given accuracy ε.

The complete listing of the package is presented in Appendix (the file st1.ma is also available on the attached diskette). The input parameters used in the program st1.ma are described in Table 2.2. The main function in the package st1.ma is

$$\text{stab1}\,[\varepsilon, n1plot, ac1, bc1, sch1].$$

It is not difficult to obtain the symbolic form expression for the amplification factor $\lambda(\kappa,\xi)$ with the aid of *Mathematica*. For this purpose let us define the following function which contains the difference equation (2.1.9) under consideration:

Table 2.2. Parameters used in program st1.ma

Parameter	Description
eps	the given accuracy of the computation of the coordinates of points of the stability region boundary
n1plot	the number of graphics pictures showing the curves $(Re[\lambda(\kappa,\xi)], Im[\lambda(\kappa,\xi)])$, $\xi = 0, 2\pi$, where $\lambda(\kappa, \xi)$ is the amplification factor; n1plot≥ 0
ac1	the left end of the interval in which the stability region boundary is to be determined
bc1	the right end of the above interval ac1$\leq \kappa \leq$ bc1
sc	the left-hand side of the difference equation sc$= 0$

```
sch1 = (u[j,n+1] - u[j,n])/dt +
                    a (u[j,n] - u[j-1,n])/h
```

Here we have denoted by dt the time step τ. We have written in this definition the subscript j and the superscript n of the difference solution value u_j^n as the arguments j , n . The definition sch1 $= expr$ specifies that whenever *Mathematica* encounters an expression sch1, it should replace sch1 by *expr*.

Note that in the case of more complex difference schemes the definition sch1 can include several expressions separated by a semicolon. In this case the final expression for the difference equation is simply the last expression in the function definition. For example, we can write the difference equation (2.1.9) with the aid of the following function:

```
sch1 =
( ut = (u[j, n+1] - u[j,n])/dt;
      ux1 =. (u[j,n] - u[j-1,n])/h; ut + ux1 )
```

In accordance with the Fourier procedure, we now must substitute the Fourier harmonic $\lambda^n e^{ij\xi}$ into the difference equation instead of u[j, n]. To do this we define the transformation rule

```
u[j_, n_] -> l^n Exp[I j xi]
```

We have denoted the amplification factor λ by l, and the variable ξ by xi. The *Mathematica* system uses the variable I for $i = \sqrt{-1}$. In what follows we will need the expressions $Re(\lambda)$ and $Im(\lambda)$. Unfortunately, the built-in *Mathematica* functions Re[z], Im[z] can extract the real part and the imaginary part of a complex number z only in the case when $Re(z)$ and $Im(z)$ are numbers rather than symbolic expressions. For example, if $z = 4 - 3 * I$, then *Mathematica* outputs Re[z] = 4, Im[z] = -3. For the symbolic manipulation of the symbolic expressions involving the imaginary unit I there is the package Algebra'ReIm'. It should be loaded before the symbolic computation of such expressions begins. The user of our program st1.ma needs not perform this loading, because it is done by our program.

The expression obtained for the amplification factor l is then transformed by using the Euler formula $e^{ix} = \cos x + i \sin x$. For this purpose we use the *Mathematica* function ComplexExpand[*expr*]. It expands the expression under the assumption that all the variables are real. In our program st1.ma this function is used as follows:

```
rel = ComplexExpand[Re[l1]]/. c1rule //Simplify;
iml = ComplexExpand[Im[l1]]/. c1rule //Simplify;
```

In this program fragment, the variables rel and iml denote $Re(\lambda)$ and $Im(\lambda)$, respectively. c1rule is here the transformation rule for the introduction of a nondimensional variable c1 = κ.

The function ParametricPlot enables the user to plot the curves given parametrically. In our case the parameter is $\xi \in [0, 2\pi]$. As a result we obtain the curves in the $(Re\,\lambda, Im\,\lambda)$ plane (see the example in Fig. 2.7). The function $Im\,\lambda = F(Re\,\lambda)$ is generally a non-single-valued function. The *Mathematica* function ListPlot, which was already used by us in the package advbac.ma, would not be applicable in the present case.

The function ParametricPlot[$\{f_x, f_y\}, \{t, tmin, tmax\}$] produces a parametric plot with x and y coordinates f_x and f_y generated as a function of t. ParametricPlot[$\{\{f_x, f_y\}, \{g_x, g_y\},...\}, \{t, tmin, tmax\}$] plots several parametric curves. In our case it is important to check whether the curves $(Re\,\lambda\,(\kappa, \xi), Im\,\lambda(\kappa, \xi))$ leave the unit disc $|\lambda| \leq 1$ in the complex plane λ. Therefore, we have introduced in our program st1.ma also the command for plotting a circle of radius 1. This closed curve may be specified parametrically with the aid of the list {Cos[t],Sin[t]}.

When calling the function ParametricPlot we have used the option

```
PlotStyle -> {Dashing[{0.04, 0.03}]
```

to produce a dashed curve in which the length of the dash is equal to 0.04, and the length of the gap between two successive gaps is equal to 0.03. These segments are repeated cyclically. The sizes 0.04 and 0.03 represent the fractions of the total width of the graph.

In order to fill the interior of the unit disc we have used the *Mathematica* function Polygon[$\{pt_1, pt_2, ...\}$]. The positions of points can be specified either in absolute coordinates as $\{x, y\}$, or in scaled coordinates as `Scaled[{x, y}]`. The boundary of the polygon is formed by joining the last point to the first one. The generation of the filled unit disc is performed in our program `st1.ma` with the aid of the commands

```
obj = Polygon[Table[{Cos[n Pi/40], Sin[n Pi/40]},{n,80}]];
mt4 = Show[Graphics[{RGBColor[0.2,0.8,1], obj}],
                    Axes -> True,
                    Ticks -> Automatic,
                    DisplayFunction -> Identity];
```

`RGBColor[red, green, blue]` is a graphics directive which specifies that the graphical objects which follow are to be displayed, if possible, in the color given. Red, green and blue color intensities outside the range 0 to 1 will be clipped. The option `Ticks -> Automatic` specifies that the tick marks are placed automatically.

To avoid a separate plotting of the filled unit disc, we have used the option `DisplayFunction -> Identity` in the *Mathematica* function `Show`. Setting `DisplayFunction -> Identity` will cause the graphics objects to be returned, but no display to be generated.

When forming the graphics object `mt3`, we want that this object be displayed; for this purpose we use in `Show` the option `DisplayFunction -> $DisplayFunction`. To show both the filled unit disc and the curves $(Re(\lambda), Im(\lambda))$ in the same picture we have used the commands

```
mt3 = Show[mt4, mt1, mt2,
    Axes -> True,
           Ticks -> Automatic,
           DisplayFunction -> $DisplayFunction];
AppendTo[mt, mt3], {j, n1plot}];
```

The *Mathematica* function `AppendTo[mt, elem]` appends `elem` to the list `mt`. The list `mt` is at first initiated as an empty list with the aid of the command `mt = {}`.

In Fig. 2.7 we present a number of curves $(Re\,\lambda(\kappa, \xi), Im\,\lambda(\kappa, \xi))$ obtained with the aid of the program `st1.ma` for the case of the difference scheme (2.1.9). To make a PostScript file for this figure we have used the *Mathematica* function `Display["file", graphics]`. The file produced by this function contains no initialisation and termination commands;

it can be converted into a regular PostScript file with the aid of the program `psfix` (UNIX) or `printps` or `rasterps` (MS-DOS).

To perform the corresponding symbolic/numeric computation, at first load the program file beginning with the line `ClearAll[fbis]` and then perform Steps 2-4 in accordance with the internal instruction of the notebook `st1.ma`. The line of the input data is

```
stab1[0.001, 4, 0.0, 1.5, sch1]
```

We can see from Fig. 2.7 that at $\kappa > 1$ some pieces of the curves lie outside the unit disc.

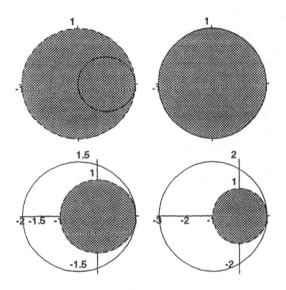

Figure 2.7: The curves $(Re\,\lambda(\kappa,\xi), Im\,\lambda(\kappa,\xi))$ for $\kappa = \kappa_j = 2j/4, j = 1,\ldots,4$

Our numerical method for determining the stability interval on the κ-axis is based on a simple idea of checking whether the curves $(Re\,\lambda(\kappa,\xi), Im\,\lambda(\kappa,\xi))$ remain inside the unit circle for a given value of κ. To have the possibility of determining the right boundary κ_{max} of the stability interval to a user-specified accuracy ε let us construct the following discontinuous function:

$$f(\kappa) = \begin{cases} 1, & |\lambda(\kappa,\xi)| \leq 1 \quad \forall \xi \in [0, 2\pi] \\ -1 & \text{otherwise} \end{cases} \qquad (2.3.18)$$

This function undergoes a jump on the stability region boundary and, in addition, it changes its sign when the κ point leaves the stability

interval. In our program st1.ma, the function (2.3.18) was implemented
as follows:

```
fbis[x_, ss1_List, cc1_List, nspg_]:=
( fbs1 = 1.0; deps = N[1.0 + 1.0/(10^8)];
      j = 1; nspg1 = nspg;
Label[loop];
rj = N[r /. {sa -> ss1[[j]], ca -> cc1[[j]], c1 -> x}];
   If[rj[[1]] > deps, nspg1 = j];
   j = j + 1; If[j <= nspg1, Goto[loop]];
   If[nspg1 < nspg, fbs1 = -1.0]; fbs1 )
```

The definition

$$\text{fbis[x\_, ss1\_List, cc1\_List, nspg\_] := } expr$$

specifies that whenever *Mathematica* encounters an expression fbis[...]
it should replace fbis[...] by *expr*. The arguments in the definition
of a function are always supplied with the _ (underline). The underline
is omitted in the text of the program, where the functions are used to
specify some expression.

Since the variables ss1 and cc1 are indeed lists, we have assigned
them the type _List in the definition of the function fbis. Then each
time when the function fbis is called it is verified whether ss1 and cc1
are the lists.

The sign-change property of the function (2.3.18) enables us to use
the bisection method for determining κ_{max}. This value is computed
numerically with the aid of arithmetic operations on floating-point numbers. *Mathematica* provides several possibilities for such operations. In
particular, the built-in *Mathematica* function N[*expr*] gives the numerical value of *expr*, see the above program fragment. It does computations
with machine precision numbers. N converts all numbers to Real form.
In our program st1.ma the bisection method is implemented with the
aid of the function

```
bisec[eps_, ac1_, bc1_, ss1_List, cc1_List, nspg_]:=
(* -- This function determines the stability region
      boundary by the bisection method --------- *)
   ( fa = N[fbis[ac1, ss1, cc1, nspg]];
       fb = N[fbis[bc1, ss1, cc1, nspg]];
```

```
       isl = Sign[fa]; isr = Sign[fb];
     ac2 = ac1; bc2 = bc1;
    If[isl!= isr, Goto[bisproc]];
    If[isr== -1, Goto[unstable]];
  Print["The scheme is stable in the interval ", ac1,
       "<= c1 <= ", bc1];
    xbis = bc1; Goto[endbis];
Label[unstable];
    Print["The scheme is unstable"];
    xbis = ac1; Goto[endbis];
Label[bisproc];
xbis = N[(ac2 + bc2)*0.5];
yy = N[fbis[xbis, ss1, cc1, nspg]];  ny = Sign[yy];
If[ny==isr, bc2=xbis, ac2=xbis];
If[N[Abs[ac2 - bc2]] >= eps, Goto[bisproc]];
Label[endbis]; xbis )
```

Here eps is the user-specified accuracy with which the value κ_{max} will be determined. ac1 and bc1 are the given values determining the interval on the κ-axis, in which the stability region boundary κ_{max} is to be computed. Since the values of $\cos\xi$ and $\sin\xi$ will be computed many times during the bisection process, it is reasonable, for the purpose of saving computer time, to compute these values once and store them in the form of tables or lists. This is done in our program by defining the lists ss1 and cc1 with the aid of the commands

```
ss1 = Table[N[Sin[(k-1)*2*Pi/nspg]], {k, nspg}];
cc1 = Table[N[Cos[(k-1)*2*Pi/nspg]], {k, nspg}];
```

The variable nspg is a positive number which specifies the number of nodes of an uniform grid in the interval $0 \le \xi < 2\pi$. The numerical value of nspg is determined automatically with the aid of the following simple iterative process. Let us denote by J some starting value for nspg; we have taken the value $J = 4$ in our program. Let us further denote by $(\kappa_{max})_J$ the value of κ_{max}, which was determined with the aid of our function bisec for a given J. As soon as the inequality

$$|(\kappa_{max})_J - (\kappa_{max})_{J+2}| \le \varepsilon$$

is satisfied for some J, we assume nspg $= J$ and terminate the iteration process. We now present the fragment of the program st1.ma, in which the value of nspg is determined.

```
(* -- Determination of nspg -------------------------- *)
  nspg0 = 4; c1sb = 10000;
    Do[nspg = nspg0 + 2*j;
  ss1 = Table[N[Sin[(k-1)*2*Pi/nspg]], {k, nspg}];
  cc1 = Table[N[Cos[(k-1)*2*Pi/nspg]], {k, nspg}];
 c1s0 = bisec[eps, ac1, bc1, ss1, cc1, nspg];
    If[j == 1, c1sb = c1s0, dabsc1 = N[Abs[c1s0-c1sb]];
  Print["nspg, dabsc1 = ", nspg, "  ", dabsc1]];
        If[dabsc1<=eps, Break[]];
 c1sb = c1s0,
                {j, 48}];
```

The numerical value of the function (2.3.18) at a specific value of κ is computed in our program st1.ma by calling the function fbis[x_, ss1_List, cc1_List, nspg_].

The *Mathematica* function

Simplify[*expr*]

can introduce trigonometric functions like $\sin(\xi/2), \cos(3\xi)$, etc., in the process of the simplifying expressions containing trigonometric functions. But we use only the functions $\cos \xi$ and $\sin \xi$ in our function fbis. To prevent the above uncontrolled simplification effect we have expressed the functions $\exp(\pm ik\xi)$ in the expression for the amplification factor λ in terms of the trigonometric functions sa $= \sin \xi$ and ca $= \cos \xi$ as soon as the symbolic expression for λ was obtained. For this purpose we have used the *Mathematica* function TrigReduce[*expr*].

Let us take, for example, the values ac1 = 0.0, bc1=1.5. We know there is a root of the equation $f(\kappa) = 0$ between ac1=0.0 and bc1=1.5, because $f(0) = 1$ and $f(1.5) = -1$. We can reduce the uncertainty by evaluating $f(\kappa)$ at the midpoint $\kappa = 0.75$. Doing this shows that $f(0.75) = 1$, so we conclude that the root lies between 0.75 and 1.5. This and two subsequent steps of the bisection method are depicted in Fig. 2.8. We continue the bisection until the length of the bracketing interval becomes less than the user-specified error tolerance eps.

For example, at eps=0.01 we have obtained the value $\kappa_{max} = 1.00195$. At eps=0.00001 the obtained result was $\kappa_{max} = 1.$, which coincides with the above analytic result (2.3.17).

If f[ac1]=f[bc1]=-1, then our function bisec prints out the message The scheme is unstable and no bisection is made. But it may happen that the chosen interval ac1$\leq \kappa \leq$bc1 contains a subinterval in

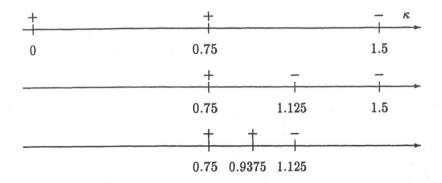

Figure 2.8: Bisection method

which the scheme under consideration is stable. To localize this subinterval we introduce a uniform mesh of 15 nodes in the interval $ac1 \leq \kappa \leq bc1$ and print out the table of values of the function (2.3.18) in these nodes. If there are positive numbers (1) in this table, we can re-specify the values of $ac1$ and $bc1$ in such a way that the function (2.3.18) has the opposite signs at $\kappa = ac1$ and $\kappa = bc1$.

2.3.3 Method of Differential Approximation

Now we want to present one more method for the stability analysis of difference schemes. This method was rigorously justified only for very limited classes of linear difference schemes for hyperbolic PDEs. [8,9] Therefore, its application is in many cases heuristic. Nevertheless it turns out that this method can become especially useful when one wants to obtain some insight into the stability of nonlinear difference schemes.

Let us explain the idea of the method for the example of the difference scheme (2.1.9). First we expand the grid values u_j^{n+1} and u_{j-1}^n into a Taylor series at to the point (x_j, t_n):

$$u_j^{n+1} = u_j^n + \left(\frac{\partial u}{\partial t}\right)_j^n \tau + \frac{\tau^2}{2}\left(\frac{\partial^2 u}{\partial t^2}\right)_j^n + O(\tau^3)$$

$$u_{j-1}^n = u_j^n - h\left(\frac{\partial^2 u}{\partial x^2}\right)_j^n + O(h^3) \qquad (2.3.19)$$

Now we substitute the right hand sides of the expressions (2.3.19) instead of u_j^{n+1} and u_{j-1}^n into the difference equation (2.1.9). As a result we obtain the PDE of the form

$$\frac{\partial u}{\partial t} + a\frac{\partial u}{\partial x} = a\frac{h}{2}\frac{\partial^2 u}{\partial x^2} - \frac{\tau}{2}\frac{\partial^2 u}{\partial t^2} + O(h^2) + O(\tau^2) \qquad (2.3.20)$$

Let us now neglect in this equation the terms of the order $O(h^2)+O(\tau^2)$. Then we obtain the following differential equation:

$$\frac{\partial u}{\partial t} + a\frac{\partial u}{\partial x} = a\frac{h}{2}\frac{\partial^2 u}{\partial x^2} - \frac{\tau}{2}\frac{\partial^2 u}{\partial t^2} \qquad (2.3.21)$$

It is possible to draw a conclusion about the stability already from this PDE, which is hyperbolic in the case $a > 0$. But we proceed further by using the *differential consequences* of the original PDE (2.1.1). These are the PDEs which are obtained from the equation (2.1.1) by differentiating the both sides of (2.1.1) with respect to the independent variables x and t. Some the differential consequences are:

$$\frac{\partial^2 u}{\partial t^2} = -a\frac{\partial^2 u}{\partial t \partial x} = -a\frac{\partial}{\partial x}\left(\frac{\partial u}{\partial t}\right) = -a\frac{\partial}{\partial x}\left(-a\frac{\partial u}{\partial x}\right) = a^2\frac{\partial^2 u}{\partial x^2} \qquad (2.3.22)$$

Replacing the second time derivative u_{tt} in (2.3.21) by the formula u_{tt} in (2.3.22), we obtain the following PDE:

$$\frac{\partial u}{\partial t} + a\frac{\partial u}{\partial x} = \frac{h}{2}a(1-\kappa)\frac{\partial^2 u}{\partial x^2} \qquad (2.3.23)$$

where $\kappa = a\tau/h$. Let us introduce

$$\nu = a\frac{h}{2}(1-\kappa) \qquad (2.3.24)$$

and rewrite equation (2.3.23) in the form

$$\frac{\partial u}{\partial t} + a\frac{\partial u}{\partial x} = \nu\frac{\partial^2 u}{\partial x^2} \qquad (2.3.25)$$

We again consider the initial condition (2.1.2). It is not difficult to show that the numerical diffusion coefficient ν in (2.3.23) should be positive to ensure the boundedness of the solution of the initial-value problem (2.3.25), (2.1.2). For this purpose we at first transform the equation (2.3.25) to the standard heat equation with the aid of new variables X and t' introduced via

$$X = x - at, \quad t' = t$$

If $\tilde{u}(X, t') = u(x, t)$, then

$$\frac{\partial u}{\partial x} = \frac{\partial \tilde{u}}{\partial X}, \frac{\partial u}{\partial t} = \frac{\partial \tilde{u}}{\partial X} \frac{\partial X}{\partial t} + \frac{\partial \tilde{u}}{\partial t'} = \frac{\partial \tilde{u}}{\partial t'} - a \frac{\partial \tilde{u}}{\partial X} \qquad (2.3.26)$$

Substituting the expressions (2.3.26) into the equation (2.3.25), we immediately obtain the heat equation

$$\frac{\partial \tilde{u}}{\partial t'} = \nu \frac{\partial^2 \tilde{u}}{\partial X^2} \qquad (2.3.27)$$

Since $X = x$ at $t = 0$, we can write the initial condition in the form

$$\tilde{u}(X, 0) = u_0(X) \qquad (2.3.28)$$

The exact solution of the initial-value problem (2.3.27), (2.3.28) is well known; it is expressed via the Poisson's integral

$$\tilde{u}(X, t') = \frac{1}{2\sqrt{\pi}} \int_{-\infty}^{\infty} \frac{1}{\sqrt{\nu t'}} \exp\left(\frac{-(X - \xi)^2}{4\nu t'}\right) u_0(\xi) d\xi$$

Therefore,

$$u(x, t) = \frac{1}{2\sqrt{\pi}} \int_{-\infty}^{\infty} \frac{1}{\sqrt{\nu t}} \exp\left(\frac{-(x - at - \xi)^2}{4\nu t}\right) u_0(\xi) d\xi \qquad (2.3.29)$$

It is easy to see that for $\nu < 0$

$$\lim_{|\xi| \to \infty} e^{\frac{-(x - at - \xi)^2}{4\nu t}} = +\infty \qquad (2.3.30)$$

at fixed x and $t > 0$. If the function $u_0(x)$ is, for example, bounded from below, so that $u_0(x) > C > 0$, where C is a constant, then it is easy to see from (2.3.29) and (2.3.30) that the solution $u(x, t)$ is unbounded for $\nu < 0$. Since it is usually assumed that stable solutions should remain bounded, negative values of the numerical diffusion coefficient in (2.3.25) must be avoided. From the inequality $\nu > 0$ we obtain with regard for (2.3.24) the inequality $\kappa \leq 1$, which coincides with the Courant-Friedrichs-Lewy condition (2.3.17).

The PDEs (2.3.21) and (2.3.23) are called the *modified equations* of the difference scheme. This heuristic method was presented for the first time by C.W. Hirt.[10] In the Russian literature the equations (2.3.21) and (2.3.23) are termed the *first differential approximations* of the difference scheme. Y.I. Shokin [8,9] has presented, in his books, the mathematical validation of the modified equation approach for certain classes of difference schemes for hyperbolic PDEs. In particular, the classes of difference schemes were found in,[4,5] for which the results of the stability analysis with the aid of the method of differential approximation coincide with the results obtained by the Fourier method.

2.3.4   The Notion of Convergence

Now we would like to discuss the notion of the *convergence* of a difference scheme to a PDE. This property also means that the solution of the finite difference equation converges to the solution of the partial differential equation as $h \to 0, \tau \to 0$. The formulation of the convergence criteria for difference initial-value problems and difference initial- and boundary-value problems is one of the important parts of the theory of difference schemes for PDEs. The convergence studies usually use the techniques of the functional analysis. We do not want to present here the exact form of any of the theoretical convergence results. We only give a general formulation of the basic result of the convergence studies.

Theorem 2.1. Let the initial-value problem for a PDE well posed and let the difference initial-value problem approximate the initial-value problem for the PDE. Then the stability of the solution of the difference initial-value problem is necessary and sufficient for the convergence of the difference solution to the solution of the initial-value problem for the given PDE.

One of the theorems of this kind is the Lax equivalence theorem.[5] It follows from these convergence results that the difference scheme chosen for the solution of an applied problem should possess the properties of the approximation and of the stability.

There are at least two practical ways for the checking for convergence. One way is to check the accuracy of the numerical solution for given sizes of the mesh steps τ and h by comparing the numerical solution with the exact analytic solution of some test problem. For the quantitative estimation of the numerical solution error, one can use a discrete analog of the L_2 norm error, which is defined as

$$\delta u^n = \| u^n - u(x, t_n) \|_{2,d} = \left[\sum_{j=0}^{M} (u_j^n - u(x_j, t_n))^2 h \right]^{1/2} \quad (2.3.31)$$

where u_j^n is the computed solution of the difference equation, and $u(x_j, t_n)$ is the value of the analytic solution computed at the same mesh point (x_j, t_n). If the error (2.3.31) is considered too large, then one can: (a) take the finer mesh steps, say $h/2$ and $\tau/2$; or (b) take another difference method, which possesses a higher convergence order that ensures for the same values of h and τ, a smaller magnitude of the error (2.3.31).

Another practical way of studying the convergence of numerical solution is as follows. Let us denote by $u(h_1, \tau_1)$ the numerical solution obtained for some chosen value of $t > 0$, and let $u(h_2, \tau_2)$ be the numerical solution obtained at smaller values of h and τ, for example, $h_2 = h_1/2, \tau_2 = \tau_1/2$. If the difference $|u(h_1, \tau_1) - u(h_2, \tau_2)|$ is sufficiently small at all nodes (j) of the spatial computing mesh, for example, the error does not exceed some value ε, then one may draw a subjective judgment on the convergence.

Exercise 2.6. Consider the case $a < 0$ in the equation (2.1.1). Take the forward differences in x and t to approximate the spatial derivative $\partial u/\partial x$ and the time derivative $\partial u/\partial t$. Determine the stability condition of the difference scheme with the aid of the Fourier method.

Hint. Use the program st1.ma. Change the line

```
(u[j,n+1] - u[j,n])/dt + a*(u[j,n] - u[j-1,n])/h
```

to the line

```
(u[j,n+1] - u[j,n])/dt + a*(u[j+1,n] - u[j,n])/h
```

The line of input data now reads

```
stab1[0.00001, 6, -1.5, 0.0,sch1]
```

Exercise 2.7. Take the difference equation obtained in the previous exercise. Plot, with the aid of *Mathematica*, the closed curves of the amplification factor $\lambda(\xi)$ for this scheme in the complex plane $\{Re(\lambda), Im(\lambda)\}$ for the values $\kappa = -1$, $\kappa = -3/4$, $\kappa = -1/2$, $\kappa = -1/4$. Show that all these curves lie inside the unit disc $|\lambda| \leq 1$. Also plot the curve for $\lambda(\xi)$ in the case $\kappa = -5/4$. Show that some points of this curve lie outside the unit disc $|\lambda| \leq 1$.

Hint. Use the program st1.ma modified as in the above Exercise 2.6. Change the line

```
c1j = Table[N[2*j/n1plot], {j, n1plot}];
```

to the line

```
c1j = Table[N[-5*j/(4*n1plot)], {j, n1plot}];
```

The line of input data should now read

```
stab1[0.00001, 5, -1.5, 0.0,sch1]
```

Exercise 2.8. For the difference equation (2.1.9) which is stable for $a > 0$, and the difference equation obtained in the Exercise 2.6, show that these two difference schemes can be written as one difference scheme involving only the central spatial differencing.

Hint. Assume that the desired difference equation has the form

$$\frac{u_j^{n+1} - u_j^n}{\tau} + a\frac{u_{j+1}^n - u_{j-1}^n}{2h} = b\frac{u_{j+1}^n - 2u_j^n + u_{j-1}^n}{h^2} \qquad (2.3.32)$$

where b is an indeterminate coefficient. Now rewrite this equation as

$$u_j^{n+1} = u_j^n - \frac{\kappa}{2}(u_{j+1}^n - u_{j-1}^n) + \frac{\tau b}{h^2}(u_{j+1}^n - 2u_j^n + u_{j-1}^n) \qquad (2.3.33)$$

First consider the case $a > 0$. Subtract from both sides of equation (2.1.11) the two sides of (2.3.33):

$$0 = (u_{j+1}^n - 2u_j^n + u_{j-1}^n)\left(\frac{\tau b}{h^2} - \frac{\kappa}{2}\right) \qquad (2.3.34)$$

Since $u_{j+1}^n - 2u_j^n + u_{j-1}^n \neq 0$ for arbitrary solution profile $u^n(x)$, we find from (2.3.34) that $b = ah/2$. Consider the case $a < 0$ in a similar way. This difference scheme is called the *upwind* difference scheme, because it performs an automatic switch from the backward spatial differencing to the forward spatial differencing in the case when the advection velocity $a = a(x,t)$ changes its sign within the spatial computational domain. Thus the upwind difference scheme automatically changes its structure depending on the behavior of the velocity of the advection, or of the "wind".

Exercise 2.9. Study the stability of the upwind scheme obtained in Exercise 2.8 with the aid of the Fourier method and obtain the stability condition of this scheme.

Exercise 2.10. Obtain the stability condition of scheme (2.2.11) with the aid of the Fourier method.

Hint. Use the program st1.ma.

Exercise 2.11. Obtain the stability condition of the upwind scheme from Exercise 2.8 with the aid of the Hirt's method.

2.4   ELEMENTARY SECOND-ORDER SCHEMES

We have already noted in Section 1.2 that the schemes with higher approximation orders (that is the schemes for which $k_1 + k_2 \geq 3$ in (2.2.5)), that is higher-order schemes, generally yield more accurate numerical solutions than the first-order schemes which we have considered in Sections 2.1-2.3. Therefore, it is desirable to be able to construct higher-order difference schemes for the numerical solution of applied problems. Another advantage of higher-order schemes is that they enable the user to get sufficiently accurate numerical results on coarser spatial grids than the grids which are used in the computer programs implementing the first-order schemes. This is especially important when the small computers are applied to the numerical solution of complex mathematical physics problems, because computer storage limitations may impose the substantial limitations for the number of spatial grid nodes which can be used.

2.4.1   Lax-Wendroff Scheme

It appears that the first higher-order difference scheme for the hyperbolic PDEs was proposed by P. Lax and B. Wendroff in 1960.[11,1,3,5] We illustrate the idea behind the construction of this type of scheme for the approximation of the advection equation (2.1.1). Let us at first approximate this equation by a difference scheme with a forward difference in time and a central difference in the space: this is the scheme (2.2.11). We already know that this scheme is unstable (see Exercise 2.10). Let us write down the modified equation of scheme (2.2.11):

$$\frac{\partial u}{\partial t} + a\frac{\partial u}{\partial x} = -\frac{\tau}{2}\frac{\partial^2 u}{\partial t^2} \tag{2.4.1}$$

The idea of Lax and Wendroff was to add to the right-hand side of the difference equation formula which cancels the first-order term in the right hand side of (2.4.1). If we directly approximate the term $(\tau/2)\partial^2/\partial t^2$, we obtain a difference equation which is relatively difficult to solve. A better way is to use the differential consequence (see (2.3.22))

$$\partial^2 u/\partial t^2 = a^2 \partial^2 u/\partial x^2 \tag{2.4.2}$$

Thus let us add the term

$$\frac{\tau}{2}a^2\frac{u_{j+1}^n - 2u_j^n + u_{j-1}^n}{h^2}$$

which approximates the term $(\tau/2)a^2\partial^2 u/\partial x^2$, to the right-hand side of equation (2.2.11). As a result we obtain the Lax-Wendroff scheme for the advection equation (2.1.1):

$$\frac{u_j^{n+1} - u_j^n}{\tau} + a\frac{u_{j+1}^n - u_{j-1}^n}{2h} = \frac{\tau}{2}a^2\frac{u_{j+1}^n - 2u_j^n + u_{j-1}^n}{h^2} \tag{2.4.3}$$

This scheme has the order of approximation $O(h^2) + O(\tau^2)$, thus this is a second-order scheme.

By using the *Mathematica* program advbac.ma from Section 2.2, it is not difficult to write a similar program implementing scheme (2.4.3). For programming, it is convenient to rewrite the difference equation (2.4.3) in the form

$$\frac{u_j^{n+1} - u_j^n}{\tau} + \frac{F_{j+1/2}^n - F_{j-1/2}^n}{h} = 0 \tag{2.4.4}$$

where

$$F_{j+1/2}^n = \frac{au_j^n + au_{j+1}^n}{2} - \frac{a\tau}{2h}(au_{j+1}^n - au_j^n) \tag{2.4.5}$$

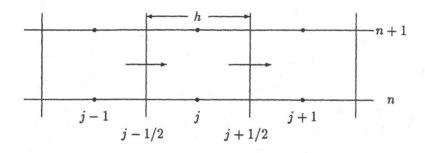

Figure 2.9: Location of (j, n) and $(j \pm 1/2, n)$ points

It is especially important to interpret the difference equation (2.4.4). If we assume that the subscript "j" refers to the geometric center of a cell of the spatial computing mesh, see Fig. 2.9, then u_j^n, u_j^{n+1} are determined at cell centers and the quantities $F_{j-1/2}^n$ and $F_{j+1/2}^n$ in (2.4.4) represent the fluxes across the left cell boundary $j - 1/2$, and across the right cell boundary $j + 1/2$, respectively.

Let us now increase the subscript j in (2.4.4) by +1:

$$\frac{u_{j+1}^{n+1} - u_{j+1}^n}{\tau} + \frac{F_{j+3/2}^n - F_{j+1/2}^n}{h} = 0 \qquad (2.4.6)$$

Comparing Eqs. (2.4.4) and (2.4.6) we can see that the flux $F_{j+1/2}^n$ across the left boundary of the $(j+1)$th cell is the flux across the right boundary of the jth cell. This fact can be used for writing a program in which the fluxes $F_{j+1/2}^n$ are computed for each cell boundary only once. As a result of such programming, the computer time is saved significantly in comparison with the case when the difference equation (2.4.3) is directly programmed.

Note that the difference scheme (2.1.9) can also be written down in the form (2.4.4), if we assume that $F_{j+1/2}^n = au_j^n$, $F_{j-1/2}^n = au_{j-1}^n$ in (2.4.4). However, we have not used this possibility when writing the program **advbac.ma**; that is we have programmed in **advbac.ma** directly the difference equation (2.1.11).

The difference equation (2.4.3) requires the specification of a *numerical* boundary condition in the right boundary cell, that is at $j = M$. We will discuss the questions of the choice of the numerical boundary conditions below in Section 2.11. For the Lax-Wendroff scheme (2.4.3) we chose the "box" scheme (2.11.13).

In order to program the scheme (2.4.3) we have replaced the lines

```
(* -- Do-loop over grid nodes on the x-axis ---------- *)
        Do[
(* -- The first-order scheme (2.1.9)
                        for equation (2.1.1)  *)
up[[j]] = uj[[j]] - cap*(uj[[j]] - uj[[j - 1]]),
        {j, 1, M}];
(*  The boundary condition on the left boundary x = a1 *)
                up[[0]] = ul;
```

in the notebook advbac.ma by the lines

```
(* -- Do-loop over grid nodes on the x-axis ---------- *)
    fxr = 0.5*((uj[[0]] + uj[[1]]) -
                        cap*(uj[[1]] - uj[[0]]));
        Do[ fxl = fxr;
(* - The Lax-Wendroff scheme (2.4.3) for Eq. (2.1.1) - *)
 fxr = 0.5*((uj[[j]] + uj[[j + 1]]) -
                        cap*(uj[[j + 1]] - uj[[j]]));
    up[[j]] = uj[[j]] - cap*(fxr - fxl),
        {j, 1, M2 }];
(*  The boundary condition on the left boundary x = a1 *)
                up[[0]] = ul;
(*  The boundary condition on the right boundary x= b1 *)
up[[M]] = ((uj[[M]] + uj[[M - 1]] - up[[M - 1]]) +
    cap*(uj[[M - 1]] + up[[M - 1]] - uj[[M]]))/(1 + cap);
```

In the above fragment, M2 = M - 1.

Let us consider a problem of the advection of a semi-ellipse pulse. For this problem the function $u_0(x)$ entering the initial condition (2.1.2) is

$$u(x,0) = u_0(x) = \begin{cases} u_l + u_r\sqrt{1 - \left(\frac{x-x_c^0}{15h}\right)^2}, & |x - x_c^0| \leq 15h \\ u_l & \text{otherwise} \end{cases} \quad (2.4.7)$$

Here u_l, u_r and x_c^0 are user-specified constants; we have set $u_l = 0.5$, $u_r = 1.0$, $x_c^0 = 20h$ in the computations whose results are presented below in Fig. 2.10. The semi-ellipse pulse features both sharp corners and a rounded top.

The initial condition (2.4.7) was programmed in the form of the function `uinit`, which has the form

```
uinit[x_, ul_, ur_, jf0_, h_]:=
( xc0 = jf0*h; y1 = (x - xc0)/(15*h);
z = If[Abs[x - xc0] <= 15*h, ul+ur*Sqrt[1-y1*y1], ul]; z )
```

The complete listing of the *Mathematica* package `lw1.ma` implementing the Lax-Wendroff scheme (2.4.3) with the initial condition (2.4.7) is available on the attached diskette.

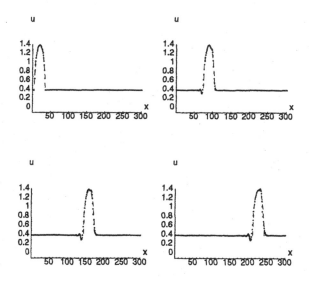

Figure 2.10: $\kappa = 0.8$

In Fig. 2.10 we present the computational results which were obtained by the Lax-Wendroff scheme (2.4.3) at $\kappa = a\tau/h = 0.8$. Note that there are oscillations of the numerical solution behind the left corner of the semi-ellipse pulse. The amplitude of these oscillations increases with time. The profile in the right lower corner of Fig. 2.10 represents the difference solution at $t = 1.37674$. By this time, the pulse has already propagated a distance of about $205h$ along the x–axis.

To perform the above computation, use the file **lw1.ma** available on the attached diskette.

The fine details of the numerical solutions shown in Fig. 2.10 are very small. This makes more difficult the comparison of numerical solutions obtained by different difference schemes. To improve the graphical resolution one can show in the graphs only the subregion of a semi-ellipse pulse. For this purpose we have modified the graphics part of our program **lw1.ma**. We have given the name **lw2.ma** to the modified program; this program is also available on the attached diskette. The graphical output is formed in **lw2.ma** with the aid of the following fragment:

```
(* -- Storing the pictures in the elements mt[[j]],
                                 j = 1,...,Npic -- *)
     umin = If[umin > 0, 0.0, umin];
     umax = If[umax < 0, 0.0, umax]; mt = {};
     Do[
         ug = ut[[j]];
         t  = tnt[[j]];
(* ----- Determining the values of mlpl and mrpl ----- *)
   Do[uk = ug[[k]];
   If[Abs[uk - ul] > 0.0001, mlpl = k; Break[]], {k, M}];
If[j==1, mlpl = 1];   xorg = mlpl; uminpl = Min[0, umin];
               mrpl = Min[mlpl + 80, M1]; xplmax = mrpl;
(* xorg = a1 + (mlpl - 1)*h;   xplmax = a1 + (mrpl - 1)*h;*)
xtab = {}; ytbnum = {}; ytbex = {}; jxmax = mrpl -mlpl + 1;
   Do[j3 = j2 + mlpl - 1; xj = j3;
     (*       xj = a1 + (j3 - 0.5)*h;            *)
     unum = ug[[j3]]; AppendTo[xtab, xj];
                              AppendTo[ytbnum, unum],
   {j2, jxmax}];
     xytab = Table[{xtab[[j]], ytbnum[[j]]}, {j, jxmax}];
     Print["The solution graph at t = ", t];
     mt1 = ListPlot[xytab,
             AxesLabel -> {"x", "u"},
             AxesOrigin -> {xorg, 0},
             PlotRange -> {{xorg, xplmax}, {uminpl, umax}},
             DisplayFunction -> Identity];
(* ----- Plotting the curves of the exact solution ----- *)
             uex = { };
     Do[xj = a1 + (j2 - 0.5)*h;
```

```
      zj = xj - a*t; uj2 = uinit[zj, ul, ur, jf0, h];
                     AppendTo[uex, uj2],
   {j2, M1}];

  Do[j3 = j2 + mlpl - 1;
   uj2 = uex[[j3]]; AppendTo[ytbex, uj2], {j2, jxmax}];
    xytab = Table[{xtab[[j]], ytbex[[j]]}, {j, jxmax}];

   mt2 = ListPlot[xytab,
           PlotJoined -> True,
           AxesOrigin -> {xorg, 0},
           PlotStyle -> {Dashing[{0.04}]},
           PlotRange -> {{xorg, xplmax}, {uminpl, umax}},
           DisplayFunction -> Identity];
  mt3 = Show[mt1, mt2, DisplayFunction -> $DisplayFunction];
   AppendTo[mt, mt3],
                        {j, nt}];
If[nt == 2,
    Show[GraphicsArray[{mt[[1]], mt[[2]]}]] ];
If[nt == 3,
   Show[GraphicsArray[{{mt[[1]], mt[[2]]}, {mt[[3]]}}]] ];
If[nt == 4,
    Show[GraphicsArray[{{mt[[1]], mt[[2]]},
                        {mt[[3]], mt[[4]]}}]] ];
```

In this fragment, we at first determine the numbers mlpl and mrpl of the grid nodes bounding the interval on the x-axis, within which the numerical solution is then plotted (see lines 8-11).

In Fig. 2.11 we present the same numerical results as in Fig. 2.10. It is easy to see that the graphical resolution in the region of the pulse has improved significantly.

In Fig. 2.12 we show the results of the application of the same scheme (2.4.3) for the solution of the same problem (2.1.1), (2.4.7), but with the Courant number $\kappa = 0.6$. It may be seen that the deviation of the numerical solution profiles from the semi-ellipse profile is, in this case, larger than in the case when $\kappa = 0.8$. This is due to the fact that the *dispersion effects*, to be discussed later, of the Lax-Wendroff scheme (2.4.3) become stronger for smaller values of κ.

In Fig. 2.13 we show the results of the numerical solution of problem (2.1.1), (2.4.7) using the first-order scheme (2.1.9). To perform the corresponding computation, use the file upw1.ma available on the attached diskette. It may be seen from Fig. 2.13 that the upwind scheme (2.1.9)

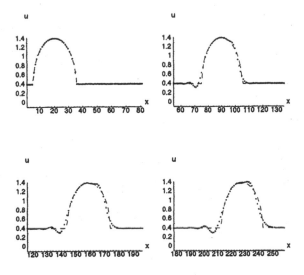

Figure 2.11: $\kappa = 0.8$

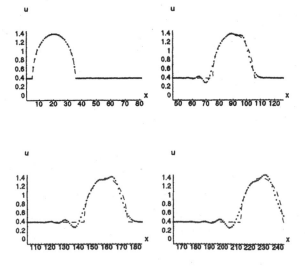

Figure 2.12: The scheme (2.4.3), $\kappa = 0.6$

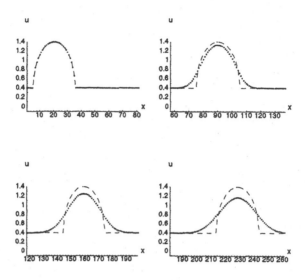

Figure 2.13: The scheme (2.1.9), $\kappa = 0.6$

produces no oscillations near the corners. But the height of the numerically computed pulse decreases as the time t increases. The corners,
which are present in the exact solution profiles, are strongly rounded.
These effects of the numerical solution by scheme (2.1.9) are caused by
the fact that the *dissipation*, or *diffusion* of this scheme increases as the
smaller values of the Courant number κ are taken.

2.4.2   The Method of Noh and Protter

Now consider the construction of the method,[12,1] which was proposed by W.F. Noh and M.H. Protter in 1963. This method may be
considered as a method for obtaining higher-order difference schemes for
hyperbolic PDEs. In accordance with the exact solution (2.2.10), the
value $u_j^{n+1} = u^*$, see Fig. 2.14.

If the Courant number $\kappa = 1$, then $a\tau = h$, and therefore $u^* = u_{j-1}^n$,
and the value u^* is thus determined exactly.

We now consider a more general situation where $\kappa \neq 1$. In Fig. 2.14,
$a\tau < h$, u^* is located between the nodes $(j-1)$ and (j). If we use linear
interpolation between the spatial nodes $(j-1)$ and (j) to estimate u^*,
we obtain that

$$u^* = u_j^n - (u_j^n - u_{j-1}^n)a\tau/h$$

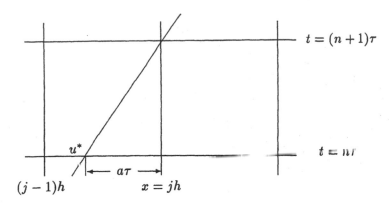

Figure 2.14: Translation of the value u^* along the characteristic $dx/dt = a$ into the point $(j, n+1)$

Setting $u_j^{n+1} = u^*$ then gives the familiar upwind difference scheme (2.1.9). If we use a quadratic polynomial fit to interpolate over the intervals $(j-1), (j)$, and $(j+1)$, we obtain the Leith's scheme

$$u_j^{n+1} = u_j^n - \frac{1}{2}\left(\frac{a\tau}{h}\right)(u_{j+1}^n - u_{j-1}^n) + \frac{1}{2}\left(\frac{a\tau}{h}\right)^2(u_{j+1}^n - 2u_j^n + u_{j-1}^n) \quad (2.4.8)$$

It is easy to see that in the case when the Leith's method is applied to the one-dimensional advection equation (2.1.1) this method coincides with the Lax-Wendroff method. The stability condition of (2.4.3) and (2.4.8) is $|\kappa| \leq 1$. This restriction can now be interpreted as a statement that u^* must be determined by *interpolation* rather than *extrapolation*.

2.4.3 Richtmyer's Scheme

Now consider the two-step version of the Lax-Wendroff scheme, which was proposed by R.Richtmyer in 1963.[13,5] This scheme is much simpler than the original Lax-Wendroff scheme, especially in multidimensional problems. For the scalar equation (2.1.1) the method is as follows.

$$u_j^{n+1} = \frac{1}{2}(u_{j+1}^n + u_{j-1}^n) - \tau a\left(\frac{u_{j+1}^n - u_{j-1}^n}{2h}\right) \quad (2.4.9)$$

$$u_j^{n+2} = u_j^n - 2\tau a\left(\frac{u_{j+1}^{n+1} - u_{j-1}^{n+1}}{2h}\right) \qquad (2.4.10)$$

The values $u_{j\pm1}^{n+1}$ used in the second step (2.4.10) are based on the $u_{j\pm1}^{n+1}$ computation in the first step (2.4.9). The first step is to be considered a provisional step, with significance attached only to the result of the second step in each sequence. This method does not look anything like the original Lax-Wendroff scheme (2.4.3), but substitution of (2.4.9) into (2.4.10) shows that the difference equation coincides with the original Lax-Wendroff scheme, if we assume that the scheme (2.4.3) is written down on a computing mesh with the steps $2h$ and 2τ.

The two-step Lax-Wendroff scheme was presented as follows:[14]

$$\tilde{u}_{j+1/2} = \frac{1}{2}(u_j^n + u_{j+1}^n) - \tau a\left(\frac{u_{j+1}^n - u_j^n}{h}\right) \qquad (2.4.11)$$

$$u_j^{n+1} = u_j^n - \tau a\left(\frac{\tilde{u}_{j+1/2} - \tilde{u}_{j-1/2}}{h}\right) \qquad (2.4.12)$$

It is easy to show that the substitution of the provisional values $\tilde{u}_{j\pm1/2}$ from (2.4.11) into (2.4.12) yields exactly the Lax-Wendroff difference equation (2.4.3).

Let us assume as in the case of the Lax-Wendroff scheme (2.4.3) that the subscript "j" by the grid values u_j^n, u_j^{n+1} refers to the geometric center of a cell of the spatial computing mesh, see Fig. 2.9. Then the equation (2.4.11) may be interpreted as the equation for the computation of the provisional values of the solution values on the left boundary $(j - 1/2)$ and on the right boundary $(j + 1/2)$ of the jth cell. And the difference equation (2.4.12) enables one to compute the new value u_j^{n+1}, which generally differs from the value u_j^n because of the difference of the fluxes $a\tilde{u}_{j+1/2}$ and $a\tilde{u}_{j-1/2}$ across the boundaries $x = (j \pm 1/2)h$ of the jth cell. In the version (2.4.11), (2.4.12) is characterized as a more common form of the use of the scheme (2.4.9), (2.4.10).[15]

2.4.4   MacCormack Scheme

Now let us present the MacCormack difference method.[16] The basic form of this two-step method as applied for the numerical solution of the advection equation has the form

$$\tilde{u}_j = u_j^n - \tau a\left(\frac{u_{j+1}^n - u_j^n}{h}\right) \qquad (2.4.13)$$

$$u_j^{n+1} = \frac{1}{2}\left[u_j^n + \tilde{u}_j - \tau a\left(\frac{\tilde{u}_j - \tilde{u}_{j-1}}{h}\right)\right] \qquad (2.4.14)$$

As we can see from these formulas, the provisional values $\tilde{u}_j$ are computed on the basis of forward spatial differences, whereas the final values u_j^{n+1} are computed with the use of backward spatial differences. If we substitute the values $\tilde{u}_j$ and $\tilde{u}_{j-1}$ from the first step (2.4.13) into the equation (2.4.14), we easily obtain the Lax-Wendroff difference equation (2.4.8). But in cases when the MacCormack method is applied for the numerical integration of nonlinear PDEs it does already not coincide with the Lax-Wendroff method.

It is possible to interchange the spatial differencing operators in (2.4.13) and (2.4.14), that is to consider instead of (2.4.13), (2.4.14) the MacCormack scheme of the form

$$\tilde{u}_j - \tau a \left(\frac{u_j^n - u_{j-1}^n}{h} \right)$$

$$u_j^{n+1} = \frac{1}{2} \left[u_j^n + \tilde{u}_j^n - \tau a \left(\frac{\tilde{u}_{j+1} - \tilde{u}_j}{h} \right) \right] \tag{2.4.15}$$

In the case of the linear equation (2.1.1), the scheme (2.4.15) also reverts, after the elimination of the provisional values, to the Lax-Wendroff scheme (2.4.8).

In applied nonlinear problems involving the propagation of shock waves, the MacCormack schemes produce better results than the Lax-Wendroff scheme.

Exercise 2.12. Derive the Lax-Wendroff scheme approximating Eq. (2.3.1).

Hint. At first show that the analogue of the differential consequence (2.4.2) has in the case of Eq. (2.3.1) the form (see also Reference 3)

$$u_{tt} = a^2 u_{xx} - a\varphi_x + \varphi_t$$

Exercise 2.13. Derive formula (2.4.8) with the aid of the parabolic interpolation and *Mathematica*.

Hint. At first find the coefficients of the parabolic interpolation function $f(x)$ of the form

$$f(x) = \alpha + \beta(x - jh) + \gamma(x - jh)^2 \tag{2.4.16}$$

where the coefficients α, β, and γ are to be determined from the requirements

$$f((j-1)h) = u_{j-1}^n, \; f(jh) = u_j^n, \; f((j+1)h) = u_{j+1}^n \tag{2.4.17}$$

The conditions (2.4.17) lead to the following algebraic system:

$$\begin{aligned} \alpha - \beta h + \gamma h^2 &= u_{j-1}^n \\ \alpha &= u_j^n \\ \alpha + \beta h + \gamma h^2 &= u_{j+1}^n \end{aligned} \tag{2.4.18}$$

Solve the system (2.4.18) in the symbolic form with using *Mathematica* and then substitute the expressions for α, β, and γ into (2.4.16). Find u_j^{n+1} by substituting the value $x = jh - a\tau$ into (2.4.16):

$$u_j^{n+1} = f(jh - a\tau)$$

Exercise 2.14. Show the equivalence of the two-step Lax-Wendroff scheme (2.4.9), (2.4.10) to the one-step Lax-Wendroff scheme (2.4.3), in which the steps h and τ are replaced formally by 2τ and $2h$.

Hint. Use *Mathematica*. For this purpose

(a) write the equation (2.4.9) in *Mathematica*;

(b) shift the subscript j by $+1$ and -1 to obtain the expressions for u_{j+1}^{n+1} and u_{j-1}^{n+1};

(c) substitute the computed expressions for u_{j+1}^{n+1} and u_{j-1}^{n+1} into equation (2.4.10);

(d) solve (2.4.3) for u_j^{n+1};

(e) replace τ and h in the equation obtained from (2.4.3) by 2τ and $2h$;

(f) compute the difference $(u_j^{n+2})_1 - (u_j^{n+2})_2$, where $(u_j^{n+2})_1$ is the expression obtained at the stage (e), and $(u_j^{n+2})_2$ is the expression obtained at the stage (c). The result should be zero.

Exercise 2.15. Show the equivalence of the MacCormack scheme (2.4.13), (2.4.14) to the Lax-Wendroff scheme (2.4.3).

Exercise 2.16. Study the stability of the one-step Lax-Wendroff scheme with the aid of the Fourier method.

Hint. Use the program st1.ma from Section 2.3.

2.5   ALGORITHM FOR AUTOMATIC DETERMINATION OF APPROXIMATION ORDER OF SCALAR DIFFERENCE SCHEMES

We now present a symbolic algorithm for the determination of the approximation order of scalar difference schemes approximating scalar PDEs of the form

$$\frac{\partial u}{\partial t} = P\left(u, \frac{\partial u}{\partial x}, \frac{\partial^2 u}{\partial x^2}, \ldots, \frac{\partial^{m_x} u}{\partial x^{m_x}}\right) \tag{2.5.1}$$

where $P(\cdot)$ is a polynomial function of its arguments, and $m_x \geq 1$. The mesh on the x-axis is assumed to be uniform. The basic steps of the algorithm are as follows.

Step 1. The data are input into *Mathematica*:

a) the difference equation under consideration;
b) the original partial differential equation;
c) the order $kt \geq m_x$ of the Taylor expansions to be used;
d) the center (xc, tc) of the Taylor expansions.

Step 2. All the grid functions u_k^m in the difference equation are expanded at point (j, n) into a truncated Taylor series up to the above specified order kt in accordance with the formula

$$u_k^m = v(x,t) + \sum_{\nu=1}^{kt} \sum_{i_1+i_2=\nu} \frac{1}{i_1! i_2!} \frac{\partial^{i_1+i_2} v}{\partial x^{i_1} \partial t^{i_2}} (kh - x_c)^{i_1} (m\tau - t_c)^{i_2} + R$$

(2.5.2)

where R is the remainder term of the Taylor formula. To distinguish between the difference solution u_j^n and the solution of the first differential approximation, or the modified equation, we have introduced the notation $v(x,t)$ for the solution of the modified equation.

Step 3. Determine m_x in (2.5.1).

Step 4. The differential consequences of the original PDE (2.5.1) are determined up to the order kt. For the case of the PDE (2.1.1) and $kt = 2$ the differential consequences are given by the formulas (cf. (2.3.22))

$$\frac{\partial^2 v}{\partial x \partial t} = -a \frac{\partial^2 v}{\partial x^2}, \quad \frac{\partial^2 v}{\partial t^2} = a^2 \frac{\partial^2 v}{\partial x^2}$$

The differential consequences may be computed symbolically with the aid of a recursive procedure. Let us assume, for example, that we have already computed the differential consequence

$$\frac{\partial^2 v}{\partial x \partial t} = P_{1,1}\left(v, \frac{\partial v}{\partial x}, \ldots, \frac{\partial^{m_x+1} v}{\partial x^{m_x+1}}\right)$$

where

$$P_{1,1}(\cdot) = \frac{\partial}{\partial x} P\left(v, \frac{\partial v}{\partial x}, \ldots, \frac{\partial^{m_x} v}{\partial x^{m_x}}\right)$$

and the polynomial $P(\cdot)$ enters the definition (2.5.1) of the original PDE. Then we have that

$$\frac{\partial^3 v}{\partial x \partial t^2} = \frac{\partial}{\partial t} P_{1,1}(\cdot) = \tilde{P}_{1,2}\left(v, \frac{\partial v}{\partial t}, \frac{\partial^2 v}{\partial x \partial t}, \ldots, \frac{\partial^{m_x+2} v}{\partial x^{m_x+1} \partial t}\right)$$

We easily find from here the differential consequence

$$\frac{\partial^3 v}{\partial x \partial t^2} = P_{1,2}\left(v, \frac{\partial v}{\partial x}, \frac{\partial^2 v}{\partial x^2}, \ldots, \frac{\partial^{2m_x+1} v}{\partial x^{2m_x+1}}\right)$$

where

$$P_{1,2}(\cdot) = \tilde{P}_{1,2}\left[v, P(\cdot\cdot), \frac{\partial}{\partial x}P(\cdot\cdot), \dots, \frac{\partial^{m_x+1}}{\partial x^{m_x+1}}P(\cdot\cdot)\right]$$

and $(\cdot\cdot) = (v, \partial v/\partial x, \dots, \partial^{m_x}v/\partial x^{m_x})$.

Step 5. The expressions for the values u_k^m from (2.5.2) without the remainder term are substituted into the difference equation under consideration. Let us denote the equation appearing as the result of this substitution by

$$H\left(v, \frac{\partial v}{\partial t}, \frac{\partial v}{\partial x}, \frac{\partial^2 v}{\partial x \partial t}, \dots, \frac{\partial^{kt}v}{\partial x^{kt-1}\partial t}, \dots, \frac{\partial^{kt}v}{\partial t^{kt}}\right) = 0 \qquad (2.5.3)$$

Step 6. Compute symbolically the difference

$$R_1 v = H\left(v, \frac{\partial v}{\partial t}, \frac{\partial v}{\partial x}, \frac{\partial^2 v}{\partial x \partial t}, \dots, \frac{\partial^{kt}v}{\partial t^{kt}}\right) - \left[\frac{\partial v}{\partial t} - \right.$$
$$\left. P\left(v, \frac{\partial v}{\partial x}, \frac{\partial^2 v}{\partial x^2}, \dots, \frac{\partial^{m_x}v}{\partial x^{m_x}}\right)\right]$$

so that

$$\frac{\partial v}{\partial t} = P(\cdot) - R_1 v$$

Step 7. The derivatives $\partial^k v/\partial x^{k-m}\partial t^m$, where $k \geq 2$ and $m = 1, \dots, kt$, are replaced in the expression for $R_1 v$ by the differential consequences $P_{k-m,m}$. As a result of this we obtain a modified equation of the form

$$\frac{\partial v}{\partial t} - P(\cdot) + R_1 v \equiv P_1\left(v, \frac{\partial v}{\partial t}, \frac{\partial v}{\partial x}, \frac{\partial^2 v}{\partial x^2}, \dots, \frac{\partial^{kt\cdot m_x}v}{\partial x^{kt\cdot m_x}}\right) = 0 \quad (2.5.4)$$

Step 8. Find the order of the remainder term $R_1 v$ with respect to the grid parameters h and τ.

We have implemented steps 1–8 in the program **appr1.ma** whose complete listing is presented in Appendix 2 (see also the attached diskette).

The derivative $\partial v(x, t)/\partial x$ is denoted in the *Mathematica* system by the command D[v[x, t], x]. So the left hand side of equation (2.1.1) may be programmed as the function

```
eqn[x_, t_] := D[v[x, t], t] + a D[v[x, t], x]
```

The Taylor expansions, in accordance with formula (2.5.2), can be conveniently programmed with the aid of the function u[k_, m_]:

```
(* --- The expansion of the difference solution into the
       Taylor series with respect to (xc, tc) point --- *)

   u[k_, m_]:= Block[{x, t, sum},
                                     sum = v[x, t];
Do[
           Do[ i2=ktj - i1;
    dr1 = If[i1==0, 1, (k*h - xc)^i1];
    dr2 = If[i2==0, 1, (m*dt - tc)^i2];
    dr = dr1*dr2/(i1!*i2!);
   sum = sum + Derivative[i1, i2][v][x, t]*dr,
             {i1, 0, ktj}],
      {ktj, kt}];
   sum ]
```

The built-in *Mathematica* function m! enables us to compute the factorials i1! and i2!.

The higher-order derivatives of functions with several arguments can be represented in *Mathematica* as:

$$\texttt{Derivative[}n_1,\ n_2,\ldots\texttt{][v][}x_1,\ x_2,\ldots\texttt{]}.$$

For example, the partial derivative $\partial^3 v(x,t)/\partial x^2 \partial t$ is represented as

$$\texttt{Derivative[2, 1][v][x, t]}.$$

Derivative[2, 1][v][x, t] is the *Mathematica* differentiation functional. The arguments in the functional Derivative[n_1, n_2,..] specify how many times to differentiate with respect to each "slot" of the function on which it acts. By using functionals to represent differentiation, *Mathematica* avoids any need to introduce explicit "dummy variables".

The determination of m_x in (2.5.1) at Step 3 is performed with the aid of the built-in *Mathematica* function Exponent[*expr, form*]. It gives the maximum power with which *form* appears in *expr*. As soon as the computed exponent is different from zero, when taking *form* as Derivative[k, 0][v][x, t], we assume m_x =k, k= 20, 19, 18, ..., 1. The value of m_x is determined with the aid of the following fragment of the program appr1.ma:

```
(* --- Determining mx, the highest order of the x-deriv-
```

```
     ative in the original partial differential equation  *)
   Do[k1 = 20- k + 1;
      mk = Exponent[pv, Derivative[k, 0][v][x, t]];
   If[mk==0, Goto[endloop], mx = k; Break[ ]];
   Label[endloop],
   {k, 20}];
   Print["The highest order of x-derivative in PDE is ", mx];
```

In our program **appr1.ma** the differential consequences $P_{\alpha,\beta}$ are stored at step 4 in the one-dimensional table **dcs**. The integer indices α and β are stored in the two-dimensional table **indcs**, so that the number **k** of the differential consequence $P_{\alpha,\beta}$ in the table **dcs** is **k=indcs[** $\alpha+1,\beta$**]**, and $P_{\alpha,\beta}$ =**dcs[[k]]**. Since the amount of computer memory needed for storage of the $P_{\alpha,\beta}$ is not known in advance, it is more convenient to fill in the table **dcs** step by step, as the next $P_{\alpha,\beta}$ is computed. For this purpose we at first initiate the list **dcs** as an empty list with the aid of the command **dcs = { }** and then use the *Mathematica* function **AppendTo[**v, elem**]**, which appends an element elem to the list v; see the program fragment

```
(* - Initialization of the table dcs for the differential
        consequences of the differential equation ------ *)
   dcs = {}; ktp = kt + 1;
   indcs = Table[99, {i, 1, ktp}, {k, 1, kt}];

(*  Symbolic computation of dcs[[i,j]] for 2<=i+j<=kt  *)
ind = 1; indcs[[1, 1]] = ind; AppendTo[dcs, pv];
ind = 2; indcs[[2, 1]] = ind;
sc2 = D[pv, x]; AppendTo[dcs, sc2];
        Print[dcs];
Do[
   Do[ nxj = ktj - ntj;
                           If[nxj==1&&ntj==1, Goto[endloop]];
If[ntj==1, Goto[ntj1]];
   k1 = mx*ntj;
   k2 = indcs[[nxj+1,ntj-1]]; vtt = D[dcs[[k2]], t];
        Do[
vtt = vtt /. Derivative[k, 1][v][x, t] ->
                D[dcs[[1]], {x, k}],
      {k, k1}];
Goto[ready];
```

```
Label[ntj1]; k= indcs[[nxj, ntj]]; vtt = D[dcs[[k]], x];
Label[ready]; ind = ind + 1; indcs[[nxj+1, ntj]] = ind;
                  AppendTo[dcs, vtt];
OutputForm[nxj] >>> ap.m;
OutputForm[ntj] >>> ap.m;
OutputForm[vtt] >>> ap.m;
Label[endloop],
                    {ntj, ktj}],
     {ktj, 2, kt}];
```

The symbolic expression for the leading term $R_1 v$ of the truncation error is computed at step 6 with the aid of the following fragment of the program appr1.ma:

```
(*  Substitution of the differential consequences into the
     modified equation ------------------------------- *)
sc1 = sch[j, n];
        sc1 = sc1 /. centexp;
                              sc1 = Simplify[sc1];
OutputForm[sc1] >>> ap.m;
     Print[sc1]; sc2 = Expand[sc1];
     r1v = sc2 - eq1;
  Do[
    Do[ nxj = ktj - ntj; k = indcs[[nxj+1, ntj]];

  r1v = r1v /. Derivative[nxj, ntj][v][x, t] -> dcs[[k]],
                  {ntj, ktj}],
     {ktj, kt}];
  r1v = r1v /. Derivative[0, 1][v][x, t] -> pv;
    r1v = Expand[r1v]; r1v = Simplify[r1v];
    sc2 = eq1 + r1v;
                    Print[sc2];
OutputForm[sc2] >>> ap.m;
                    r1v = Expand[r1v];
  Print[r1v];
  If[r1v===0,
  Print["The modified equation of scheme coincides with ",
       "the original PDE at kt = ",kt, "."];
  k1 = kt +1;
  Print["Please take the value kt = ",k1," and repeat the
                              computation"];
```

```
Print["with this new value of kt."]; Interrupt[ ]];
```

To determine the order of the remainder term R_1v at Step 8 with respect to the grid parameters h and τ, we make use of the function `Coefficient[Rv, dt, k]`. The built-in *Mathematica* function

$$\text{Coefficient}[poly, \, expr, \, n]$$

finds the coefficient of $expr\char`^n$ in *poly*. After that our program `appr1.ma` prints out the message `The approximation order of scheme is` $O(h^{k_1})+ O(\tau^{k_2})$, where k_1 and k_2 are the integers found as the result of the execution of Step 8:

```
(* -- Determination of the local approximation order
        of the difference scheme --------------------- *)
k1 = mx*kt; ntx = 0;

Do[ ak = Coefficient[r1v, h, k];
 If[ak===0, Goto[nextk]]; nxord = k;
    ntx= Exponent[ak, dt]; If[ntx > 0, Goto[nextk]];
      Break[ ];
Label[nextk],
     {k, k1}];

 ntx = 0;
Do[ ak = Coefficient[r1v, dt, k];
 If[ak===0, Goto[nextk]]; ntord = k; ntx= Exponent[ak,h];
If[ntx > 0, Goto[nextk]]; Break[ ];
Label[nextk],
     {k, k1}];

Print["The approximation order of scheme is O(h^",
      nxord, ") + O(dt^", ntord, ")"];
napmax = Max[nxord, ntord];

    If[napmax > 1,
  Do[ak = Coefficient[r1v, dt, k];
   If[ak=!=0, ntx1 = k; ntx2 = 0;
   Do[bj = Coefficient[ak, h, j];
     If[bj=!=0, ntx2 = j; Break[ ]],
     {j, napmax}]];
```

```
If[ntx2!=0&&ntx1 + ntx2 == napmax,

Print["+ O(dt^", ntx1, "*h^", ntx2, ")"]],
{k, napmax}]];
```

To perform the computation discribed in steps 1–8 given above, by the program appr1.ma, one must at first specify the positive integer value of kt (see line 1 of the above listing). Then one must input the difference scheme as the body of the function sch[j_, n_] and the original differential equation as the body of the function eqn[x_,t_]. Then one can perform a computation by the program appr1.ma, see the internal instruction to the notebook appr1.ma. The results of symbolic computations are stored in the file ap.m. These are:

$$\alpha \quad \beta \quad P_{\alpha,\beta} \quad H(\cdot) \quad P_1(\cdot\cdot)$$

Example 1. Let us consider the difference scheme (2.1.9) approximating the PDE (2.1.1). In this case the left-hand side $H(\cdot)$ of equation (2.5.3) was computed symbolically by the program appr1.ma with kt = 2 as

```
                        (0,2)
  (0,1)           dt v     [x, t]        (1,0)
v       [x, t] + ---------------- + a v       [x, t] -
                        2


        (2,0)
  a h v     [x, t]
  -----------------
         2
```

The expression for $P_1(\cdot\cdot)$ (see equation (2.4.7)) was found to be

```
                                      2      (2,0)
  (0,1)                  (1,0)      a  dt v     [x, t]
v       [x, t] + a v       [x, t] + ------------------- -
                                            2
```

```
        (2,0)
   a h v     [x, t]
   ----------------
          2
```

We can easily see that the terms of the order $O(h)$ and $O(\tau)$ have the same form as in the modified equation (2.3.23). And finally the following message was printed out:

```
    The approximation order of scheme is O(h^1) + O(dt^1)
```

Example 2. Let us take the difference scheme (2.2.11). Our program appr1.m has printed out in this case the message

```
    The approximation order of scheme is O(h^2) + O(dt^1)
```

Example 3. We finally take the difference scheme (2.4.3). Let us at first perform a symbolic computation with the value $kt = 2$ in (2.5.2). In this case our program appr1.m prints out the messages

```
The modified equation of scheme coincides with the original
PDE at kt = 2. Please take the value kt = 3 and repeat the
computation with this new value of kt.
```

The reason for this message is that the second-order derivatives are cancelled in the modified equation corresponding to the Lax-Wendroff scheme for equation (2.1.1). The expansion order kt = 2 is thus too small for the difference scheme under consideration. In this connection, the computation is interrrupted with the aid of the *Mathematica* command Interrupt[ ]. Let us now input the value kt = 3 and perform the symbolic computation with this new value of kt. As a result the program appr1.ma has produced the following expression for $P_1(\cdot\cdot)$, see (2.5.4):

```
                                 3    2  (3,0)
   (0,1)               (1,0)    a  dt  v      [x, t]
 v      [x, t] + a v      [x, t] - ------------------- +
                                          6

      2  (3,0)
   a h  v      [x, t]
   ------------------
           6
```

From this it follows that the modified equation corresponding to the Lax-Wendroff scheme (2.4.3) has the form

$$\frac{\partial v}{\partial t} + a\frac{\partial v}{\partial x} = \frac{1}{6}(a^3\tau^2 - ah^2)\frac{\partial^3 v}{\partial x^3}$$

so that the scheme (2.4.3) is a second-order approximation both in space and time.

2.6 MONOTONICITY PROPERTY OF DIFFERENCE SCHEMES

We now consider the following important property of the solution of the initial-value problem (2.1.1), (2.1.2): if the function $u_0(x)$ is a nondecreasing (nonincreasing) function, then the solution $u(x,t)$ at any fixed $t \geq 0$ will be a nondecreasing (nonincreasing) function of x. This property follows from the exact solution (2.2.10). Therefore, it is reasonable to require that the difference solution also possesses a similar property.

Let us write a finite difference scheme approximating Eq. (2.1.1) in the form

$$u^{n+1} = Su^n \qquad (2.6.1)$$

where S is called the step operator of the scheme.

Definition 2.2. The grid function u^n is called *monotone increasing* if $u_j^n \leq u_{j+1}^n \; \forall j$, and *monotone decreasing* if $u_j^n \geq u_{j+1}^n \; \forall j$.

Definition 2.3. The finite difference scheme (2.6.1) is called *monotone*, if the step operator S translates a monotone grid function u^n into a monotone grid function u^{n+1} with the same direction of increase or decrease of its values.

Now consider a linear difference scheme

$$u_j^{n+1} = \sum_{k=-p}^{q} \alpha_k u_{j+k}^n \qquad (2.6.2)$$

where $p \geq 0$, $q \geq 0$, $p + q > 0$.

Theorem 2.2 (S.K.Godunov[17,18]). The difference scheme (2.6.2) is monotone if and only if all the coefficients α_k are nonnegative.

Proof. Let $\alpha_k \geq 0$ and the sequence $\{u_j\}$ is monotone. Let us assume for definiteness that $\{u_j\}$ is an increasing sequence, that is $u_j - u_{j-1} \geq 0 \; \forall \; j$. Then

$$u_j^n - u_{j-1}^n = \sum_k \alpha_k u_{j+k}^n - \sum_k \alpha_k u_{j+k-1}^n =$$

$$\sum_k \alpha_k(u_{j+k}^n - u_{j+k-1}^n) \geq 0$$

Thus we have proved the sufficiency of the condition $\alpha_k \geq 0 \ \forall \ k$.

Let us now prove the necessity of this condition. Assume that $\alpha_{k_0} < 0$, and set

$$u_j^n = \left\{ \begin{array}{ll} 1, & j \geq k_0 \\ 0, & j \leq k_0 - 1 \end{array} \right.$$

Then $u_j^n - u_{j-1}^n \geq 0 \ \forall j$. On the other hand, we obtain from (2.6.2) at $j = k_0$ and at $j = k_0 - 1$ that $u_{k_0}^{n+1} - u_{k_0-1}^{n+1} = \alpha_0 < 0$, which contradicts our assumption that the scheme translates the monotone sequence $\{u^n\}$ into a monotone sequence $\{u^{n+1}\}$ with the same growth direction. The proof of the theorem is completed.

Thus the criterion for the monotonicity of the linear difference scheme (2.6.2) has the form

$$\alpha_k \geq 0, \quad k = -p, -p+1, \ldots, q-1, q \tag{2.6.3}$$

We also require of the scheme (2.6.2) that it reproduces the exact result in the case when $u^n = const$:

$$\sum_{k=-p}^{q} \alpha_k = 1 \tag{2.6.4}$$

Example 1. Consider the upwind scheme (2.1.11). In this case $p = 1$, $q = 0$ in (2.6.2), and

$$\alpha_{-1} = \kappa, \quad \alpha_0 = 1 - \kappa \tag{2.6.5}$$

The monotonicity conditions (2.6.3) in this case lead to the inequalities $0 \leq \kappa \leq 1$, which coincide with the stability conditions of scheme (2.1.11).

Example 2. In the case of the Lax-Wendroff scheme (2.4.3) we have that in (2.6.2)

$$p = q = 1, \quad \alpha_{-1} = \frac{\kappa}{2}(1+\kappa), \quad \alpha_0 = 1 - \kappa^2, \quad \alpha_1 = \frac{\kappa}{2}(\kappa - 1)$$

We can see from here that $\alpha_{-1} > 0$, $\alpha_0 > 0$ if $0 < \kappa < 1$. But $\alpha_1 < 0$ if $0 < \kappa < 1$. This means that the Lax-Wendroff scheme is a nonmonotone difference scheme.

It has been shown[17] that, among the linear second-order difference schemes for equation (2.1.1), there is no scheme satisfying the monotonicity condition. Thus the problem of the construction of a monotone linear higher-order difference scheme on a fixed stencil unfortunately has no solution.

The definition of the monotonicity can be extended to the case of nonlinear difference schemes in the following way. Let us consider the difference scheme

$$u_j^{n+1} = H(u_{j-p}^n, u_{j-p+1}^n, \ldots, u_{j+q}^n) \tag{2.6.6}$$

Definition 2.4 (A. Harten et al.[19]). The finite difference scheme (2.6.6) is said to be *monotone* if H is a monotone increasing function of each of its arguments.

Let us consider the difference

$$\begin{aligned} u_j^{n+1} - u_{j-1}^{n+1} &= H(u_{j-p}^n, u_{j-p+1}^n, \ldots, u_{j+q}^n) - \\ & H(u_{j-p-1}^n, u_{j-p}^n, \ldots, u_{j+q-1}^n) = \\ \sum_{k=-p}^{q} & \frac{\partial H(X_{-p}, \ldots, X_q)}{\partial u_{j+k}^n}(u_{j+k}^n - u_{j+k-1}^n) \end{aligned} \tag{2.6.7}$$

where $X_k = u_{j+k-1}^n + \Theta_k(u_{j+k}^n - u_{j+k-1}^n)$, $0 < \Theta_k < 1$, $k = -p, \ldots, q$ in accordance with the mean value theorem (the Lagrange theorem). It follows from (2.6.7) that when the inequalities

$$\frac{\partial H(u_{j-p}, u_{j-p+1}, \ldots, u_{j+q})}{\partial u_{j+k}} \geq 0, \; k = -p, \ldots, q \tag{2.6.8}$$

are satisfied for any values of $u_{j-p}, \ldots, u_{j+q}$, then the difference scheme is monotone. It is easy to see that in the particular case of the linear scheme (2.6.2) the inequalities (2.6.8) coincide with the inequalities (2.6.3).

Note that the analogue of the condition (2.6.4) has, in the case of scheme (2.6.6), the form

$$H(u, u, \ldots, u) = u$$

Exercise 2.17. Obtain the monotonicity condition for the upwind scheme from Exercise 2.8.

2.7 TVD SCHEMES

2.7.1 The Properties of TVD Schemes

In the foregoing sections we have considered a number of elementary difference schemes for the advection equation (2.1.1). They are either too diffusive or too dispersive, as can be seen by computational examples presented above in Figs. 2.12, 2.13. The higher-order schemes generate the spurious, or parasitic oscillations (see Figs. 2.11, 2.12).

For the reduction of the amplitude of parasitic oscillations of numerical solutions some averaging operators as well as the "artificial viscosity terms" were introduced previously into the higher-order difference schemes. However, the problem of the averaging or smoothing of functions has no unique solution. Therefore, the above mentioned methods for the regularization of numerical solutions require the choice of certain empirical constants which ensure the simultaneous suppression of the oscillations and the minimal dissipation of the difference solution.

However, it should be noted that the higher-order difference schemes can preserve the monotonicity of a *specific* solution. This observation has led the researchers to the idea of constructing schemes having a variable stencil on different solution intervals to ensure both a high order approximation and the preservation of the monotonicity. Based on this idea, many efficient finite difference schemes have been developed: the hybrid schemes, the FCT ("Flux-Corrected Transport") schemes, the TVD ("Total Variation Diminishing") schemes, the ENO ("Essentially Non-Oscillatory") schemes, and a number of other schemes. A detailed bibliography of the relevant works may be found in.[15,20] In this Section we restrict ourselves to the description of the construction of TVD schemes, which have now gained a widespread acceptance.

The theory of the TVD schemes was proposed by A. Harten.[21] We consider this theory for the example of equation (2.1.1). We first note the following property of the initial-value problem (2.1.1), (2.1.2). Take a smooth function $u_0(x)$ with a bounded *total variation*:

$$TV[u_0(x)] = \int_{-\infty}^{\infty} |\partial u_0(x)/\partial x| dx \leq C < \infty$$

Then it is easy to see from the exact solution for (2.2.10), that at any moment of time $t = t^* > 0$

$$TV[u(x, t^*)] = TV[u_0(x)]$$

Let us construct an explicit, generally nonlinear, difference scheme approximating Eq. (2.1.1) on an uniform mesh:

$$u_j^{n+1} = u_j^n - H(u_{j-p}^n, \ldots, u_j^n, \ldots, u_{j+q}^n) \qquad (2.7.1)$$

where $p \geq 0$, $q \geq 0$, $p + q > 0$. We require that the difference operator H in the scheme (2.7.1) satisfies the condition

$$H(u, \ldots, u, \ldots, u) = 0 \qquad (2.7.2)$$

that is the difference approximation of a derivative of a constant function is equal to zero.

Again write Eq. (2.7.1) in the form

$$u^{n+1} = Su^n \tag{2.7.3}$$

where S is the step operator. We require for scheme (2.7.3) that the total variation of the numerical solution u^n does, at least, not increase when passing from the nth time level to the $(n+1)$th time level.

We will call the scheme (2.7.1) a *TVD scheme* (the scheme which satisfies the condition that the total variation of the numerical solution does not increase)[21], if

$$TV^h(u^{n+1}) = TV^h(Su^n) \leq TV^h(u^n)$$

where

$$TV^h(u^n) = \sum_{j=-\infty}^{\infty} |\Delta_{j+1/2} u^n|$$

$$\Delta_{j+1/2} u^n = u^n_{j+1} - u^n_j$$

The following theorem establishes the relation between the TVD schemes and the monotonous schemes[21]:

Theorem 2.3.
1. A TVD scheme preserves the monotonicity.
2. A monotone scheme satisfies the TVD condition.

Proof.
1. Note that for the computation of the value of u^{n+1}_j by the explicit scheme (2.7.1) we need to know the values of the grid function at the nth time level only in the $(p+q+1)$ nodes, therefore, it is sufficient to prove the theorem for a monotone grid function with the bounded total variation of the following form:

$$u^n_j = \begin{cases} u_l = const, & j \leq j^- \\ monotone, & J^- \leq j \leq J^+ \\ u_r = const, & j > J^+ \end{cases}$$

where $J^- < J^+$, and $J^+ - J^- > p+q+1$. It is obvious that $TV(u^n) = |u_l - u_r|$. Assume that u^{n+1} is non-monotone. Then there will appear in u^{n+1} at least one local minimum $u^{n+1}_{min} = Min$ and one local maximum $u^{n+1}_{max} = Max$, consequently,

$$TV^h(u^{n+1}) \geq |u_l - u_r| + |Max - Min| > TV^h(u^n)$$

but the scheme (2.7.1) will then not satisfy the TVD condition.
2. Assuming that the difference equation (2.7.1) is linear, let us rewrite it in the form (2.6.2). Since the difference scheme (2.6.2) is

assumed to be monotone, all the coefficients α_k in (2.6.2) should be nonnegative. Then we obtain from (2.6.2) that

$$\Delta_{j+1/2}u^{n+1} = \alpha_{-p}\Delta_{j-p+1/2}u^n + \cdots + \alpha_0\Delta_{j+1/2}u^n + \cdots + \alpha_q\Delta_{j+q+1/2}u^n$$

where all the coefficients α_i are nonnegative, and their sum is equal to unity. Therefore, the inequality

$$|\Delta_{j+1/2}u^{n+1}| \le \alpha_{-p}|\Delta_{j-p+1/2}u^n| + \cdots + \alpha_0|\Delta_{j+1/2}u^n| + \cdots + \alpha_q|\Delta_{j+q+1/2}u^n|$$

is valid, and consequently

$$TV^h(u^{n+1}) = \sum_{j=-\infty}^{\infty} |\Delta_{j+1/2}u^{n+1}| \le \alpha_{-p} \sum_{j=-\infty}^{\infty} |\Delta_{j-m+1/2}u^n| +$$

$$\cdots + \alpha_0 \sum_{j=-\infty}^{\infty} |\Delta_{j+1/2}u^n| + \cdots + \alpha_q \sum_{j=-\infty}^{\infty} |\Delta_{j+q+1/2}u^n| = TV^h(u^n)$$

that was to be proved.

Let us now proceed to the construction of a specific TVD scheme for Eq. (2.1.1). Write the scheme (2.7.1) in the form

$$u_j^{n+1} = u_j^n + C_{j+1/2}^n\Delta_{j+1/2}u^n \qquad (2.7.4)$$

if $a < 0$, or in the form

$$u_j^{n+1} = u_j^n - D_{j-1/2}^n\Delta_{j-1/2}u^n \qquad (2.7.5)$$

if $a > 0$, where

$$C_{j+1/2}^n = C_{j+1/2}^n(u_{j-p}^n, \ldots, u_j^n, \ldots, u_{j+q}^n)$$

$$D_{j-1/2}^n = D_{j-1/2}^n(u_{j-p}^n, \ldots, u_j^n, \ldots, u_{j+q}^n)$$

and

$$C_{j+1/2}^n \ge 0, \qquad D_{j-1/2}^n \ge 0 \qquad (2.7.6)$$

$$C_{j+1/2}^n \le 1, \qquad D_{j-1/2}^n \le 1 \qquad (2.7.7)$$

Then the scheme (2.7.4)-(2.7.7) can be shown to be a TVD scheme.

Let us prove this fact for the case $a > 0$. Using Eq. (2.7.5) we can write that

$$\Delta_{j+1/2}u^{n+1} = D_{j-1/2}^n\Delta_{j-1/2}u^n + (1 - D_{j+1/2}^n)\Delta_{j+1/2}u^n$$

It follows from the conditions (2.7.6) and (2.7.7) that

$$|\Delta_{j+1/2}u^{n+1}| \leq D_{j-1/2}^n|\Delta_{j-1/2}u^n| + (1 - D_{j+1/2}^n)|\Delta_{j+1/2}u^n|$$

Then

$$TV^h(u^{n+1}) = \sum_{j=-\infty}^{\infty} |\Delta_{j+1/2}u^{n+1}| \leq \sum_{k=-\infty}^{\infty} D_{k+1/2}^n|\Delta_{k+1/2}u^n|$$

$$+ \sum_{i=-\infty}^{\infty} (1 - D_{i+1/2}^n)|\Delta_{i+1/2}u^n| = \sum_{j=-\infty}^{\infty} |\Delta_{j+1/2}u^n| = TV^h(u^n)$$

The proof for the case $a < 0$ is carried out in a similar way.

Thus for certain limitations for the Courant number the scheme (2.7.1) will meet the requirement of the non-increasing total variation if the value of the difference approximation of a derivative computed with the aid of the operator H will have the same sign as the approximate value of the derivative computed with the aid of the corresponding first-order divided difference. This criterion enables one to construct various TVD schemes with an adaptive approximation of the derivatives that depends on the solution behavior.

2.7.2   The Scheme of Osher and Chakravarthy

As an example let us consider the second-order TVD scheme due to S. Chakravarthy and S. Osher:[22]

$$\frac{u_j^{n+1} - u_j^n}{\tau} + a\frac{\hat{u}_{j+1/2}^n - \hat{u}_{j-1/2}^n}{h} = 0 \qquad (2.7.8)$$

where

$$\hat{u}_{j+1/2}^n = u_j^n + \frac{1+\beta}{4}\overline{\Delta_{j+1/2}u^n} + \frac{1-\beta}{4}\overline{\overline{\Delta_{j-1/2}u^n}} \qquad (2.7.9)$$

for $a > 0$ in (2.1.1), and

$$\hat{u}_{j+1/2}^n = u_{j+1}^n - \frac{1+\beta}{4}\overline{\Delta_{j+1/2}u^n} - \frac{1-\beta}{4}\overline{\overline{\Delta_{j+3/2}u^n}} \qquad (2.7.10)$$

for $a < 0$. Here

$$\overline{\Delta_{j+1/2}u^n} = minmod[\Delta_{j+1/2}u^n, b\Delta_{j-1/2}u^n] \qquad (2.7.11)$$

$$\overline{\overline{\Delta_{j-1/2}u^n}} = minmod[\Delta_{j-1/2}u^n, b\Delta_{j+1/2}u^n] \qquad (2.7.12)$$

The function $minmod(r, m)$ is defined as follows:[15-17]

$$minmod[r, m] = sign(r)\max\{0, \min[|r|, m sign(r)]\} \qquad (2.7.13)$$

The weight parameter β in (2.7.9)-(2.7.10) is chosen to be in the interval $-1 \le \beta \le 1$. If the bars over $\Delta_{j+1/2}u^n$, $\Delta_{j-1/2}u^n$, $\Delta_{j+3/2}u^n$ in (2.7.9) and (2.7.10) are absent, the scheme (2.7.8)-(2.7.10) approximates Eq. (2.1.1) with the order $O(\tau) + O(h^3)$ for $\beta = 1/3$, and with the order $O(\tau) + O(h^2)$ for other values of β. The functions (2.7.11) and (2.7.12) limit the variation of the difference approximation of the spatial derivative; these functions are called the *flux limiters.*

Let us show that the scheme (2.7.8)-(2.7.13) is a TVD scheme provided that the following conditions are satisfied:

$$\kappa = a\tau/h \le \frac{4}{5 - \beta + b(1 + \beta)}, \quad 1 \le b \le \frac{3 - \beta}{1 - \beta}$$

Consider the case when $a < 0$. The minmod function can be written in the form

$$minmod[r, m] = sign(r)\Phi(r, m)$$

Then for $r \ne 0$ the equation

$$minmod[r, m] = r\Phi(1, \frac{m}{r}) \qquad (2.7.14)$$

To prove the property (2.7.14), let us make use of the obvious inequalities:

$$\min(a, b) \le a \qquad (2.7.15)$$

$$\min(a, b) \le b \qquad (2.7.16)$$

When $\Delta_{j+1/2}u^n = 0$, we obtain, by virtue of the flux limiters (2.7.11) and (2.7.12), that $u_j^{n+1} = u_j^n$ and consequently the conditions (2.7.6), (2.7.7) are satisfied. Now assume that $\Delta_{j+1/2}u^n \ne 0$. Introduce the notation

$$Z_{j+1} = \Delta_{j+3/2}u^n / \Delta_{j+1/2}u^n \qquad (2.7.17)$$

Then, using Eqs. (2.7.14) and (2.7.17), we can rewrite the scheme (2.7.8) in the form

$$u_j^{n+1} = u_j^n - \kappa W_{j+1/2}\Delta_{j+1/2}u^n \qquad (2.7.18)$$

where

$$W_{j+1/2} = 1 - \frac{1 - \beta}{4}Z_{j+1}\Phi(1, b\frac{1}{Z_{j+1}}) - \frac{1 + \beta}{4}\Phi(1, bZ_{j+1})$$

For scheme (2.7.18) to satisfy the condition (2.7.6) it is necessary that $W_{j+1/2} \geq 0$. Since $\Phi(r, m) \geq 0$, and $\Phi(r, m) = 0$ when $rm \leq 0$, it is easy to obtain, using (2.7.15) and (2.7.16), that

$$W_{j+1/2} \geq 1 - \frac{1-\beta}{4}b - \frac{1+\beta}{4} \geq 0,$$

and consequently the inequality

$$b \leq (3 - \beta)/(1 - \beta)$$

should be satisfied. On the other hand, in accordance with the approximation condition, the parameter b cannot be less than unity.

The limitation for the Courant number $\kappa = a\tau/h$ follows from the condition (2.7.7): $|a|(\tau/h)W_{j+1/2} \leq 1$. It is not difficult to show that

$$W_{j+1/2} \leq 1 + \frac{1-\beta}{4} + \frac{1+\beta}{4}b$$

and consequently

$$\kappa \leq 4/(5 - \beta + b(1 + \beta)) \qquad (2.7.19)$$

Let us find the maximum value of κ allowed by the stability condition (2.7.19) of the TVD scheme under consideration. Let us introduce the function

$$\kappa_{\max}(b, \beta) = 4/(5 - \beta + b(1 + \beta))$$

Since $\beta \geq -1$, the maximum of $\kappa_{\max}(b, \beta)$ at fixed β is achieved at the minimal value of b, that is at $b = 1$. Then we obtain that $\kappa_{\max}(1, \beta) = 2/3$. This means that the TVD scheme (2.7.8)-(2.7.13) imposes a more severe limitation on the time step than the upwind scheme and the Lax-Wendroff schemes, which were considered in the previous sections.

The TVD scheme has either the second or the third order of approximation only on the intervals of smooth monotone variation of the solution. On the intervals where the sharp changes of the derivatives occur as well as in the regions of the local solution extrema the TVD scheme transforms into a first-order upwind biased scheme. The larger is the value of b, the larger the solution intervals are which are calculated by the method of a higher approximation order.

Note that for $b = 1$, the TVD scheme has the second approximation order on monotone solutions, which does not depend on the choice of the parameter β. In this case we obtain the scheme of the minimal derivative values, which was proposed by V.P.Kolgan[23] long before the publication of the works von TVD schemes.[21,22]

2.7.3   The Program tvd1.m

We have implemented the TVD scheme (2.7.8)-(2.7.13) in the program tvd1.ma. With the aid of this program one can solve numerically the test problem (2.1.1), (2.4.7). We have constructed the program tvd1.ma on the basis of the program lw2.ma from Section 2.4. The input parameters used in the program tvd1.ma are described in Table 2.3. The main function in the file tvd1.ma is

$$\text{tvd}[a, a_1, b_1, j_{f0}, u_l, u_r, M, \kappa, \beta, b, Npic]$$

Table 2.3. Parameters used in program tvd1.ma

Parameter	Description
a	the advection speed a in eqn (2.1.1), $a > 0$
a1	the abscissa a_1 of the left end of the interval on the x-axis
b1	the abscissa b_1 of the right end of the above interval, $a_1 < b_1$
jf0	the number of the grid node j_{f0}, $0 < j_{f0} < M$, such that the maximum of the initial semi-ellipse pulse at $x = j_{f0}h$
ul,ur	are the constants u_l and u_r entering the initial condition (2.4.7), the value u_l corresponds to the constant background, and $u_l + u_r$ is the extremal value of the initial function $u(x, 0)$
M	the number M of the cell whose right boundary coincides with the right boundary $x = b_1$ of the spatial integration interval
cap	the Courant number $\kappa = a\tau/h$, where τ is the time step of the difference scheme, and h is the step for the uniform grid in the interval $a_1 \leq x \leq b_1$
beta	the weight parameter β in the TVD scheme, $-1 \leq \beta \leq 1$
bfl	the coefficient b entering the flux limiters, see (2.7.11) and (2.7.12), $1 \leq b \leq (3 - \beta)/(1 - \beta)$
Npic	the number of the pictures of the difference solution graphs, Npic > 1

Since the complete listing of the program tvd1.ma may be found on the attached diskette, we present only the part of the program, which implements the scheme (2.7.8)-(2.7.13):

```
(* ------ Do-loop over grid nodes on the x-axis ------- *)
      fxr = ul;
```

```
        Do[ fxl = fxr; unj = uj[[j]];
(* The TVD scheme (2.7.8)-(2.7.13) for equation (2.1.1) *)
    djpun = uj[[j+1]] - unj; djmun = unj - uj[[j - 1]];
      dlun = minmod[djpun, bfl*djmun];
        d2un = minmod[djmun, bfl*djpun];
fxr = unj + ((1 + beta)*dlun + (1 - beta)*d2un)/4;
        up[[j]] = uj[[j]] - cap*(fxr - fxl),
        {j, 1, M2 }];
(*  The boundary condition on the left boundary x = a1  *)
              up[[0]] = ul;
(*  The boundary condition on the right boundary x= b1  *)
up[[M]] = ((uj[[M]] + uj[[M - 1]] - up[[M - 1]]) +
    cap*(uj[[M - 1]] + up[[M - 1]] - uj[[M]]))/(1 + cap);
```

The identifier bfl in the above fragment is used to denote the coefficient b entering the flux limiter in (2.7.11) and (2.7.12). The function minmod[r, m] was programmed in accordance with its definition (2.7.13) as

```
minmod[r_,m_]:= Sign[r]*Max[0, Min[Abs[r], m*Sign[r]]]
```

Let us illustrate the influence of the flux limiters (2.7.11)-(2.7.13) on the accuracy of the difference scheme (2.7.8) for solution of the problem (2.1.1), (2.4.7). In Figs. 2.15 and 2.16 we present the computational results obtained with the aid of the program tvd1.ma. To perform the computation for obtaining Fig. 2.15, first compile the file tvd[...] and then click anywhere in the line of input data

```
tvd[1.0, 0.0, 2.0, 20, 0.5, 1.0, 300, 0.4, 1.0, 2.0, 4]
```

A similar line in the case of Fig. 2.16 has the form

```
tvd[1.0, 0.0, 2.0, 20, 0.5, 1.0, 300, 0.4, 1/3, 4.0, 4]
```

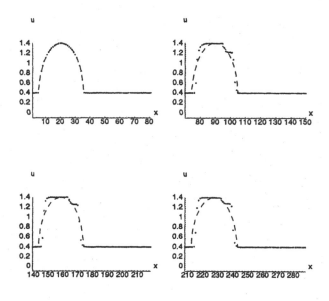

Figure 2.15: $\beta = 1$, $b = 2$, $\kappa = 0.4$

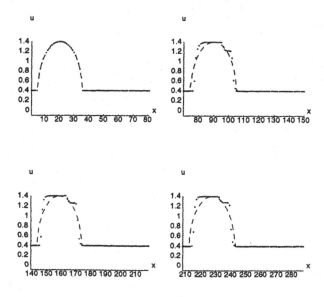

Figure 2.16: $\beta = 1/3$, $b = 4$, $\kappa = 0.4$

Comparing Figs. 2.15 and 2.16 with Fig. 2.11 we can see that the TVD scheme (2.7.8)-(2.7.13) has a much smaller dispersion than the Lax-Wendroff scheme (2.4.3). At the same time, the accuracy of the TVD scheme is much better than in the case of the first-order scheme (2.1.9) (compare Figs. 2.15 and 2.16 with Fig. 2.13). The height of the semi-ellipse pulse is preserved in the TVD solution with a remarkable accuracy. There is a clear trend to the formation of a staircase profile in the TVD solution as the time increases.

It should be noted that the TVD scheme (2.7.8)-(2.7.13) requires, at the same value of the Courant number κ, much more computing time than the Lax-Wendroff scheme (2.4.3). This is not surprising, because the scheme (2.7.8)-(2.7.13) involves much more complex formulas than the Lax-Wendroff scheme (2.4.3).

Exercise 2.18. Perform a series of computations using the program **tvd1.ma** to study the effect of the constants β and b on the accuracy of the numerical solution.

Exercise 2.19. Modify the program **tvd1.ma** by taking the initial condition $u(x,0) = u_0(x)$ corresponding to the square wave test from Section 2.2.2 (see the formula (2.2.6) for $u_0(x)$). Perform the computation by the program in order to see whether the speed of the propagation of a discontinuity is reproduced correctly in the numerical solution.

2.8 THE CONSTRUCTION OF DIFFERENCE SCHEMES FOR SYSTEMS OF PDEs

Mathematical models of many mathematical physics problems involve systems of partial differential equations or systems of integro-differential equations. Finite difference schemes are applicable also for the numerical solution of such equation systems.

A system of difference equations may also arise in the case when the method of finite differences is applied to the solution of scalar PDEs. One such situation arises when the scalar PDE contains higher derivative terms. As an example let us consider the *wave equation*

$$\partial^2 u/\partial t^2 = c^2 \partial^2 u/\partial x^2 \qquad (2.8.1)$$

where $c^2 = const > 0$, and assume that the initial conditions

$$u(x,0) = f_1(x), \ \partial u(x,0)/\partial t = f_2(x) \qquad (2.8.2)$$

are given for (2.8.1). Let us now introduce two new dependent variables $u_1 = \partial u/\partial t, u_2 = c\partial u/\partial x$. Then we may rewrite the scalar equation

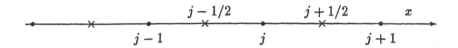

Figure 2.17: Staggered mesh system

(2.8.1) as the following PDE system:

$$\begin{cases} \partial u_1/\partial t = c\,\partial u_2/\partial x \\ \partial u_2/\partial t = c\,\partial u_1/\partial x \end{cases} \qquad (2.8.3)$$

Let us now assume that the function $f_1(x)$ is differentiable. Then we obtain from the formula $u_2(x,t) = c\,\partial u/\partial x$ that $u_2(x,0) = cf_1'(x)$. Thus we may formulate the following initial conditions for equation system (2.8.3):

$$u_1(x,0) = f_2(x); \ u_2(x,0) = cf_1'(x), \quad -\infty < x < \infty \qquad (2.8.4)$$

Now let us assume that the discrete values u_{1j} are computed at the points $(jh, n\tau)$ and that the values $u_{2j\pm 1/2}$ are computed at the points $((j\pm 1/2)h, n\tau)$. Thus we want to use two mesh systems (see Fig. 2.16). Such a grid is called *staggered*, or hybrid. We can write the following difference scheme to approximate the system (2.8.3):

$$\begin{cases} \dfrac{u_{1j}^{n+1}-u_{1j}^{n}}{\tau} = c\dfrac{u_{2j+1/2}^{n}-u_{2j-1/2}^{n}}{h} \\ \dfrac{u_{2j+1/2}^{n+1}-u_{2j+1/2}^{n}}{\tau} = c\dfrac{u_{1j+1}^{n+1}-u_{1j}^{n+1}}{h} \end{cases} \qquad (2.8.5)$$

It is possible to write down the scheme (2.8.5) in a vector-matrix form. For this purpose let us introduce the column vector $\mathbf{u}_j^n$ and the 2×2 matrix A via the formulas

$$\mathbf{u}_j^n = \begin{pmatrix} u_{1j}^n \\ u_{2j-1/2}^n \end{pmatrix}, \qquad A = \begin{pmatrix} 0 & c \\ c & 0 \end{pmatrix} \qquad (2.8.6)$$

Then we can write the difference scheme (2.8.5), in vector form, as

$$\frac{\mathbf{u}_j^{n+1} - \mathbf{u}_j^n}{\tau} = A\frac{\mathbf{u}_{j+1}^n - \mathbf{u}_j^n}{h} \qquad (2.8.7)$$

Note that the use of the staggered meshes for the construction of difference schemes for the systems of PDEs is not necessary. It is usually possible to construct schemes which are simpler in their computer implementation when only the points $(jh, n\tau)$ are used. For this purpose the vector matrix representation of a PDE system is used. For example, in the case of the system (2.8.3) we can rewrite this system as

$$\frac{\partial \mathbf{u}}{\partial t} = A \frac{\partial \mathbf{u}}{\partial x} \qquad (2.8.8)$$

where

$$\mathbf{u} = \begin{pmatrix} u_1 \\ u_2 \end{pmatrix}, \qquad A = \begin{pmatrix} 0 & c \\ c & 0 \end{pmatrix} \qquad (2.8.9)$$

The MacCormack scheme can be written for the vector equation (2.8.8) as

$$\tilde{\mathbf{u}}_j = \mathbf{u}_j^n - \tau A \left(\frac{\mathbf{u}_{j+1}^n - \mathbf{u}_j^n}{h} \right) \qquad (2.8.10)$$

$$\mathbf{u}_j^{n+1} = \frac{1}{2} \left[\mathbf{u}_j^n + \tilde{\mathbf{u}}_j - \tau A \left(\frac{\tilde{\mathbf{u}}_j - \tilde{\mathbf{u}}_{j-1}}{h} \right) \right] \qquad (2.8.11)$$

compare with Eqs. (2.4.13), (2.4.14). We have assumed in (2.8.10) and (2.8.11) that

$$\mathbf{u}_j^n = \begin{pmatrix} u_{1j}^n \\ u_{2j}^n \end{pmatrix}, \qquad \tilde{\mathbf{u}}_j = \begin{pmatrix} \tilde{u}_{1j} \\ \tilde{u}_{2j} \end{pmatrix}$$

Comparing formulas (2.8.10), (2.8.11) with the formulas (2.4.13), (2.4.14), we can see that the generalization of the difference schemes to a scalar equation to the case of a PDE system is quite straightforward: the scalar values u_j^n, $\tilde{u}_j$ are replaced by the vector values $\mathbf{u}_j^n$, $\tilde{\mathbf{u}}_j$, and the scalar advection speed a is replaced by a matrix A.

Note that the use of the same type of spatial differencing, for example, the central differencing as in the scheme (2.8.5), is not necessary. There are, in the literature, difference schemes for PDE systems, in which the individual difference equations have different stencils, but they will not be discussed here.

Now consider another situation where the system of difference equations may be used to represent the difference scheme for a scalar PDE. Let us assume that equation (2.1.1) is approximated by the following *multi-level* difference scheme:

$$b_q u_j^{n+q} + b_{q-1} u_j^{n+q-1} + \ldots + b_0 u_j^n = 0 \qquad (2.8.12)$$

where $b_q, \ldots, b_0$ denote finite difference operators. Let us assume that equation (2.8.12) has a unique solution for u^{n+q} and that the solution

u^{n+q} continuously depends on $u^{n+q-1}, \ldots, u^n$. This means that equation (2.8.12) is equivalent to the equation

$$u_j^{n+q} = c_{q-1}u^{n+q-1} + \ldots + c_0 u^n \qquad (2.8.13)$$

where

$$c_j = -b_q^{-1}b_j, \; j = 0, \ldots, q - 1$$

Now introduce $q - 1$ new dependent variables

$$u^{(1),n} = u^{n-1}, \; u^{(2),n} = u^{n-2}, \ldots, u^{(q-1),n} = u^{n-(q-1)} \qquad (2.8.14)$$

Then equation (2.8.13) can be replaced by the system of equations

$$\begin{cases} u^{n+1} = c_{q-1}u^n + c_{q-2}u^{(1),n} + \ldots + c_0 u^{(q-1),n} \\ u^{(1),n+1} = u^n \\ u^{(2),n+1} = u^{(1),n} \\ \qquad \ldots \\ u^{(q-1),n+1} = u^{(q-2),n} \end{cases} \qquad (2.8.15)$$

It was assumed, in the derivation of the above system (2.8.15), that the involved grid functions $u^{n+1}, u^n, \ldots, u^{(q-1),n}$ have the same subscript "j", therefore, it has been omitted in (2.8.15). The difference equation (2.8.12) involves the grid values u_j^m on $(q + 1)$ time levels $t = m\tau$, $m = n + 0, n + 1, \ldots, n + q$. By introducing the new variables $u^{(1),n}, \ldots, u^{(q-1),n}$, via formulas (2.5.14), we have reduced the $(q+1)$st-level difference scheme (2.8.12) to a two-level difference scheme (2.8.15), because the superscripts in (2.8.15) take only the values n and $n + 1$.

The initial data for equation (2.1.1) are assumed to be given at $t = 0$. This enables us to compute the values u_j^0, $j = 0, \pm 1, \pm 2, \ldots,$. But the difference equation (2.8.12) also needs the values u_j^m for $m = 1, 2, \ldots, q - 1$ in order that we can start a calculation by the formula (2.8.13). These values must be determined with the aid of approximations using the initial data to be discussed later.

Let us now present an example of the derivation of the two-level representation (2.8.15) for the three-level difference scheme

$$u_j^{n+1} - 2u_j^n + u_j^{n-1} = (c\tau/h)^2(u_{j+1}^n - 2u_j^n + u_{j-1}^n) \qquad (2.8.16)$$

which approximates the wave equation (2.8.1). In accordance with Eq. (2.8.12) we at first write the difference equation

$$u_j^{n+2} - 2u_j^{n+1} + u_j^n = (c\tau/h)^2(u_{j+1}^{n+1} - 2u_j^{n+1} + u_{j-1}^{n+1}) \qquad (2.8.17)$$

which is obtained from (2.8.16) by shifting the superscript n by $+1$. It may be seen from (2.8.17) that $q = 2$ in this example. Therefore, we introduce only one new dependent variable $u_j^{(1),n+1} = u_j^n$. Then we may rewrite the difference equation (2.8.17) as the following system of two two-level equations:

$$\begin{cases} u_j^{n+1} = 2u_j^n - u_j^{(1),n} + (c\tau/h)^2(u_{j+1}^n - 2u_j^n + u_{j-1}^n) \\ u_j^{(1),n+1} = u_j^n \end{cases} \quad (2.8.18)$$

Let us introduce the shift operators

$$T_1 \equiv T_{+1}, \quad T_0 = t_{+0}; \quad T_{+1}u(x,t) = u(x \pm h, t)$$

$$T_{+0}u(x,t) = u(x,t+\tau); \quad T_1^m u(x,t) = u(x+mh,t) \quad (2.8.19)$$

$$T_0^m u(x,t) = u(x,t+m\tau), \quad m \in R, \, Iu(x,t) \equiv u(x,t)$$

The operator I is called the identity operator. Using the operators (2.8.19) we can write the system (2.8.18) in the form

$$\begin{cases} u_j^{n+1} = 2u_j^n - u_j^{(1),n} + (c\tau/h)^2(T_1 - 2I + T_{-1})u_j^n \\ u_j^{(1),n+1} = u_j^n \end{cases} \quad (2.8.20)$$

Now it is easy to write this system in the vector matrix form

$$\mathbf{u}_j^{n+1} = C\mathbf{u}_j^n \quad (2.8.21)$$

where

$$\mathbf{u}_j^n = \begin{pmatrix} u_j^n \\ u_j^{(1)n} \end{pmatrix}, \, C = \begin{pmatrix} 2 + (c\tau/h)^2(T_1 - 2I + T_{-1}) & -1 \\ 1 & 0 \end{pmatrix} \quad (2.8.22)$$

Thus we can see that at least some entries of the matrix C are the finite difference operators. Therefore, the matrix C is indeed a finite difference operator, which acts on the difference solution $\mathbf{u}_j^n$ to give the difference solution $\mathbf{u}_j^{n+1}$ for the next time level.

It is not difficult to write a *Mathematica* program for computing the above matrix C for the arbitrary linear scalar multi-level difference equation (2.8.12). Let us at first describe the steps of the corresponding symbolic algorithm.

Step 1. Determine q in (2.8.12).

Step 2. Reduce the superscripts in (2.8.12) by the amount $(q-1)$. As a result we obtain the difference equation

$$b_q u_j^{n+1} + b_{q-1} u_j^n + \ldots + b_0 u_j^{n-(q-1)} = 0 \quad (2.8.23)$$

Step 3. Introduction of the shift operators T[k], such that u[j+k, m] = T[k]*u[j, m], and then substitute these operator representations of the grid values into the difference equation (2.8.23).

Step 4. Solve (2.8.23) for u_j^{n+1} to get the equation

$$u_j^{n+1} = c_{q-1}u^{n-1} + \ldots + c_0 u^{n-(q-1)} \qquad (2.8.24)$$

Step 5. Generate the matrix C in (2.8.21) on the basis of (2.8.24). This is a square $q \times q$ matrix of the form

$$C = \begin{pmatrix} c_{q-1} & c_{q-2} & \cdots & \cdots & \cdots & c_0 \\ 1 & 0 & \cdots & \cdots & \cdots & 0 \\ 0 & 1 & 0 & \cdots & \cdots & 0 \\ \cdots & \cdots & \cdots & \cdots & \cdots & \cdots \\ 0 & 0 & 0 & \cdots & 1 & 0 \end{pmatrix} \qquad (2.8.25)$$

so that if $C = \| c_{jk} \|_1^q$, then

$$c_{1,1} = c_{q-1}; \; c_{1,2} = c_{q-2}; \ldots c_{1,q} = c_0$$

$$c_{j,k} = \begin{cases} 1, & k+1 = j \\ 0 & \text{otherwise} \end{cases}, j = 2, \ldots, q; \; k = 1, \ldots, q$$

Let us now describe how we have implemented the above symbolic algorithm in *Mathematica*. Let us illustrate our algorithm on the example of the difference equation (2.8.16). This equation is input in to our *Mathematica* program multlev.ma in the form of the function

```
sch[j_, n_]:= u[j, n+1] - 2*u[j, n] + u[j, n - 1] -
    (c*dt/h)^2*(u[j+1,n] - 2*u[j, n] + u[j - 1, n])
```

so that the difference equation is represented by equation

```
sch[j, n] = 0
```

For the determination of the number of time levels q it is convenient to use the transformation rule

```
utrt = u[k_, m_] -> T0[m - n]
```

The underlines after k and m show that the transformation rule utrt vill act on u[k, m] for any k and m. The application of the rule utrt o the difference equation under consideration converts it to the form

```
'[1] - 2 T[0] + T[-1] = 0
```

Basing on this simple equation it is easy to determine q with the aid of the corresponding Do-loop, see the subsequent listing of the program multlev.ma.

The shift operators T[k] can be conveniently introduced with the aid of the transformation rule

utrx = u[k_, m_] -> T[k - j]*v[j, m]

In the result of the application of the shift operators T[k] the difference equation (2.8.12) is rewritten in the operator form, which is convenient both for the representation of the multi-level difference equation in the matrix form (2.8.21) and for the stability analysis of difference scheme (2.8.12) by the Fourier method.

Note that with the aid of the transformation rule

shift = T[k_] v[j_, m_] -> v[j + k, m]

we can eliminate the shift operators T[k] from the difference equation and return to its initial form (2.8.16). As an example let us consider the operator difference equation

sch = a T[1] v[j, n] + b T[2] v[j, n];

Now apply the transformation rule shift to sch:

sch = sch /. shift;

The result obtained is

sch = a v[1 + j, n] + b v[2 + j, n]

Mathematica thus provides the utilities for the various operations on the operator difference equations.

The new dependent variables (2.8.14) are denoted, in our program multlev.ma, by w[j], $j = 1,\ldots,q$, so that

$$w[1] = u_j^n, \quad w[2] = u_j^{(1),n}, \ldots, w[q] = u_j^{(q-1),n}.$$

The capital letter C cannot be used in a *Mathematica* program to denote the matrix C, because this letter is reserved by *Mathematica* 0for naming the undetermined coefficients C[1], C[2], etc., which may appear as a result of the use of the *Mathematica* function DSolve[eqns, y[x], x], which solves a differential equation for y[x], taking x as the independent variable. Instead, we have used the variable CC to denote the matrix C. It is convenient to initialize the matrix CC as a $q \times q$ identity matrix. This can be done using the *Mathematica* function IdentityMatrix[q]. Then the entries of the identity matrix can be replaced by the true expressions corresponding to the difference equation under consideration using the *Mathematica* function ReplacePart[CC, elem, {j, k}]. After generating the matrix CC, our program multlev.ma prints out the expressions for the elements of the matrix C. For the case of scheme (2.8.16) they were computed symbolically as

```
              2        2              2
C[[1,1]] = {2 - 2 cp  + cp  T[-1] + cp  T[1]}
C[[1,2]] = {-1}
C[[2,1]] = 1
C[[2,2]] = 0
```

Here the variable cp is used to denote the dimensionaless quantity $c\tau/h$. Comparing the above expressions for C[[j,k]] with the matrix C in (2.8.22), we can see that the results coincide.

Now we present the listing of our program multlev.ma.

```
sch[j_, n_]:= u[j, n+1] - 2*u[j, n] + u[j, n - 1] -
    (c*dt/h)^2*(u[j+1,n] - 2*u[j, n] + u[j - 1, n])
(* -- The formula for the nondimensional variable entering
        the difference equation: ------------------------ *)
        rulecp = dt -> h*cp/c
sc1 = sch[j, n];
Print["The difference equation"];
Print[sc1, " = 0"];
(* -- Introduction of the variable cp into the difference
                                          scheme -- *)
 sc1 = sc1 /. rulecp;
OutputForm[sc1] >> tscr.m;
(* --- Determination of the number q of time layers -- *)
        utrt = u[k_, m_] -> T0[m - n]
        sc2 = sc1;
        sc2 = sc2 /. utrt;
OutputForm[sc2] >>> tscr.m;
 nmin = 10; nmax = -10;
   Do[aj = Coefficient[sc2, T0[j]];
        If[aj===0, Goto[nexj]];
If[j < nmin, nmin = j];
  If[j > nmax, nmax = j];
    Label[nexj],
    {j, -10, 10, 1}];
 q = nmax - nmin;
Print["q = ",q, "; nmin = ", nmin, "; nmax = ", nmax];
(* -- Solution of the difference equation
```

```
                        with respect to u[j,n + q] -- *)
    utrt1 = u[k_, m_] -> u[k, m -nmin]
      sc3 = sc1 /. utrt1;
                         OutputForm[sc3] >>> tscr.m;
(* --- Introduction of the shift operators T[k] along the
      x-axis, such that T[k]*u[j,m] = u[j + k, m] ---- *)
  utrx = u[k_, m_] -> T[k - j]*v[j, m]
    sc4 = sc3 /. utrx;
                    sol = Solve[sc4 == 0, v[j, n + q]];
    up = v[j, n + q] /. sol;
      up = up /. T[0] -> 1;
                         OutputForm[up] >>> tscr.m;
    utrt2 = v[j_, m_] -> V[j, m - q + 1]
      up = up /. utrt2;
                         OutputForm[up] >>> tscr.m;
    utrt3 = V[j_, m_] -> w[n - m + 1]
      up1 = up /. utrt3; up1 = Simplify[up1];
      Do[Collect[up1, w[j]], {j,q}];
                         OutputForm[up1] >>> tscr.m;
(* ------- Generation of the qxq matrix C ------------ *)
    CC = IdentityMatrix[q];
  Do[aj = Coefficient[up1, w[j]];
    CC = ReplacePart[CC, aj, {1, j}],
    {j,q}];
  celem[j_, k_] := If[k + 1 == j, 1, 0]
        Do[
          Do[  CC[[j,k]] = 0;
              CC[[j, k]] = celem[j, k],
                {k, q}],
        {j, 2, q}];
Print["The expressions for the elements of the matrix C"];
  Do[
      Do[aj = CC[[j, k]];
  Print["C[[", j, ",", k, "]] = ", aj],
      {k, q}],
    {j, q}];
```

To perform a computation using this program, one must at first specify the difference equation by typing the body of the function sch[j_, n_], see line 1 in the above listing of multlev.ma. Then one must specify the transformation rule rulecp for the introduction of some nondimensional variable involving the mesh steps h and τ. In the case of the difference scheme (2.8.16) this rule has the form `rulecp = dt -> h*cp/c`, with the aid of which the nondimensional variable `cp` $= c\tau/h$ is introduced. Then one can perform a computation using the program `multlev.ma`. The intermediate results of symbolic computations are stored in the file `tscr.m`.

Exercise 2.20. In (2.8.5), make the substitutions

$$u_{1j}^n = (u_j^n - u_j^{n-1})/\tau, \ u_{2j+1/2}^n = (u_{j+1}^n - u_j^n)/h$$

Show that the resulting difference equation coincides with the difference equation (2.8.16).

2.9   IMPLICIT DIFFERENCE SCHEMES

Let us again consider the difference scheme (2.1.9). We may rewrite this scheme in the form (2.1.11). The grid values u_j^n and u_{j-1}^n, which are present in the right hand side of this equation, are assumed to be known as the result of a computation at the previous time level $t = t_n = n\tau$. Therefore, the equation (2.1.11) gives an *explicit* formula for the computation of the grid values u_j^{n+1} of the solution at the next time level $t = t_{n+1} = (n+1)\tau$. The difference schemes (2.4.3), (2.4.9)-(2.4.10)), (2.4.11)-(2.4.12), (2.4.13)-(2.4.14), (2.4.15), (2.8.5), (2.8.10)-(2.8.11), (2.8.16) possess the same property. Such schemes are called *explicit* difference schemes.

Let us now consider the following difference scheme for the advection equation (2.1.1):

$$\frac{u_j^{n+1} - u_j^n}{\tau} + a\frac{(u_{j+1}^{n+1} - u_{j-1}^{n+1}) + (u_{j+1}^n - u_{j-1}^n)}{4h} = 0 \qquad (2.9.1)$$

$$j = 0, \ldots, M; \quad n = 0, 1, \ldots$$

The difference approximation of the derivative $\partial u/\partial x$ represents here the arithmetic mean of the central differences at $t = n\tau$ and $t = (n+1)\tau$. This difference equation cannot be resolved explicitly with respect to the known values u_j^n, $u_{j\pm1}^n$. The reason for this is that the equation (2.9.1) contains not only the unknown value u_j^{n+1}, but also the unknown values u_{j+1}^{n+1} and u_{j-1}^{n+1}. Therefore, the determination of the solution of equation (2.9.1) is not so easy as in the case of explicit difference schemes. Such difference schemes are called the *implicit* difference schemes. The stencil of scheme (2.9.1) contains six points, shown in Fig. 2.18. Note that the

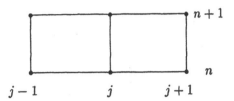

Figure 2.18: The stencil of scheme (2.9.1)

stencil of explicit difference schemes usually contains only one point at the $(n + 1)$st time level.

The specific difference scheme (2.9.1) is called the *Crank-Nicolson* scheme.[24,1,2,3,25] It is easy to determine the approximation order of scheme (2.9.1) with the aid of the program appr1.ma from Section 2.5. To make use of this program we at first type in the difference equation (2.9.1) as the body of the function sch[j_, n_]:

```
( ut1 = (u[j, n+1] - u[j,n])/dt;

  ux1 = a (u[j+1, n+1] - u[j-1,n+1] +

          u[j+1,n] - u[j-1,n])/(4 h); ut1 + ux1 )
```

Let us now specify the center of Taylor expansion as in the case of the Lax-Wendroff scheme (2.4.3). In this case the program appr1.ma produces the following expression for $P_1(\cdot\cdot)$, see (2.5.4):

```
 (0,1)              (1,0)
v       [x, t] + a v       [x, t] +
```

```
   2   2      2  (3,0)
a (a   dt  + 2 h ) v      [x, t]
--------------------------------
              12
```

It may be seen from this formula that the scheme has the order of approximation $O(\tau^2) + O(h^2)$ with respect to the $(jh, n\tau)$ point. From this it follows that the modified equation corresponding to the Crank-Nicolson scheme (2.9.1) has the form

$$\frac{\partial v}{\partial t} + a\frac{\partial v}{\partial x} = -\frac{a}{12}(a^2\tau^2 + 2h^2)\frac{\partial^3 v}{\partial x^3}$$

Let us now investigate the stability of scheme (2.9.1) with the aid of the Fourier method. For this purpose we substitute the particular solution of the form

$$u_j^n = \lambda^n e^{ij\alpha h}, \quad i = \sqrt{-1}$$

into the equation (2.9.1):

$$\frac{\lambda - 1}{\tau} + \frac{a}{4h}\left[\lambda(e^{i\alpha h} - e^{-i\alpha h}) + (e^{i\alpha h} - e^{-i\alpha h})\right] = 0 \qquad (2.9.2)$$

Using the formula $e^{\pm i\alpha h} = \cos\alpha h \pm i\sin\alpha h$, we can transform the equation (2.9.2) to the form

$$\frac{\lambda - 1}{\tau} + \frac{a}{4h}(2\lambda i\sin\xi + 2i\sin\xi) = 0 \qquad (2.9.3)$$

where $\xi = \alpha h$. From this we find the following expression for the amplification factor λ:

$$\lambda = (1 - ib)/(1 + ib) \qquad (2.9.4)$$

where $b = (\kappa/2)\sin\xi$, $\kappa = a\tau/h$. It follows from (2.9.4) that $|\lambda| = 1$ for any values of κ and ξ. This means that the difference scheme (2.9.1) remains stable for all values of the mesh steps τ and h. The difference schemes, which are stable for any values of the mesh parameters, are called *unconditionally* or *absolutely stable* difference schemes. Thus the Crank-Nicholson scheme (2.9.1) represents an example of an absolutely stable difference scheme. If we expand the grid values in (2.9.1) into Taylor series with respect to the point $(jh, (n + 1/2)\tau)$, then we find that the order of approximation of this scheme is $O(\tau^2) + O(h^2)$.

Using the shift operators (2.8.19) we can rewrite the equation (2.9.1) in the operator form

$$\Lambda_0 u_j^{n+1} = \Lambda_1 u_j^n \qquad (2.9.5)$$

where

$$\Lambda_0 = I + (\kappa/4)(T_1 - T_{-1}), \quad \Lambda_1 = I - (\kappa/4)(T_1 - T_{-1})$$

Let us assume that the inverse operator Λ_0^{-1} exists. Then we can formally write the solution of the difference equation (2.9.5) as

$$u_j^{n+1} = \Lambda_0^{-1} \Lambda_1 u_j^n$$

Now consider the question of the solution of the difference equation (2.9.1). Let us rewrite the difference equation (2.9.1) in the form

$$a_j u_{j-1}^{n+1} + b_j u_j^{n+1} + c_j u_{j+1}^{n+1} = f_j^n, \ 0 < j < M \qquad (2.9.6)$$

$$u_0^{n+1} = g(t_{n+1}) \qquad (2.9.7)$$

$$u_{M-1}^{n+1} = u_M^{n+1} \qquad (2.9.8)$$

where

$$a_j = -\kappa/4, \ b_j = 1, c_j = \kappa/4, \ f_j = u_j^n - (\kappa/4)(u_{j+1}^n - u_{j-1}^n)$$

The function $g(t)$ in (2.9.7) is assumed to be given. The condition (2.9.8) is used as a boundary condition at the right boundary of the spatial computational domain. Note that in the case when $a > 0$ in (2.1.1) the condition (2.9.8) is not needed to obtain the solution of the original PDE by virtue of the behavior of the characteristics, see Fig. 2.4. Nevertheless the difference scheme (2.9.1) requires some boundary condition at the right end. Thus (2.9.8) is a purely difference boundary condition. Let us write the equations (2.9.6)-(2.9.8) sequentially for $j = 0$, $j = 1, \ldots, j = M$:

$$\begin{cases} u_0^{n+1} = g(t_{n+1}) \\ a_1 u_0^{n+1} + b_1 u_1^{n+1} + c_1 u_2^{n+1} = f_1^n \\ a_2 u_1^{n+1} + b_2 u_2^{n+1} + c_2 u_3^{n+1} = f_2^n \\ \cdots\cdots\cdots \\ a_{M-1} u_{M-2}^{n+1} + b_{M-1} u_{M-1}^{n+1} + c_{M-1} u_M^{n+1} = f_{M-1}^n \\ u_{M-1}^{n+1} - u_M^{n+1} = 0 \end{cases} \qquad (2.9.9)$$

We can see that the equations (2.9.9) represent a system of linear algebraic equations for determining the unknown values $u_0^{n+1}, \ldots, u_M^{n+1}$. If we introduce the vectors

$$U = \begin{pmatrix} u_0^{n+1} \\ u_1^{n+1} \\ . \\ . \\ . \\ u_M^{n+1} \end{pmatrix}, \quad b = \begin{pmatrix} g(t_{n+1}) \\ f_1^n \\ . \\ . \\ . \\ f_{M-1}^n \\ 0 \end{pmatrix}$$

and the $(M+1) \times (M+1)$ matrix

$$A = \begin{pmatrix} 1 & 0 & 0 & \ldots & \ldots & & \ldots & 0 \\ a_1 & b_1 & c_1 & 0 & \ldots & & \ldots & 0 \\ \ldots & \ldots & \ldots & \ldots & \ldots & & \ldots & \ldots \\ 0 & \ldots & \ldots & 0 & a_{M-1} & & b_{M-1} & c_{m-1} \\ 0 & \ldots & \ldots & \ldots & 0 & & 1 & -1 \end{pmatrix}$$

then we can write the system (2.9.9) in the vector matrix form as

$$AU = b \qquad (2.9.10)$$

where A is a tridiagonal matrix. The banded tridiagonal structure of the matrix A in (2.9.10) enables us to use an efficient variant of the Gauss elimination method for the solution of the system (2.9.10).

Let us write the equation $u_0^{n+1} = g(t_{n+1})$ of the system (2.9.9) in the form

$$u_0^{n+1} = X_0 u_1^{n+1} + Y_0$$

where $X_0 = 0, Y_0 = g(t_{n+1})$. From the second equation of system (2.9.9) we have that

$$a_1(X_0 u_1^{n+1} + Y_0) + b_1 u_1^{n+1} + c_1 u_2^{n+1} = f_1^n \qquad (2.9.11)$$

that is we have eliminated the value u_0^{n+1}. Let us rewrite (2.9.11) in the form

$$u_1^{n+1} = X_1 u_2^{n+1} + Y_1 \qquad (2.9.12)$$

where

$$X_1 = (-c_1/b_1), \ Y_1 = (f_1^n - a_1 Y_0)/b_1$$

We can make use of the relationship (2.9.12) to eliminate u_1 from the equation

$$a_2 u_1^{n+1} + b_2 u_2^{n+1} + c_2 u_3^{n+1} = f_2^n$$

We can write the result of this elimination in the form

$$u_2^{n+1} = X_2 u_3^{n+1} + Y_2$$

The above elimination process can be continued for $n = 3, 4, \ldots$.

Substituting the relationship

$$u_{j-1}^{n+1} = X_{j-1} u_j^{n+1} + Y_{j-1}$$

into the equation (2.9.6), we obtain

$$u_j^{n+1} = \frac{-c_j}{b_j + a_j X_{j-1}} u_{j+1}^{n+1} + \frac{f_j^n - a_j Y_{j-1}}{b_j + a_j X_{j-1}}$$

It may be seen from here that the coefficients of the relationships

$$u_j^{n+1} = X_j u_{j+1}^{n+1} + Y_j \qquad (2.9.13)$$

obtained in the process of elimination are calculated by the recurrence formulas

$$X_j = \frac{-c_j}{b_j + a_j X_{j-1}}, \quad Y_j = \frac{f_j^n - a_j Y_{j-1}}{b_j + a_j X_{j-1}} \qquad (2.9.14)$$

The last of the relationships (2.9.13) has the form

$$u_{M-1}^{n+1} = X_{M-1} u_M^{n+1} + Y_{M-1} \qquad (2.9.15)$$

On the other hand we have the equation (2.9.8). Solving the system of two equations (2.9.8) and (2.9.15) with respect to u_{M-1}^{n+1} and u_M^{n+1}, we can find the value u_{M-1}^{n+1} as

$$u_{M-1}^{n+1} = Y_{M-1}/(1 - X_{M-1})$$

After that the values $u_{M-2}^{n+1}, u_{M-3}^{n+1}$, etc. are determined from the equalities

$$u_{M-2}^{n+1} = X_{M-2} u_{M-1}^{n+1} + Y_{M-2}, \quad u_{M-3}^{n+1} = X_{M-3} u_{M-2}^{n+1} + Y_{M-3}$$

respectively, until the value u_1^{n+1} is determined.

Let us now determine the number of arithmetic operations, which are needed for the solution of the system (2.9.9) by the above variant of the Gauss elimination method. For the computation of the pair of coefficients X_j and Y_j by the formulas (2.9.14) we need 2 additions and subtractions, 2 multiplications and 2 divisions. The computation of the value u_j^{n+1} by (2.9.13) requires 1 multiplication and 1 addition. Thus the overall number of arithmetic operations needed for the computation of the values $u_1^{n+1}, \ldots, u_M^{n+1}$ is proportional to the number M of the nodes of a spatial mesh. Note that in the case of the solution of an arbitrary linear system of M equations with M unknowns by the elimination method, one usually has to perform $O(M^3)$ operations. It is clear that the substantial reduction of the number of arithmetic operations in the case of system (2.9.9) became possible owing to the specifics of this system.

The general algebraic system (2.9.10) is called *well conditioned*, if the small perturbations of the coefficients and of the right hand sides of this system lead to small perturbations of the solution $\mathbf{U}$. Otherwise the system (2.9.10) is called *ill-conditioned*.

The source of the perturbations introduced in the coefficients of (2.9.10) is the use of the machine arithmetic of floating-point numbers. It can be shown that the particular system (2.9.9) with a tridiagonal matrix is well-conditioned, if the condition of the dominance of the diagonal elements

$$|b_j| \geq |a_j| + |c_j| + \delta \qquad (2.9.16)$$

is satisfied, where δ is a small positive number. The condition (2.9.16) also ensures that the denominators $b_j + a_j X_{j-1}$ in formulas (2.9.14) do not vanish.

Let us, for example, check the condition (2.9.16) in the case of the Crank-Nicolson scheme (2.9.1):

$$1 \geq |-\kappa/4| + |\kappa/4| + \delta$$

or $|\kappa/2| + \delta \leq 1$. Therefore, if κ satisfies the inequality $|\kappa| \leq 2$, the condition (2.9.16) will be satisfied for some small $\delta > 0$.

The extension of the implicit difference schemes to the case of systems of PDEs may be performed in a straightforward way. For example, the Crank-Nicolson scheme for the system (2.8.8) may be written as:

$$\frac{\mathbf{u}_j^{n+1} - \mathbf{u}_j^n}{\tau} = A \frac{(\mathbf{u}_{j+1}^{n+1} - \mathbf{u}_{j-1}^{n+1}) + (\mathbf{u}_{j+1}^n - \mathbf{u}_{j-1}^n)}{4h} \qquad (2.9.17)$$

We will consider in subsequent sections a number of other implicit difference schemes.

Exercise 2.21. Write the implicit scheme (2.9.17) in the form

$$B_1 \mathbf{u}^{n+1} = B_0 \mathbf{u}^n$$

where B_1 and B_0 are matrix difference operators.

2.10 VON NEUMANN STABILITY ANALYSIS IN THE CASE OF SYSTEMS OF DIFFERENCE EQUATIONS

2.10.1 Fourier Transform of Difference Equations

In Section 2.3 and in a number of subsequent sections, we have already used the Fourier method for the stability analysis of difference schemes approximating the scalar PDEs. This method can be extended to the case of the systems of difference equations approximating an initial-value problem for PDEs.

We will assume that the difference equations are given in the two-level form. The multi-level difference schemes can be reduced to the two-level difference schemes with the aid of a procedure described in Section 2.8. We will further assume that the number of spatial variables $x_1, \ldots, x_L$ is $L \geq 1$. Denote by $h_1, \ldots, h_L$ the steps of a uniform spatial grid along the axes $x_1, \ldots, x_L$, respectively. Let $\mathbf{x} = (x_1, \ldots, x_L)$ and let

$$\mathbf{u}(\mathbf{x}, t_n) = \{u^{(1)}(\mathbf{x}, t_n), \ldots, u^{(m)}(\mathbf{x}, t_n)\}, \ m \geq 1$$

be the approximation for $\mathbf{u}(\mathbf{x}, t)$ obtained from the system of difference equations for the moment of time $t = t_n = n\tau$, $n = 1, 2, \ldots,$. Similarly to (2.8.19) let us define the shift operators $T_j^{l_j}$ as

$$T_j^{l_j}\mathbf{u}(\mathbf{x}) = \mathbf{u}(x_1, \ldots, x_j + h_j l_j, \ldots, x_L), \quad 1 \leq j \leq L \qquad (2.10.1)$$

We shall denote by T^l the operator

$$T^l = T_1^{l_1} \ldots T_L^{l_L}$$

where l is a multi-index, $l = (l_1, \ldots, l_L)$. Now consider a system of difference equations of the form

$$B_1\mathbf{u}^{n+1} = B_0\mathbf{u}^n \qquad (2.10.2)$$

where B_1 and B_0 are linear difference matrix operators involving the shift operators. We further assume that the system of difference equations (2.10.2) is uniquely solvable for $\mathbf{u}^{n+1}$, that is, we assume that there exists the unique inverse operator B_1^{-1}. Then we may solve the system (2.10.2) and obtain the solution $\mathbf{u}^{n+1}$ as

$$\mathbf{u}^{n+1} = S\mathbf{u}^n(\mathbf{x}), \ n = 0, 1, \ldots \qquad (2.10.3)$$

where

$$S = B_1^{-1}B_0$$

The operator S in (2.10.3) is called the *step operator* of the difference scheme. Let us now substitute into (2.10.3) the solution of the form

$$\mathbf{u}^n(\mathbf{x}) = \lambda^n \mathbf{U}_0 \exp\{i(k_1 x_1 + \ldots + k_L x_L)\}$$

where $k_1, \ldots, k_L$ are real wave numbers, $\mathbf{U}_0$ is a constant vector, and λ is a complex number. This gives the equation

$$\lambda \mathbf{U}_0 \exp\{i(k_1 x_1 + \ldots + k_L x_L)\} = G\mathbf{U}_0 \exp\{i(k_1 x_1 + \ldots + k_L x_L)\}$$

where the matrix $G(k_1, \ldots, k_L, h_1, \ldots, h_L, \tau)$ is obtained as a result of the action of the difference operator S on the function of the form $\exp\{i(k_1 x_1 + \ldots + k_L x_L)\}$. The matrix $G(\mathbf{k}, \mathbf{h}, \tau)$, where $\mathbf{k} = (k_1, \ldots, k_L)$, $\mathbf{h} = (h_1, \ldots, h_L)$, is called the *amplification matrix* of scheme (2.10.2). It is possible to determine the stability of this scheme by studying the behavior of certain norm of the amplification matrix G. For this purpose we first introduce the set $S_\infty(R_L)$ of such functions

$$\mathbf{v}(\mathbf{x}) = \{v^{(1)}(\mathbf{x}), \ldots, v^{(m)}(\mathbf{x})\}$$

which rapidly decrease as $|\mathbf{x}| \to \infty$ and which are defined on all of

$$R_L = \{-\infty < x_j < \infty, \ j = 1, \ldots, L\}$$

Let us now use the mean square, L_2 norm

$$\| \mathbf{v} \| = \left(\int_{R_L} |\mathbf{v}(\mathbf{x})|^2 dx \right)^{1/2}$$

where

$$|\mathbf{v}(\mathbf{x})|^2 = \sum_{j=1}^{m} |v^{(j)}(\mathbf{x})|^2$$

$$dx = dx_1 \ldots dx_L$$

The direct and inverse Fourier transforms are given by the formulas

$$F_{\mathbf{k}}(\mathbf{v}) = (2\pi)^{-L/2} \int_{R_L} e^{-i\mathbf{k}\mathbf{x}} \mathbf{v}(\mathbf{x}) dx \qquad (2.10.4)$$

$$\mathbf{v}(\mathbf{x}) = (2\pi)^{-L/2} \int_{R_L} e^{i\mathbf{k}\mathbf{x}} F_{\mathbf{k}}(\mathbf{v}) dk$$

where $\mathbf{k}\mathbf{x} = k_1 x_1 + \ldots + k_L x_L$. Let us now assume that the function $\mathbf{v}(\mathbf{x})$ is such that it satisfies the Parseval equality

$$\int_{R_L(\mathbf{x})} |\mathbf{v}(\mathbf{x})|^2 dx = \int_{R_L(\mathbf{k})} |F_{\mathbf{k}}(\mathbf{v})|^2 dk \qquad (2.10.5)$$

This equation means that the norm of the function has not changed in the result of the Fourier transform. We have, in particular, the equality

$$\| S \| = \| G \| \qquad (2.10.6)$$

where S is the step operator of the difference scheme. The scheme (2.10.2) is called *stable*, if for any initial data $\mathbf{u}^0(\mathbf{x})$ the estimate

$$\| \mathbf{u}^n \| \leq M \| \mathbf{u}^0 \|, \ n = 1, 2, \ldots$$

is satisfied, where M is a constant which does not depend on τ, $\mathbf{h}$, and n. Since the step operator S does not depend on n, the stability is equivalent to a uniform boundedness of the powers of the step operator:

$$\| S^n \| \leq M, \ n = 1, 2, \ldots \qquad (2.10.7)$$

Under the above choice of the norm this condition is equivalent (see (2.10.6)) to the condition

$$\| G^n \| \leq M, \ n = 1, 2, \ldots \qquad (2.10.8)$$

The difference operator S in (2.10.2) contains the shift operators T_j. Let us compute the Fourier transform of the expression (2.10.1). In accordance with the definition (2.10.4) of the direct Fourier transform we have that

$$F_{\mathbf{k}}(T_j^{l_j}\mathbf{u}) = (2\pi)^{-L/2}\int_{R_L} e^{-i\mathbf{k}\mathbf{x}}\mathbf{u}(x_1, \ldots, x_j + h_j l_j, \ldots, x_L)d\mathbf{x} \ \ (2.10.9)$$

Let us introduce the new variables

$$y_k = \begin{cases} x_k, \ k \neq j \\ x_j + h_j l_j, \ k = j \end{cases}$$

and let $\mathbf{y} = (y_1, \ldots, y_L)$. It is obvious that $dx_k = dy_k$, $k = 1, \ldots, L$, and that $x_j = y_j - h_j l_j$. Therefore, we can rewrite the formula (2.10.9) as follows:

$$F_{\mathbf{k}}(T_j^{l_j}\mathbf{u}) = (2\pi)^{-L/2}\int_{R_L} e^{-i(\mathbf{k}\mathbf{y}-k_j h_j l_j)}\mathbf{u}(\mathbf{y})d\mathbf{y} =$$

$$e^{ik_j h_j l_j}(2\pi)^{-L/2}\int_{R_L} e^{-i\mathbf{k}\mathbf{y}}\mathbf{u}(\mathbf{y})d\mathbf{y} = e^{ik_j h_j l_j}F_{\mathbf{k}}(\mathbf{u}) \qquad (2.10.10)$$

It follows from (2.10.10) that the wave numbers k_j enter the amplification matrix G only in the form of the products $k_j h_j$, $j = 1, \ldots, L$. Let us introduce the $\xi_j = k_j h_j, j = 1, \ldots, L$ and $\boldsymbol{\xi} = (\xi_1, \ldots, \xi_L)$. Then we may write that $G = G(\boldsymbol{\xi}, \mathbf{h}, \tau)$. The eigenvalues λ_j, $j = 1, \ldots, m$, of the matrix G are the zeroes of the *characteristic equation*

$$f_m(\lambda) = \det(G - \lambda I) = 0 \qquad (2.10.11)$$

where I is the $m \times m$ identity matrix. The coefficients of the equation (2.10.11) prove to be the functions of $\boldsymbol{\xi}$ and of $\boldsymbol{\kappa} = (\kappa_1, \ldots, \kappa_\nu)$, $\nu \geq 1$, where $\kappa_1, \ldots, \kappa_\nu$ are certain nondimensional variables (for example, the Courant number). Therefore, we may write that

$$f_m(\lambda) = \det(G - \lambda I) = \sum_{j=0}^{m} c_j(\boldsymbol{\kappa}, \boldsymbol{\xi}) \lambda^{m-j} = 0 \qquad (2.10.12)$$

Since $e^{i\xi_k l_k} = \cos(\xi_k l_k) + i\sin(\xi_k l_k)$, we obtain that the coefficients $c_j(\boldsymbol{\kappa}, \boldsymbol{\xi})$ are the periodic functions of ξ_k with the periods $2\pi/l_k$, $k = 1, \ldots, L$.

2.10.2 Von Neumann Necessary Stability Condition

The Fourier transform enables one to reduce the stability of the difference initial-value problems with constant coefficients to a purely algebraic problem of obtaining the conditions of boundedness of the amplification matrix powers. Define $\| G \|$ by the formula

$$\| G \| = \max_{\mathbf{v} \neq 0} |G\mathbf{v}| / |\mathbf{v}|$$

Let

$$R(\boldsymbol{\xi}, \mathbf{h}, \tau) = \max_{1 \leq l \leq m} |\lambda_l(G)| \qquad (2.10.13)$$

where $\lambda_l(G)$, $1 \leq l \leq m$ is the lth eigenvalue of the matrix G and $\boldsymbol{\xi} = (k_1 h_1, \ldots, k_L h_L)$, $\mathbf{h} = (h_1, \ldots, h_L)$. The quantity (2.10.13) is called the *spectral radius of the matrix* G. The matrix G is called *normal*, if the equality

$$GG^* = G^*G \qquad (2.10.14)$$

holds. The matrix G^* is defined as follows: if $G = \| a_{jk} + ib_{jk} \|_1^m$, then $G^* = \| a_{kj} - ib_{kj} \|_1^m$. In the case of normal matrices G, the equality $R(\boldsymbol{\xi}, \mathbf{h}, \tau) = \| G \|$ holds. In the general case

$$R(\boldsymbol{\xi}, \mathbf{h}, \tau)^n \leq \| G(\tau, \mathbf{k})^n \| \leq \| G(\tau, \mathbf{k}) \|^n \qquad (2.10.15)$$

Taking into account (2.10.8), we obtain from (2.10.15) the following necessary stability condition:

$$R(\boldsymbol{\xi}, \mathbf{h}, \tau)^n \leq M \quad \forall \, \mathbf{k}$$

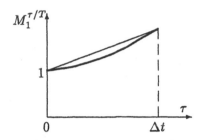

Figure 2.19: The bound for a piece of the exponential curve

If $M < 1$, then we can take another constant $M_1 \geq 1$, for which the inequality $R(\xi, \mathbf{h}, \tau)^n \leq M_1$ is valid. Therefore,

$$R(\xi, \mathbf{h}, \tau) \leq M_1^{1/n}, \ 0 < n \leq T/\tau$$

where $0 \leq t \leq T$ is a given interval on the time axis, in which the difference solution is to be determined. We have, in particular, that $R(\xi, \mathbf{h}, \tau) \leq M_1^{\tau/T}$. For the value of τ from the interval $0 < \tau < \Delta t$, the quantity $M_1^{\tau/T}$ is bounded by a linear expression of the form $1 + C_2 \tau$ (the straight line in Fig. 2.19). Thus $R(\xi, \mathbf{h}, \tau) \leq 1 + C_2 \tau$ for $0 < \tau < \Delta t$. From the definition of the spectral radius we obtain the condition

$$|\lambda_i| \leq 1 + O(\tau) \ for \ \begin{cases} 0 < t < \Delta t, \\ i = 1, 2, \dots, m \\ \text{and all } \mathbf{k} \in R_L(\mathbf{k}) \end{cases} \qquad (2.10.16)$$

where $\lambda_1, \lambda_2, \dots, \lambda_m$ are the eigenvalues of the amplification matrix $G(\xi, \mathbf{h}, \tau)$. The condition (2.10.16) is called the *von Neumann necessary stability condition*.

J.C. Strikwerda[3] distinguishes between the two forms of the von Neumann condition: the general von Neumann condition (2.10.16) and the *restricted* von Neumann condition

$$|\lambda_i| \leq 1, \quad i = 1, \dots, m \qquad (2.10.17)$$

He proves that one must apply the general condition (2.10.16), if the eigenvalues λ_i depend explicitly on $\mathbf{h}$ and τ, otherwise one may apply the restricted von Neumann condition (2.10.17).

Example 1. Let us find the necessary stability condition for the difference scheme (2.8.17). Let us make use of the vector matrix representation (2.8.21), (2.8.22) of this scheme. First, we find the entries of the amplification matrix G. We note that

$$(T_1 - 2I + T_{-1})e^{ijk_1h} = (2\cos k_1h - 2)e^{ijk_1h} = -2\sin^2(\xi/2)e^{ij\xi}$$

where $\xi = k_1h$. Therefore,

$$\det(G - \lambda I) = \begin{vmatrix} 2 - 4\kappa^2\sin^2\frac{\xi}{2} - \lambda & -1 \\ 1 & -\lambda \end{vmatrix} = 0$$

This determinant is easy to compute, and we obtain the following characteristic equation for determining λ:

$$\lambda^2 - \lambda(2 - a^2) + 1 = 0$$

where $a = (2c\tau/h)\sin(\xi/2)$. The roots $\lambda_{1,2}$ of this quadratic equation are

$$\lambda_{1,2} = (2 - a^2)/2 \pm (a/2)\sqrt{a^2 - 4}$$

If $a^2 \leq 4$, then $|\lambda_1| = |\lambda_2| = 1$. Therefore, we can see that the inequality

$$c\tau/h \leq 1$$

is the von Neumann necessary stability condition for the scheme (2.8.17).

Example 2. Now consider the MacCormack scheme (2.8.10), (2.8.11), where the matrix A is given by the formula (2.8.9). Using the shift operators (2.8.19) we at first rewrite the difference equation (2.8.10) in the form

$$\tilde{u}_j = [I - \tau A(T_1 - I)/h]u_j^n \qquad (2.10.18)$$

Substituting the expression (2.10.18) into the equation (2.8.11), we eliminate the intermediate values $\tilde{u}_j$:

$$u_j^{n+1} = \frac{1}{2}\Big[u_j^n + u_j^n - \tau A\Big(\frac{T_1 - I}{h}\Big)u_j^n -$$

$$\tau A\Big(\frac{I - T_{-1}}{h}\Big)\Big(I - \tau A\Big(\frac{T_1 - I}{h}\Big)\Big)u_j^n\Big] =$$

$$u_j^n - \tau A\Big(\frac{T_1 - T_{-1}}{2h}\Big)u_j^n + \frac{\tau^2 A^2}{2h^2}(T_1 - 2I + T_{-1})u_j^n$$

We easily obtain from this formula the expression for the amplification matrix G:

$$G = I - (\tau/h)i\sin\xi A + (\tau/h)^2(\cos\xi - 1)A^2 \qquad (2.10.19)$$

We can see that the matrix G is a polynomial in the matrix A. Let μ_1 and μ_2 be the egenvalues of the matrix G, and let λ_1 and λ_2 be the eigenvalues of the matrix A. Then it follows from (2.10.19) that

$$\mu_j = 1 - (\tau/h)i\sin\xi\lambda_j - (\tau/h)^2 \cdot 2\sin^2(\xi/2)\lambda_j^2$$

We have from (2.8.9) that $\lambda_1 = -\lambda_2 = c$. Therefore,

$$|\mu_j|^2 = (1 - 2\kappa^2\sin^2(\xi/2))^2 + \kappa^2\sin^2\xi =$$
$$1 - 4\kappa^2(1 - 4\kappa^2(1 - \kappa^2))\sin^4(\xi/2)$$

where $\kappa = c\tau/h$. We can see that the eigenvalues do not depend explicitly on h and τ. The steps h and τ can be varied in such a way that the quantity κ remains constant. Therefore, we may apply the restricted von Neumann condition (2.10.17). It can easily be shown that this condition is satisfied, if

$$c\tau/h \leq 1$$

Exercise 2.22. Show that the difference scheme (2.9.17) is absolutely stable, that is the condition (2.10.17) imposes no limitation on the stepsize τ.

2.11 DIFFERENCE INITIAL- AND BOUNDARY-VALUE PROBLEMS

2.11.1 The Notion of Numerical Boundary Conditions

In the vast majority of cases, mathematical physics problems are solved in the spatial domains of finite size. The requirement of the solution uniqueness, as well as the physics of the problem, usually lead to a necessity of imposing a boundary condition, at least on some pieces of the boundary of the spatial domain. The difference discretization of the initial and boundary conditions, as well as a difference approximation of the PDE, at the interior points of a spatial/temporal domain give rise to a *difference initial- and boundary-value problem*.

In this section we would like to discuss the problem of the construction of stable and accurate difference discretizations to initial- and boundary-value problems for the hyperbolic PDEs.

Let us at first consider a relatively simple initial- and boundary-value problem for the PDE (2.1.1). Let us set

$$u(x,0) = u_0(x),\ 0 \leq x \leq l \tag{2.11.1}$$

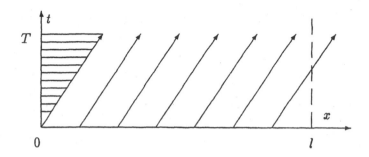

Figure 2.20: The characteristics in the problem (2.1.1), (2.11.1)

$$u(0,t) = g(t), \ 0 \le t \le T \qquad (2.11.2)$$

where $0 < l < \infty$, $0 < T < \infty$ and we have assumed that $a = const > 0$ in (2.1.1). The need for the formulation of the boundary condition (2.11.2) follows from Fig. 2.20. In the region dashed by horizontal lines the solution $u(x,t)$ is indeterminate in the case when the data on the line $x = 0$ are not specified.

In order to avoid a discontinuity in the solution $u(x,t)$ along the characteristic $x = at$, we will assume that the matching condition $u_0(0) = g(0)$ is satisfied. The solution of the initial- and boundary-value problem (2.1.1), (2.11.1), (2.11.2) may easily be found in the spatial/temporal domain and has the form

$$u(x,t) = \begin{cases} g(t - x/a), \ 0 \le x < at \\ u_0(x - at), \ at \le x \le l \end{cases}$$

Let us now consider the following difference approximation of the problem (2.1.1), (2.11.1), (2.11.2):

$$\frac{u_j^{n+1} - u_j^n}{\tau} + a\frac{u_j^n - u_{j-1}^n}{h} = 0 \qquad (2.11.3)$$

$$j = 1, \ldots, M, \ n = 0, 1, \ldots, [T/\tau]$$

$$u_j^0 = u_0(jh), \quad j = 0, 1, \ldots, M \qquad (2.11.4)$$

$$u_0^n = g(n\tau), \quad n = 0, 1, \ldots, [T/\tau] \tag{2.11.5}$$

where $hM = l$ and $[a]$ is the integral part of the number a. We can see from the stencil of the difference equation (2.11.3) that no boundary condition is needed to compute the value u_M^{n+1} on the right boundary $x = l$. The original formulation (2.11.1), (2.11.2) does also not contain a boundary condition on the right boundary $x = l$ of the spatial interval. Thus we can determine an approximate solution u^n with the aid of the formulas (2.11.3), (2.11.4) and (2.11.5) in the overall spatial/temporal domain $D = \{(x,t) : \ 0 \leq x \leq l, \ 0 \leq t \leq T\}$. The stability of the difference initial- and boundary-value problem (2.11.3)-(2.11.5) is obviously determined by the stability of the difference scheme (2.11.3). We have already found in the foregoing sections that the scheme (2.11.3) is stable provided that $0 \leq a\tau/h \leq 1$.

But such a favourable situation occurs relatively seldom. In the case when a higher-order difference schemes are used, the problem of *additional boundary conditions* arises. In Reference 25 these boundary conditions were termed *extraneous* boundary conditions. J.C. Strikwerda uses the term *numerical boundary conditions*.[3]

To illustrate the need for numerical boundary conditions let us consider the Lax-Wendroff scheme (2.4.3) for the PDE (2.1.1). Solving equation (2.4.3) for u_j^{n+1}, we obtain

$$u_j^{n+1} = u_j^n - \frac{\kappa}{2}(u_{j+1}^n - u_{j-1}^n) + \frac{\kappa^2}{2}(u_{j+1}^n - 2u_j^n + u_{j-1}^n) \tag{2.11.6}$$

where $\kappa = a\tau/h$. Since the value $u_0^n = g(n\tau)$ is given, we can compute the values $u_1^{n+1}, u_2^{n+1}, \ldots, u_{M-1}^{n+1}$ using (2.11.6). But when we try to compute the value u_M^{n+1}, we need the unknown value u_{M+1}^n in accordance with (2.11.6). Thus the Lax-Wendroff scheme (2.4.3) requires a numerical boundary condition at the right end $x = l$ of the spatial integration interval. This boundary condition is not needed for the original initial- and boundary-value problem (2.1.1), (2.11.1), (2.11.2), because the characteristics translate the information to the line $x = l$ from the line $t = 0$ (see Fig. 2.20).

The numerical boundary conditions for the given difference scheme may have different forms. The requirements for these boundary conditions are as follows:

(i) The numerical boundary conditions should not deteriorate the stability of the difference initial- and boundary-value problem very much.

(ii) The approximation order of the difference initial- and boundary-value problem should be preserved.

The stability of difference initial- and boundary-value problems can be studied theoretically by a number of methods. We mention here two most widespread methods:

(i) the method of energy inequalities;[27]

(ii) the method based on the GKS-theory.[28]

Each of these methods has its merits and shortcomings. We would like to mention the shortcomings of these methods.

The step operators of difference schemes for hyperbolic equations prove to be non-self-adjoint in the vast majority of cases. In addition, the checking of the remaining conditions of the theorems on the stability in the energy norm is often difficult. A success of the stability investigation by the method of energy inequalities depends substantially on the choice of an energy norm. Unfortunately there are no mathematically formalized criteria for the choice of the energy norm. The implementation of the energy method for a specific difference problem is an art.

The GKS theory is better formalized than the method of energy inequalities. It enables one to investigate the stability of difference initial- and boundary-value problems in the L_2 norm. But the GKS theory has the following shortcomings: (i) it was developed only for the case of one spatial variable; (ii) the practical stability studies by the GKS theory are possible indeed only for the homogeneous boundary condition, i.e. when $g(n\tau) \equiv 0$ in (2.11.5); (iii) the procedure of the GKS stability analysis becomes very difficult in the cases of the systems of difference equations.

Therefore, we will not present here the GKS theory. We only present sets of numerical boundary conditions. The choice of a stable difference boundary formula can be made with the aid of a trial-and-error procedure.

2.11.2   Forms of Numerical Boundary Conditions for the Right End

We consider at first the case when one has to use a numerical boundary condition at the right end $j = M$ of the spatial integration interval.

I. Space extrapolation formulas.

$$(T_1^{-1} - I)^k u^{n+1}(x_M) = 0 \qquad (2.11.7)$$

where k is a natural number, and T_1^{-1} is the shift operator, $T_1^{-1}u(x,t) = u(x - h,t)$. In particular, for $k = 1$ and $k = 2$ we have the following difference boundary formulas from (2.11.7):

$$u_M^{n+1} = u_{M-1}^{n+1}, \; k = 1 \qquad (2.11.8)$$

$$u_M^{n+1} = 2u_{M-1}^{n+1} - u_{M-2}^{n+1}, \; k = 2 \qquad (2.11.9)$$

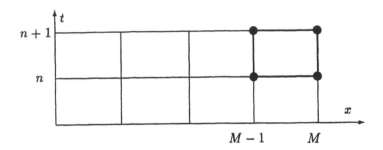

Figure 2.21: The stencil of the box scheme (2.11.3) is shown by dots

II. Space-time extrapolation.

$$(T_1^{-1}T_0^{-1} - I)^k u^{n+1}(x_M) = 0 \qquad (2.11.10)$$

where k is a natural number, $T_0^{-1}u(x,t) = u(x, t - \tau)$. For example, at $k = 2$ we get, from (2.11.10), that

$$u_M^{n+1} = 2u_{M-1}^n - u_{M-2}^{n-1} \qquad (2.11.11)$$

III. One-sided difference approximation. At the boundary node $x = Mh$, a one-sided difference scheme of the first order of approximation is used (compare with (2.11.3)):

$$u_M^{n+1} = u_M^n - (a\tau/h)(u_m^n - u_{M-1}^n) \qquad (2.11.12)$$

IV. "Box" scheme is used, which has the approximation order $O(\tau^2) + O(h^2)$ and is unconditionally stable:[26,9]

$$\frac{u_{M-1}^{n+1} + u_M^{n+1} - u_{M-1}^n - u_M^n}{2\tau} + a\frac{u_M^{n+1} + u_M^n - u_{M-1}^{n+1} - u_M^n}{2h} = 0 \qquad (2.11.13)$$

The stencil of this scheme is a box in the (x, t) plane, see Fig. 2.21.

V. The condition $\partial^2 u(x_M, t^n)/\partial x \partial t = 0$ is approximated by:[28]

$$u_M^{n+1} = u_{M-1}^{n+1} + u_M^n - u_{M-1}^n \qquad (2.11.14)$$

VI. The quantity $u_M^{n+1} - u_M^{n-1}$ is extrapolated by using the known grid values:

$$u_M^{n+1} - u_M^{n-1} = 2(u_{M-1}^{n+1} - u_{M-1}^{n-1}) - (u_{M-2}^{n+1} - u_{M-2}^{n-1}) \qquad (2.11.15)$$

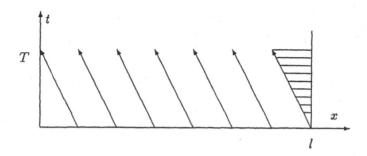

Figure 2.22: The characteristics in the problem (2.1.1), (2.11.17) (the case $a < 0$ in (2.1.1))

VII. A three-level scheme is considered:

$$u_M^{n+1} = u_M^{n-1} - 2\kappa(u_M^n - u_{M-1}^n) - c\kappa(u_M^{n+1} - 2u_M^n + u_M^{n-1}) \quad (2.11.16)$$

where $\kappa = a\tau/h$, and c is a free parameter which is chosen from the stability requirement.

Our practical experience shows that the "box" scheme (2.11.13) both for the one-step Lax-Wendroff scheme (2.4.3) and for the two-step Lax-Wendroff scheme (2.4.11)-(2.4.12) works well at least for moderate solution gradients at the right boundary $x = Mh$.

The extrapolation formula (2.11.9) works well in combination with the MacCormack scheme (2.4.13)-(2.4.14).

2.11.3   Forms of Numerical Boundary Conditions for the Left End

Let us now consider the case when $a = const < 0$ in (2.1.1). Then we may formulate the following initial and boundary values for the PDE (2.1.1):

$$u(x,0) = u_0(x), \quad 0 \le x \le l \quad (2.11.17)$$

$$u(l,t) = g_1(t), \quad 0 \le t \le T \quad (2.11.18)$$

where $0 < l < \infty$, $0 < T < \infty$. The need for the boundary condition (2.11.18) is illustrated in Fig. 2.22. In this case many difference schemes require numerical boundary conditions at the left end $x = 0$ of the spatial integration interval. Let us now present the counterparts of formulas (2.11.7)-(2.11.16) for the left boundary $j = 0$.

I'. Space extrapolation formulas.

$$(T_1 - I)^k u^{n+1}(0) = 0 \qquad (2.11.19)$$

where k is a natural number. In particular, at $k = 1$ and $k = 2$ we have the following difference boundary formulas from (2.11.19):

$$u_o^{n+1} = u_1^{n+1}, \ k = 1 \qquad (2.11.20)$$

$$u_0^{n+1} = 2u_1^{n+1} - u_2^{n+1}, \ k = 2 \qquad (2.11.21)$$

II'. Space-time extrapolation.

$$(T_1 T_0^{-1} - I)^k u_0^{n+1} = 0 \qquad (2.11.22)$$

where k is a natural number. For example, at $k = 2$ we get from (2.11.22) that

$$u_0^{n+1} = 2u_1^n - u_2^{n-1} \qquad (2.11.23)$$

III'. One-sided difference approximation.

$$u_0^{n+1} = u_0^n - (a\tau/h)(u_1^n - u_0^n) \qquad (2.11.24)$$

IV'. " Box " scheme

$$\frac{u_0^{n+1} + u_1^{n+1} - u_0^n - u_1^n}{2\tau} + a\frac{u_1^{n+1} + u_1^n - u_0^{n+1} - u_0^n}{2h} = 0 \qquad (2.11.25)$$

V'. The condition $\partial^2 u(0, t^n)/\partial x \partial t = 0$ is approximated:

$$u_0^{n+1} = u_1^{n+1} + u_0^n - u_1^n \qquad (2.11.26)$$

VI'. The quantity $u_0^{n+1} - u_0^{n-1}$ is extrapolated by using the known grid values:

$$u_0^{n+1} - u_0^{n-1} = 2(u_1^{n+1} - u_1^{n-1}) - (u_2^{n+1} - u_2^{n-1}) \qquad (2.11.27)$$

VII'. A three-level scheme is considered:

$$u_0^{n+1} = u_0^{n-1} - 2\kappa(u_1^n - u_0^n) - c_1\kappa(u_0^{n+1} - 2u_0^n + u_0^{n-1}) \qquad (2.11.28)$$

where $\kappa = a\tau/h$, and c is a free parameter which is chosen from the stability requirement.

VIII. A three-level scheme of the form

$$u_0^{n+1} = u_0^{n-1} - \frac{11}{6}\kappa\mu(u_0^{n+1} - 2u_0^n + u_0^{n-1})$$

$$+ \frac{\kappa}{3}(2u_3^n - 9u_2^n + 18u_1^n - 11u_0^n)$$

is used, where μ is a free parameter.

Let us now discuss the question of the accuracy of the numerical solution in the case when the first-order difference equation (2.11.12) is used at $j = M$ in combination with the second-order Lax-Wendroff scheme (2.4.3). Since Eq. (2.11.12) has only first order accuracy, there arises the question whether the overall accuracy of the numerical solution u_j^n, $1 \leq j < M$, will also deteriorate. Let us investigate this question. Let us consider the difference problem

$$\begin{cases} u_0^{n+1} = g((n+1)\tau) \\ u_j^{n+1} = u_j^n - \frac{\kappa}{2}(u_{j+1}^n - u_{j-1}^n) + \frac{\kappa^2}{2}(u_{j+1}^n - 2u_j^n + u_{j-1}^n) \\ \qquad 1 \leq j \leq M-1 \\ u_M^{n+1} = (1-\kappa)u_M^n + \kappa u_{M-1}^n \end{cases} \quad (2.11.29)$$

The last equation in (2.11.29) introduces a disturbance. Let us investigate the structure of this disturbance. Following Reference 29 let us consider the particular solution

$$u = \alpha + \beta(x - at) + \gamma(x - at)^2 \qquad (2.11.30)$$

of the PDE (2.1.1), where α, β and γ are constants. The Lax-Wendroff scheme (2.4.3) is a second order approximation, therefore, the function (2.11.30) also proves to be an exact solution of the difference equation (2.4.3). The last equation in (2.11.29) is also exact for the function (2.11.30), but only in the case $\gamma = 0$. The term $(x - at)^2$ introduces a disturbance. Therefore, let us take the solution $u(x,t) = (x-at)^2$ of the PDE (2.1.1) and substitute into the difference equation (2.11.29). For this purpose, we assume $u_j^n = (jh - an\tau)^2$ in (2.11.29) and compute the residual:

$$u_M^{n+1} - (1-\kappa)u_M^n - \kappa u_{M-1}^n =$$

$$(1-\kappa)(Mh - a((n+1)\tau)^2 - (1-\kappa)(Mh - an\tau)^2 -$$

$$\kappa[(M-1)h - an\tau]^2 = a\tau(a\tau - h) \qquad (2.11.31)$$

Let us rewrite equation (2.11.31) in the form

$$u_M^{n+1} = (1-\kappa)u_M^n + \kappa u_{M-1}^n + \underline{a\tau(a\tau - h)}$$

The underlined term is the residual. We now try to cancel this residual. For this purpose we consider the system

$$\begin{cases} v_0^{n+1} = g((n+1)\tau) \\ v_j^{n+1} = v_j^n - \frac{\kappa}{2}(v_{j+1}^n - v_{j-1}^n) + \frac{\kappa^2}{2}(v_{j+1}^n - 2v_j^n + v_{j-1}^n) \\ \qquad\qquad 1 \le j \le M-1 \\ v_M^{n+1} = (1-\kappa)v_M^n + \kappa v_{M-1}^n - a\tau(a\tau - h) \end{cases} \quad (2.11.32)$$

Let us find the particular solution of the system (2.11.32) in the form $v_j^n = v_j + u_0^n$. We immediately obtain that $v_0 = 0$. From (2.11.32) we obtain the following difference equations for determining v_j:

$$\begin{cases} 0 = -\frac{\kappa}{2}(v_{j+1} - v_{j-1}) + \frac{\kappa^2}{2}(v_{j+1} - 2v_j + v_{j-1}) \\ 0 = -\kappa(v_M - v_{M-1}) - a\tau(a\tau - h) \end{cases} \quad (2.11.33)$$

Let us divide both sides of the first equation of (2.11.33) by $\kappa/2$:

$$(\kappa - 1)v_{j+1} - 2\kappa v_j + (1+\kappa)v_{j-1} = 0, \ 1 \le j \le M-1 \quad (2.11.34)$$

We now search for the solution of (2.11.34) in the form

$$v_j = c\lambda^{j-M} \quad (2.11.35)$$

Substituting (2.11.35) into (2.11.34), we obtain the following quadratic equation for λ:

$$(\kappa - 1)\lambda^2 - 2\kappa\lambda + (1+\kappa) = 0$$

This equation has the zeros $\lambda_1 = 1$, $\lambda_2 = (\kappa+1)/(\kappa-1)$. Therefore, the equation (2.11.34) has the solution

$$v_j = c_1 + c_2 \left(\frac{\kappa+1}{\kappa-1}\right)^{j-M} \quad (2.11.36)$$

At $j = 0$ we have that

$$v_0 = c_1 + c_2(-1)^M \left(\frac{\kappa+1}{1-\kappa}\right)^{-M}$$

The stability condition of the Lax-Wendroff scheme is expressed by the inequalities $0 < \kappa \le 1$. Therefore, the quantity $((\kappa+1)/(1-\kappa))^{-M} \to 0$ as $M \to \infty$. For this reason we take the value $c_1 = 0$ in (2.11.36). Let us use the second equation of the system (2.11.33) to determine the constant c_2:

$$-c_2[1 - (\kappa-1)/(\kappa+1)] - (a\tau - h)h = 0$$

From this equation we find that

$$c_2 = h(h - a\tau)(\kappa + 1)/2 = h^2(1 - \kappa^2)/2$$

Now let us construct the following grid function:

$$\bar{u}_j^n = u_j^n + v_j^n = (jh - an\tau)^2 + \frac{h^2(1 - \kappa^2)}{2}\left(\frac{\kappa + 1}{\kappa - 1}\right)^{j-M}, \quad 1 \leq j \leq M$$

$$(2.11.37)$$

It is easy to see that the function (2.11.37) satisfies the difference equations (2.4.3). We can also see that the exact solution $(x - at)^2 = (jh - an\tau)^2$ is disturbed by the term of the order $O(h^2)$. This means that the overall second-order accuracy of the numerical solution is preserved despite the fact that the first-order difference equation is used in the boundary node $j = M$. The same effects indeed take place also for other difference schemes and for different hyperbolic differential equations. However, in the case of complex equations it becomes difficult to carry out such an investigation. B.Gustafsson[30] has investigated this question for certain classes of difference schemes, and his result is as follows: if the approximation order of the explicit difference scheme

$$u^{n+1}(x) = \sum_{j=-r}^{p} b_j u^n(x + jh) \qquad (r, p > 0)$$

is $O(h^k)$, and the boundary conditions are globally approximated with the order $O(h^{k-1})$, then we have the overall approximation order $O(h^k)$.

Exercise 2.23. Take the notebook lw1.ma from subsection 2.4.1 and modify it to use the initial function

$$u_0(x) = \begin{cases} 0, & a_1 \leq x < x_1 \\ k(x - x_1), & x_1 \leq x \leq b_1 \end{cases}$$

where $[a_1, b_1]$ is the integration interval on the x-axis, and $a_1 < x_1 < b_1$, $k > 0$. Consider the different forms of numerical boundary conditions at the right end $x = b_1$ (see subsection 2.11.3) and study their effect on the accuracy and stability of the numerical solution.

2.12   CONSTRUCTION OF DIFFERENCE SCHEMES FOR MULTIDIMENSIONAL HYPERBOLIC PROBLEMS

The mathematical physics problems, which are the most interesting from the point of view of practical applications, involve two or three

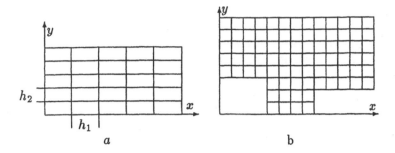

Figure 2.23: (a) rectangular spatial domain; (b) the spatial domain is a union of rectangles

spatial variables. Such problems of mathematical physics are called multidimensional problems. The addition of one spatial dimension, denoted for example by y, to the spatial variable x, leads to a drastic increase of the variety of the numerical methods for the solution of problems governed by the PDEs. The major part of these methods have been presented in Reference 2.

The scalar convection equation has in the multidimensional case the form

$$\frac{\partial u}{\partial t} + \sum_{j=1}^{L} v_j \frac{\partial u}{\partial x_j} = 0 \qquad (2.12.1)$$

where $v_1, \ldots, v_L$ are the components of the velocity vector $\mathbf{v}$ along the axes $x_1, \ldots, x_L$ ($L \geq 1$), respectively. The components $v_1, \ldots, v_L$ may generally be the functions of the spatial coordinates $x_1, \ldots, x_L$ and the dependent variable u. The particular case of Eq. (2.12.1) is the two-dimensional convection equation

$$\frac{\partial u}{\partial t} + A \frac{\partial u}{\partial x} + B \frac{\partial u}{\partial y} = 0 \qquad (2.12.2)$$

Before writing the difference equations we must define the spatial grid on which the difference equation system will be solved. Throughout this section we will assume that the spatial grid is rectangular and uniform. It is convenient to use such a grid in cases when the computational domain in the (x, y) plane is a rectangle or can be represented as a union of rectangles, see Fig. 2.23. Let us denote by h_1 and h_2 the sizes of the sides of an individual cell of the spatial grid along the x- and y-axes, respectively (see Fig. 2.23).

2.12.1   The MacCormack Scheme

In this subsection we will assume that the coefficients A and B in Eq. (2.12.2) are constant. Let us write the MacCormack scheme[16] for Eq. (2.12.2) (cf. (2.4.13), (2.4.14)):

$$\tilde{u}_{ij} = u_{ij}^n - \tau A \Delta_1^+ u_{ij}^n - \tau B \Delta_2^+ u_{ij}^n$$

$$u_{ij}^{n+1} = (1/2)(u_{ij}^n + \tilde{u}_{ij} - \tau A \Delta_1^- \tilde{u}_{ij} - \tau B \Delta_2^- \tilde{u}_{ij}) \qquad (2.12.3)$$

where the spatial differencing operators $\Delta_1^{\pm}$ and $\Delta_2^{\pm}$ are defined by the formulas

$$\Delta_1^+ u_{ij} = (u_{i+1,j} - u_{ij})/h_1, \quad \Delta_2^+ u_{ij} = (u_{i,j+1} - u_{ij})/h_2$$
$$\Delta_1^- u_{ij} = (u_{ij} - u_{i-1,j})/h_1, \quad \Delta_2^- u_{ij} = (u_{ij} - u_{i,j-1})/h_2 \qquad (2.12.4)$$

The first equation in (2.12.3) is called the predictor while the second is called the corrector MacCormack[16] pointed out that (2.12.3) is only one of four methods of second-order accuracy. Three other variants of the MacCormack method are obtained, if instead of first using two forward spatial differences and the two backward differences, the reverse procedure could be followed, or one forward and one backward difference could be followed by corresponding backward and forward difference. Let us denote the specific MacCormack scheme (2.12.3) by the formula $u^{n+1} = M_{--}^{++} u^n$, where $\tilde{u}_{ij}$ has been eliminated. Then all the four variants of the MacCormack scheme for the two-dimensional case are obtained as the set of the step operators

$$M_{--}^{++}, \ M_{++}^{--}, \ M_{-+}^{+-}, \ M_{+-}^{-+} \qquad (2.12.5)$$

We note that the MacCormack scheme with operator M_{--}^{++} has gained a wider acceptance than the remaining schemes in (2.12.5).

Substituting the function

$$u_{jk}^n = \lambda^n \exp(ijk_1 h_1 + ijk_2 h_2), \quad i = \sqrt{-1}$$

into the scheme (2.12.3) we can obtain the characteristic equation in the form

$$\lambda + b_1(\kappa, \xi) + ib_2(\kappa, \xi) = 0 \qquad (2.12.6)$$

where

$$b_1(\kappa, \xi) = -1 + \kappa_1^2(1 - \cos \xi_1) + \kappa_2^2(1 - \cos \xi_2)$$
$$+ \kappa_1 \kappa_2[(1 - \cos \xi_1)(1 - \cos \xi_2) + \sin \xi_1 \sin \xi_2] \qquad (2.12.7)$$

$$b_2(\kappa, \xi) = \kappa_1 \sin \xi_1 + \kappa_2 \sin \xi_2$$

$$\kappa_1 = A\tau/h_1, \quad \kappa_2 = B\tau/h_2 \qquad (2.12.8)$$

$$\xi_1 = k_1 h_1, \ \xi_2 = k_2 h_2, \ \kappa = (\kappa_1, \kappa_2), \ \xi = (\xi_1, \xi_2)$$

The von Neumann condition $|\lambda| \leq 1$ translates, for the characteristic equation (2.12.6), into the inequality

$$b_1^2 + b_2^2 - 1 \leq 0 \qquad (2.12.9)$$

The study of the inequality (2.12.9) by analytic methods requires intensive calculations. Therefore, we present only the final result: the von Neumann condition (2.12.9) is satisfied for any $\xi_1, \xi_2 \in [0, 2\pi]$, if

$$\begin{array}{c} \kappa_1 \geq 0, \ \kappa_2 \geq 0, \ \kappa_1 + \kappa_2 \leq 1 \\ \text{or} \\ \kappa_1 \leq 0, \ \kappa_2 \leq 0, \ \kappa_1 + \kappa_2 \geq -1 \end{array} \qquad (2.12.10)$$

Since the difference scheme (2.12.3) is a two-level scheme, the necessary stability condition (2.12.10) is also the sufficient stability condition for this scheme in the L_2 norm. In Fig. 2.24 we show the stability region of the MacCormack scheme (2.12.3) in the plane of variables (2.12.8). Thus this scheme proves to be unstable in the second and fourth quadrants of the (κ_1, κ_2) plane, in which the velocity components A and B in (2.12.2) have the opposite signs.

2.12.2 The Program mc2d.ma

To illustrate the quality of the numerical solutions obtained by using the MacCormack scheme (2.12.3), which is second-order accurate both in space and time, we have carried out the solution of a problem on a plane parallel propagation of a square wave. Let us first formulate the initial condition $u(x, y, 0) = u_0(x, y)$, which corresponds to the square wave test. In this test, the discontinuity front is represented by a straight line in the (x, y) plane. Introduce a system of Cartesian rectangular coordinates (n_1, n_2), whose n_2-axis is directed along a normal to the line of the discontinuity front at $t = 0$, and the coordinate origin is placed in an arbitrary point (x_f^0, y_f^0) lying on $\mathcal{L}_d$, see Fig. 2.25. The square wave propagates with increasing t in the direction of the normal n_2 to the discontinuity front, thus the fluid flow in the coordinate system (n_1, n_2) is one-dimensional. Taking this into account it is easy to obtain,

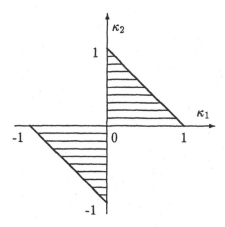

Figure 2.24: The stability region of the MacCormack scheme (2.12.3)

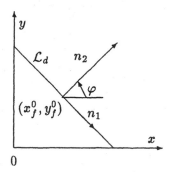

Figure 2.25: The directions of the n_1- and n_2-axes in the problem on a temporal evolution of a step.

by analogy with (2.2.6) and (2.2.10), the exact solution of the problem under consideration for any $t \geq 0$:

$$n_2 = (x - x_f^0)\cos\varphi + (y - y_f^0)\sin\varphi$$

$$u(x, y, 0) = \tilde{u}(n_2, 0) = \begin{cases} u_l, & n_2 \leq 0 \\ u_r, & n_2 > 0 \end{cases} \qquad (2.12.11)$$

at $t > 0$

$$u(x, y, t) = \begin{cases} u_l, & n_2 - Ut \leq 0 \\ u_r, & n_2 - Ut > 0 \end{cases} \qquad (2.12.12)$$

where u_l, u_r, x_f^0, y_f^0 are given constant values with $u_l \neq u_r$ and $U^2 = A^2 + B^2 \geq 0$.

Table 2.4. Parameters used in the program mc2d.ma

Parameter	Description
U	the modulus of the convection speed, $U = (A^2 + B^2)^{1/2}$, A and B are the constant components of the convection velocity along the x- and y-axes, respectively; $U > 0$
phi	the angle φ between the convection velocity vector and the positive direction of the x-axis, $0 \leq \varphi \leq \pi/2$
xf0	the abscissa x_f^0 of a point at the discontinuity front at $t = 0$ on the upper boundary $y = yb$, $x_f^0 \geq 0$
ul	the value u_l of $u(x, y, 0)$ behind the discontinuity front
ur	the value u_r of $u(x, y, 0)$ ahead of the discontinuity front
xa	the size x_a of the rectangular spatial computational domain along the x-axis, $x_a > 0$
yb	the size y_b of the rectangular spatial domain along the y-axis, $y_b > 0$
ix	the number i_x of cells of an uniform rectangular spatial grid along the x-axis, $i_x > 1$
jy	the number j_y of cells of an uniform rectangular spatial grid along the y-axis, $j_y > 1$
Saf	the safety factor in the formula for the computation of the time step τ; $0 < Saf \leq 1$ to ensure a stable computation
Npic	the number of the pictures of the difference solution surfaces $u = u(x, y, t)$ for Npic moments of time, Npic > 1
xview, yview, zview	the coordinates of the viewpoint for the surface $u = u(x, y, t)$. At least one of the coordinates xview, yview, zview should be greater than 1 in modulus

We now describe the *Mathematica* package mc2d.ma, which was written to solve numerically the test problem (2.12.2), (2.12.11) using the MacCormack difference scheme (2.12.3). The input parameters used in the program mc2d.ma are described in Table 2.4. The main function of the file mc2d.ma is

$$\text{mc2d}\,[U, \varphi, x_f^0, u_l, u_r, x_a, y_b, i_x, j_y, Saf, Npic, xview, yview, zview]$$

The organization of the computations in this program is similar to that of the program advbac.m from Section 2.2. The numerical solution is determined in a rectangular spatial region

$$D = \{(x, y)\mid 0 \le x \le x_a, \ 0 \le y \le y_b\} \qquad (2.12.13)$$

where x_a and y_b are user-specified positive constants; they are denoted in our file mc2d.ma by xa and yb, respectively. The cells of a rectangular uniform grid in the region (2.12.13) are numbered from 1 to ix in the direction of the x-axis and from 1 to jy in the direction of the y-axis. The exact solution (2.12.12) is computed with the aid of the function uex:

```
uex[x_, y_, t_] :=
(* --- The function to compute the exact solution of the
       problem on a plane parallel convection of the
       substance property u(x, y, t) ---------------- *)
( xfn = xf1 - (y - yf1)*ckf + U1*t/cphi;
  z = If[x < xfn, ul1, ur1]; z )
```

The predictor equation for $\tilde{u}_{ij}^n$ in the MacCormack scheme (2.12.3) requires the specification of the numerical boundary conditions for the computation of $\tilde{u}_{ij}$ in the upper horizontal row of cells $j = j_y$ and in the right vertical row of cells $i = i_x$. The extrapolation

$$\tilde{u}_{i,j_y+1} = 2\tilde{u}_{i,j_y} - \tilde{u}_{i,j_y-1}, \ i = 1, \ldots, i_x$$

(see also Eq. (2.11.11)) has led to the development of instability of the numerical solution u^{n+1} in the upper row of cells and in the neighboring horizontal rows. In this connection we have specified the predictor values $\tilde{u}_{i,j_y+1}$ from the exact solution (2.12.12) at $t = n\tau$.

The corrector equation for u_{ij}^{n+1} of scheme (2.12.3) requires the specification of the numerical boundary conditions for the computation of $u_{i,j}^{n+1}$ at $j = -1$, $i = 1, \ldots, i_x$ and at $i = -1$, $j = 1, \ldots, j_y$. For this purpose we again use the exact solution (2.12.12), but with $t = (n+1)\tau$. This technique of specifying the numerical boundary conditions for the MacCormack scheme (2.12.3) has enabled us to considerably reduce the numerical solution oscillations arising, in some cases, near the boundaries of the region (2.12.13). It should be noted that the construction of stable numerical boundary conditions for the two-dimensional difference schemes requires a separate detailed study. Some considerations of this question may be found in the book. [31]

The MacCormack scheme (2.12.3) was programmed similarly to the one-dimensional Lax-Wendroff scheme (see Eqs. (2.4.4) and (2.4.5)). Let us at first consider the predictor equation of the scheme. Rewrite this equation in the form

$$\tilde{u}_{ij} = u_{ij}^n - \tau \frac{F_{i+1/2,j}^n - F_{i-1/2,j}^n}{h_1} - \tau \frac{G_{i,j+1/2}^n - G_{i,j-1/2}^n}{h_2} \qquad (2.12.14)$$

where the expressions for the fluxes $F_{i\pm1/2,j}^n$ and $G_{i,j\pm1/2}^n$ are given by the formulas

$$F_{i+1/2,j}^n = A u_{i+1,j}^n, \quad F_{i-1/2,j}^n = A u_{ij}^n$$
$$G_{i,j+1/2}^n = B u_{i,j+1}^n, \quad G_{i,j-1/2}^n = B u_{ij}^n$$

Then the flux $F_{i+1/2,j}^n$ across the right boundary of the (i,j) cell will be the flux across the left boundary of the $(i+1,j)$ cell. We can observe in a similar way that the flux $G_{i,j+1/2}^n$ across the upper boundary of the (i,j) cell will be the flux across the lower boundary of the $(i, j+1)$ cell. Therefore, we can store the computed values for the fluxes $F_{i+1/2,j}^n$ and $G_{i,j+1/2}^n$ to use them for the computation of the predictor values $\tilde{u}$ in the next cells. The values of $G_{i,j+1/2}^n$ are stored in the one-dimensional list of values fb[[i]] in mc2d.ma.

The corrector equation of scheme (2.12.3) may be written, similarly to Eq. (2.12.14), in the form

$$u_{ij}^{n+1} = \frac{1}{2}\left(u_{ij}^n + \tilde{u}_{ij} - \tau\frac{\tilde{F}_{i+1/2,j} - \tilde{F}_{i-1/2,j}}{h_1} - \tau\frac{\tilde{G}_{i,j+1/2} - \tilde{G}_{i,j-1/2}}{h_2} \right)$$

where

$$\tilde{F}_{i+1/2,j} = A\tilde{u}_{i,j}, \quad \tilde{F}_{i-1/2,j} = A\tilde{u}_{i-1,j}$$
$$\tilde{G}_{i,j+1/2} = B\tilde{u}_{i,j}, \quad \tilde{G}_{i,j-1/2} = B\tilde{u}_{i,j-1}$$

Therefore, the computation by the corrector equation of scheme (2.12.3) can be optimized in the same way as in the case of the predictor equation (2.12.14).

This implementation requires the computation of only two fluxes, $F_{i+1/2,j}$ and $G_{i,j+1/2}$, for each cell of a spatial grid. The computer time needed for the overall computation by the difference scheme thus reduces by a factor of two in comparison with the case when all the four fluxes $F_{i\pm1/2,j}$ and $G_{i,j\pm1/2}$ are computed for each (i,j) cell. The achieved savings in the computer time are very significant when the Euler or Navier-Stokes equations of compressible fluids are to be solved numerically.

The numerical solution $u^n = u(x, y, n\tau)$ represents a surface in the three-dimensional Euclidean space of (x, y, u) points. The values of u^n are computed at the centers of the (i, j) cells of a spatial uniform grid. This means that we have a two-dimensional list of the values u[[i,j]].

The numerical computation by the MacCormack scheme was implemented in the following fragment of the program mc2d.ma:

```
(* ------- Begin of the Do-loop for time steps ------- *)
       Do[

(* --- Computation of the fluxes across the lower boundary
                   y = 0 ------------------- *)
   fb = Table[ul, {i, ix}]; y1 = 0.0;
   Do[
        fb[[i]] = u[[i,1]], {i, ix}];
(*  The computation of the predictor values up[[i, j]]  *)
(* --------- Do-loop over the (i, j) nodes ------------- *)
     Do[ fr = u[[1, j]];
          Do[ fl = fr; uz = u[[i, j]];
       fr = If[i == ix, uz, u[[i + 1, j]] ];
If[j == jy, x1 = (i - 0.5)*hx; y1 = (jy + 0.5)*hy;
    gt = uex[x1, y1, tn], gt = u[[i, j + 1]] ];
    gb = fb[[i]]; fb[[i]] = gt;
  up[[i, j]] = uz - cp1*(fr - fl) - cp2*(gt - gb), {i, ix}],
  {j, jy}];

(* --- Computation of the corrector values u[[i, j]] --- *)
   tn = tn + dt; umax = 0.0; umax1 = 0.0; y1 = -0.5*hy;
   Do[ x1 = (i - 0.5)*hx; fb[[i]] = uex[x1, y1, tn],
                                    {i, ix}];
    Do[ x1 = -0.5*hx; y1 = (j - 0.5)*hy;
   fr = uex[x1, y1, tn];
        Do[ fl = fr; uz = up[[i, j]];
            fr = uz; gt = uz; gb = fb[[i]];  fb[[i]] = gt;
```

```
u[[i, j]] = 0.5*(u[[i, j]] + uz - cp1*(fr - fl)
           - cp2*(gt - gb));
y1 = Abs[u[[i,j]] ]; umax1 = If[y1 > umax1, y1, umax1];
y = Abs[u[[i, j]] - um]; umax = If[y > umax, y, umax],
           {i, ix}], {j, jy}];
```

The *Mathematica* system has a number of convenient functions for making a three-dimensional plot of the array of heights $u(x, y)$. We have used in our program mc2d.ma the function ListPlot3D[u]. In Fig. 2.26

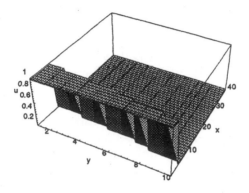

Figure 2.26: The initial surface $u = u(x, y, 0)$ using the default view point (1.3,-2.4,2)

we show the initial surface $u = u(x, y, 0)$, which is obtained at $\varphi = \pi/6$ in (2.12.11). To obtain Fig. 2.26 load first the file mc2d.ma and then click in the line of input data

```
mc2d[1.0, Pi/6, 0.2, 1.0, 0.2, 2.0, 0.5, 40, 10, 0.8, 2,
     1.3, -2.4, 2.0]
```

The appearance of a surface depends strongly on a view point (x_{view}, y_{view}, z_{view}) from which the surface is observed. If this point is not specified explicitly by the user, the *Mathematica* system uses the default view point (1.3, -2.4, 2). It should be noted that these coordinates are measured in a special coordinate system whose origin (0,0,0) is placed at the center of a box containing the three-dimensional graphics object.

The special coordinates are scaled so that the longest side of the box corresponds to one unit. For a cubical box, therefore, each of the special coordinates runs from -1/2 to 1/2 across the box. Note that the view point must always lie outside the box. With the default setting **Boxed** -> **True**, *Mathematica* draws the edges of this box explicitly (see Fig. 2.26).

The picture of Fig. 2.26 does not enable us to see, in sufficient detail, the region of the discontinuity front in the initial piecewise constant distribution $u(x, y, 0)$. Therefore, it is desirable to choose another view point. The function **ListPlot3D** can be used together with the option

$$\text{ViewPoint} \rightarrow \{\text{xview, yview, zview}\}.$$

The typical choices for the **ViewPoint** option are as follows:

{1.3,-2.4,2}	default view point
{0,-2,0}	directly in front
{0,-2,2}	in front and up
{0,-2,-2}	in front and down
{-2,-2,0}	left-hand corner
{2,-2,0}	right-hand corner
{0,0,2}	directly above

We have chosen the point

$$\text{xview} = 2.0, \ \text{yview} = 2.0, \ \text{zview} = 2.0$$

The picture of the initial surface $u = u(x, y, 0)$ with such a view point choce is shown in Fig. 2.27. Comparing Figs. 2.26 and 2.27 we can see

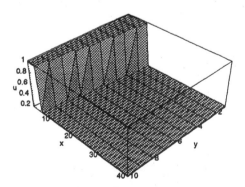

Figure 2.27: The initial surface $u = u(x, y, 0)$ using the view point (2,2,2)

that Fig. 2.27 allows us to better see the details of the initial distribution

$u = u(x, y, 0)$. Therefore, the view point with the coordinates (2,2,2) is preferable for the test problem under consideration.

The time step τ allowed by the stability of scheme (2.12.3) can be computed with regard for the stability condition (2.12.10) by the formula

$$\tau = \theta/(|A|/h_1 + |B|/h_2) \qquad (2.12.15)$$

where θ is a safety factor, $0 < \theta \leq 1$ to ensure a stable computation; h_1 and h_2 are the steps of an uniform spatial grid along the x- and y-axes, respectively (see Eqs. (2.12.3), (2.12.4)). To obtain Fig. 2.28 load the

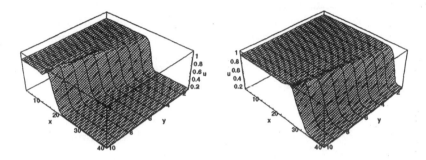

Figure 2.28: The numerical solution surfaces $u = u(x, y, t)$ obtained at $\theta = 0.8$ in (2.12.15). (a) $t = 19\tau = 0.556359$; (b) $t = 38\tau = 1.11272$. The mesh of 40×10 cells

file mc2d.ma and then click in the cell of input data

```
mc2d[1.0, Pi/6, 0.2, 1.0, 0.2, 2.0, 0.5, 40, 10, 0.8, 2,
     2, 2, 2]
```

We note that we have used a relatively crude grid of 40×10 cells for the spatial grid. This small number of grid points allows us to rapidly obtain the difference solution even on a small personal computer. In the general case, the computer time needed for obtaining the numerical solution with the aid of a difference scheme on a rectangular grid is proportional to the overall number of cells of a spatial grid. If we, for example, increase the values of i_x and j_y by a factor of two, then the needed computer time will increase by the factor of four. A crude spatial grid is usually sufficient at the stage of the debugging. But the accuracy requirements at the stage of obtaining the final solution may impose, in the case of complex real-life problems, the use of mesh of one million cells or more.

It can be seen from Fig. 2.28 that the step in the surface $u = u(x, y, t)$ propagates with increasing t in the direction of the vector with the co-ordinates $(\cos(\pi/6), \sin(\pi/6))$. At the same time, is smeared the discontinuity in the difference solution over several cells of a spatial grid takes place. The amplitude of the parasitic oscillations of the numerical solution in the region behind the discontinuity front proves to be relatively small at $\theta = 0.8$ in Eq. (2.12.15).

In many applied problems of the mathematical physics it is important to know the local behavior of the solution in any small subregion of a spatial integration domain. *Mathematica* gives the possibility of assigning different colors to the cells of an uniform spatial grid depending on the numerical solution value in an individual cell. Let us at first illustrate this technique with the following example. Consider the function

$$v(x, y) = |\sin 2x \cos 2y| \qquad (2.12.16)$$

in the square $0 \le x, y \le 2\pi$. This function has its minimum $v = 0$ along the lines $x = 0, \pi/2, 3\pi/2, 2\pi$ and $y = \pi/4, 3\pi/4, 5\pi/4, 7\pi/4$. The function (2.12.16) has its maximum $v = 1$ at points with the abscissas $x_j = (2j - 1)\pi/4$, $j = 1, 2, 3, 4$ and with the ordinates $y_k = (k - 1)\pi/2, k = 1, 2, 3, 4, 5$.

The *Mathematica* function Hue[h] is a graphics directive which specifies that graphics objects which follow are to be displayed, if possible, in a color corresponding to hue h. The parameter h must be between 0 and 1. Values of h outside this range are treated cyclically. As h varies from 0 to 1, the color corresponding to Hue[h] runs through red, yellow, green, cyan, blue, magenta, and back to red again.

The *Mathematica* function DensityGraphics[v] produces a density plot, in which certain color is assigned to each spatial cell depending on the value of v in the cell. By default, each cell is assigned a gray level, running from black to white as the value of v in the cell increases. The option ColorFunction allows the user to specify a function which is applied to each cell value to find the color of the cell. The cell values v_{jk} are scaled so as to run between 0 and 1 in a particular density plot. The DensityGraphics can be displayed using Show. In the generated picture, the boundaries of cells are shown by black lines. If the spatial grid is dense, these grid lines can obscure the local colors. In such cases it is reasonable to omit these black lines. This can be done by specifying the option Mesh -> False for the function Show.

In Fig. 2.29 we show the density graphics of the function (2.12.16) on a mesh of 40×40 cells. We have used the option

ColorFunction -> (cl2[#]&)

for this purpose, where the function cl2[f_] has the form cl2[f_] := Hue[f]. The picture of Fig. 2.29 has the following two shortcomings:

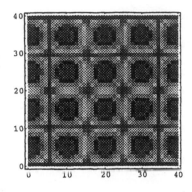

Figure 2.29: The density graphics on a 40 × 40 mesh

(a) because of a relatively crude mesh, the effect of the quantization of the color over the rectangular cells is very appreciable; (b) the red color is present both in the regions of local minima and of local maxima.

The quantization effect can be reduced by using a more dense grid in the spatial region. Our experience shows that the 100 × 100 grid already yields the color pictures, in which the effect of the color quantization over the cells is practically imperciptible. Ideally, the best choice would be to take an even more dense grid, in which the sizes of each cell are exactly equal to the sizes of a pixel (picture element) of the computer monitor. Then one would have to use a 480 × 480 mesh. However, remember that the computing time needed for obtaining a difference solution on a mesh of N^2 cells is the quantity of the order $O(N^2)$ in the case of an explicit difference scheme. Therefore, it is reasonable to apply the *interpolation* to produce a color distribution on a mesh of 100 × 100 cells. The *Mathematica* system has a built-in function Interpolation[*data*]. In the case of two spatial variables the data can have the form $\{\{x_1, y_1, v_1\}, \{x_2, y_2, v_2\}, \ldots\}$. Interpolation works by fitting polynomial curves between successive data points. The degree of the polynomial curves is specified by the option InterpolationOrder. The default setting is InterpolationOrder -> 3. However, as our experience shows, with this setting the function Interpolation works much slower than in the case of a linear interpolation specified by the option InterpolationOrder -> 1. So we have used linear interpolation.

The color function can be specified in such a way that the red color corresponds to the maximum value of the solution $v(x, y)$, and the blue color corresponds to the minimum value of $v(x, y)$. For this purpose it is sufficient to specify the color function cl1[*f*_] of the form cl1[*f*_]:= Hue[h0*(1 - f)] with h0 = 0.5. In Fig. 2.30 we show the density

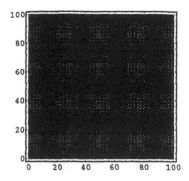

Figure 2.30: The density graph on a 100×100 mesh

graphics obtained with the aid of a linear interpolation of grid values v_{jk} from a 40×40 mesh onto an uniform grid of 100×100 cells. It can be seen that the effects of the quantization of the color over the small cells are insignificant. In addition, the regions of the local maxima and minima now have the correct colors.

The file `col.ma` is available on the attached diskette. To obtain Fig. 2.30 one must load this file and then click in the cell of input data.

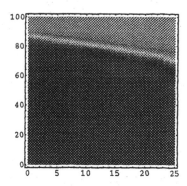

Figure 2.31: The density graph corresponding to the numerical solution $u = u(x, y, t)$ at $\theta = 0.8$ in (2.12.15) and $t = 38\tau = 1.11272$. The initial mesh has 40×10 cells

We have added the calculation of the density graphics to the program `mc2d.ma`, the corresponding file has the name `mc2dc.ma` and is available on the attached diskette. In Fig. 2.31 we show the density graphics corresponding to Fig. 2.28-b. In the corresponding color map, which is obtained on a color monitor, we can clearly distinguish between the high solution values behind the shock front and lower solution values before the shock front. The transition zone is shown in several colors.

Let us now consider the regime of an unstable computation by the MacCormack scheme (2.12.3), (2.12.4). Let us take the value $\theta = 1.1$ in (2.12.15). In this case the stability condition (2.12.10) is violated, therefore, the solution errors show growth without bound in time. This effect is indeed observed, see Fig. 2.32. To obtain Fig. 2.32 compile the

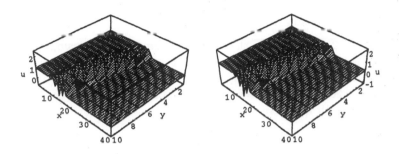

Figure 2.32: The numerical solution surfaces $u = u(x, y, t)$ obtained for $\theta = 1.1$ in (2.12.15). (a) $t = 14\tau = 0.563679$; (b) $t = 15\tau = 0.603942$. The mesh of 40×10 cells

file mc2d.ma and then click anywhere in the cell of input data

```
mc2d[1.0, Pi/6, 0.2, 1.0, 0.2, 2.0, 0.5, 40, 10, 1.1, 2,
     2, 2, 2]
```

At $t = 15\tau$ the maximum height $u = u(x, y, 15\tau)$ exceeds 3.0. The relative error $|(u^n - u_{ex})/u_{ex}|$, where u_{ex} is the exact solution, thus exceeds 200%.

In the program mc2d.ma, as in the program advbac.ma, the computation by the difference scheme is interrupted as soon as the quantity

$$\max_{i,j} |u_{ij}^n - u_{\max}|,$$

where $u_{\max} = \max(u_l, u_r)$, exceeds a certain threshold. This simple technique of computation control prevents machine overflow in the case of an unstable computation.

Exercise 2.24. Study the stability of the MacCormack scheme $u^{n+1} = M_{-+}^{+-}$ for the equation (2.12.2).

Hint. Derive the characteristic equation in the form $\lambda + b_1' + ib_2' = 0$. Then make, in the expressions for b_1' and b_2', a change of variables $\kappa_1' = -\kappa_1$, $\xi_1' = 2\pi - \xi_1$, $\kappa_2' = \kappa_2$, $\xi_2' = \xi_2$ and compare the expressions with the formulas (2.12.7) for b_1 and b_2.

Exercise 2.25. Show that, for constant coefficients A and B in (2.12.2), $M_{--}^{++} = M_{++}^{--}$ and $M_{-+}^{+-} = M_{+-}^{-+}$.

Exercise 2.26. Show that there exist, in the case of three spatial variables x_1, x_2 and x_3, eight variants of the MacCormack scheme for the advection equation (2.12.1).

Exercise 2.27. Perform the following computations using the program `col.ma` with different values of `m1int`, `m2int`: (a) `m1int = m2int = 60`; (b) `m1int = m2int = 80`; (c) `m1int = m2int = 100`; (d) `m1int = m2int = 120`, in order to see the effect of the diminution of the color quantization with increasing `m1int`, `m2int`.

Exercise 2.28. As in Exercise 2.27 use the program `col.ma` to perform computations for different valus of `h0` from the interval `0.5 <= h0 < 1.0` to find the limiting value of `h0`, which still provides a "cold" color corresponding to the lowest value of the numerical solution.

2.13 DETERMINATION OF PLANAR STABILITY REGIONS

There are many difference schemes whose stability region is a region in the plane of some nondimensional variables κ_1 and κ_2 (see, in particular, Fig. 2.24 earlier). In many cases there is no *a priori* information about the geometric form, or the "geometry" of the stability region. There are examples of simply connected stability regions, whose boundary has a complex geometric form. In addition, a stability region may be multiply connected, that is it may contain unstable "holes".

In this connection, we want to present a symbolic-numerical method for the determination of arbitrary planar multiply connected stability regions. We have implemented this method in the *Mathematica* notebook `st2.ma`. This notebook enables us:

- to obtain the expression for the amplification factor $\lambda(\kappa, \xi)$ of a scalar two-level difference scheme in the case of one or two spatial variables;

- to detect all the contours of a (generally multiply connected) stability region in the (κ_1, κ_2) plane to the user-specified accuracy ε;

- to show the stability region on the screen of the computer monitor.

The listing of the program `st2.ma` is relatively long, its length exceeds 500 lines. Therefore, we present the complete listing in the Appendix.

In what follows we will explain the structure of important functions used by the program stab2, which is the main program in the file st2.ma.

The program stab2 may be subdivided into two parts:

Part 1. Symbolic computations with the purpose of obtaining the expressions for the coefficients of the characteristic equation of a difference scheme.

Part 2. Numerical computations for the purpose of determining the coordinates of the points of all the contours of the stability region boundary.

Part 1 of the program stab2 generalizes, to the case of two spatial variables, the symbolic algorithm of the program st1.ma, see Section 2.3.

Let us assume that the coefficients of the characteristic equation (2.10.12) are periodic functions of the variables ξ_j, $j = 1, \ldots, L$, with equal periods 2π. Let us define the cube

$$\Pi = \{(\xi_1, \ldots, \xi_L) | \ 0 \le \xi_j \le 2\pi, \ j = 1, \ldots, L\} \qquad (2.13.1)$$

and introduce the function

$$f(\kappa) = \begin{cases} 1, & |\lambda_j(\kappa, \xi)| \le 1, \ j = 1, \ldots, m \ \forall \xi \in \Pi \\ 0 & \text{otherwise} \end{cases} \qquad (2.13.2)$$

(see also (2.3.18)). This function takes the value $f(\kappa) = 1$ at any point κ, at which the difference scheme under consideration is stable. At the unstable points κ the function (2.13.2) takes the value 0. A function, which takes only the values 1 or 0, is termed a binary function in the theory of digital image processing.[32] Let us now specify the rectangle

$$P: \ \{(\kappa_1, \ \kappa_2) | \ \kappa_{jl} \le \kappa_j \le \kappa_{jr}, j = 1, 2\} \qquad (2.13.3)$$

in the plane of the nondimensional variables $(\kappa_1, \ \kappa_2)$, where the quantities κ_{jl} and κ_{jr} are specified by the user on the basis of *a priori* information on the possible location of the stability region in the $(\kappa_1, \ \kappa_2)$ plane. The computer can compute the values of the function (2.13.2) only at a finite number of (κ_1, κ_2) points. Therefore, we have to introduce some grid in the rectangle (2.13.3). The simplest grid is an uniform rectangular grid with the steps $\Delta \kappa_1$ and $\Delta \kappa_2$ along the κ_1- and κ_2- axis, respectively. Let us denote by $(\kappa_{1j}, \ \kappa_{2k})$ the coordinates of the geometric center of the $(i, \ j)$ cell of an uniform rectangular grid, that is

$$\kappa_{1j} = \kappa_{1l} + (j - 0.5)\Delta \kappa_1, \ j = 1, \ldots, J_1$$
$$\kappa_{2k} = \kappa_{2l} + (k - 0.5)\Delta \kappa_2, \ k = 1, \ldots, J_2$$
$$\Delta \kappa_m = (\kappa_{mr} - \kappa_{ml})/J_m, \ m = 1, 2, \quad J_{1,2} > 1$$

Let us now define the binary, or bilevel, digital image $\{f_{j,k}\}$ as a two-dimensional array of the values $f_{j,k} = f(\kappa_{1j}, \kappa_{2k})$, $j = 1,\ldots,J_1$, $k = 1,\ldots,J_2$.

Similarly to the program st1.ma from Section 2.3 we have introduced an uniform rectangular grid G_ξ in the cube (2.13.1). The number nspg of grid points along each of the coordinate axes ξ_j, $j = 1,\ldots,L$, is assumed to be the same in our program stab2. The numerical value of nspg is determined in stab2, with the aid of the following simple iterative process. Let us denote by J some start value for nspg; we have taken the value $J = 4$ in our program. Since the stability region may consist, in a general case, of a number of stable islands, which in turn can contain the unstable holes, we have used the following quantitative measure for the determination of J. Let us denote by $A(J)$ the area of the rectangle (2.13.3), which is occupied by the (j, k) cells with $f_{jk} = 1$. As soon as the equality

$$A(J) = A(J + 2)$$

is satisfied at some J, we assume nspg$= J+2$ and terminate the iteration process.

Notice that the area $A(J)$ is proportional to the area $\Delta\kappa_1 \cdot \Delta\kappa_2$ of an individual cell. Therefore, we can write that $A(J) = N(J) \cdot \Delta\kappa_1 \cdot \Delta\kappa_2$. In this connection we have indeed checked in our program the condition $N(J) = N(J + 2)$ instead of the above condition $A(J) = A(J + 2)$.

The final value of nspg strongly depends on the difference scheme under study and also on the number of cells $J_1 \cdot J_2$ introduced in the region (2.13.3). For many schemes, which were analyzed with the aid of the program st2.ma, the value of nspg was in the range 12 $\leq$nspg$\leq$ 30.

In what follows we present the fragment of the program stab2, which implements the above iteration procedure for the determination of nspg:

```
(* - Determination of nspg, the number of grid points along
      the axis of each spectral variable xi1, xi2 ------- *)
    fbin = Table[0, {i, 1, ix}, {j, 1, jy}];
        nspg0 = 2; npold = 0; dnabs = 1000;
  Print["Determination of nspg"];
    Do[nspg = nspg0 + 2*j;
    ss1 = Table[N[Sin[(k-1)*2*Pi/nspg]], {k, nspg}];
    cc1 = Table[N[Cos[(k-1)*2*Pi/nspg]], {k, nspg}];
  {npnew, fbin} = bingen;
      If[j == 1, npold = npnew, dnabs = N[Abs[npnew-npold]]];
Print["nspg, pointsnew, pointsold, dnabs = ", npnew, "   ",
      npold, " ",nsp1, " ",dnabs];
```

```
        If[dnabs == 0, Break[]];
    npold = npnew,
                {j, 48 }];
```

In the above program fragment, the integer variables npnew and npold have the following meaning: npold= $N(J)$, npnew= $N(J + 2)$. If it turns out that the final value of npold $= 0$, then this means that in the region (2.13.3) there are no stable points of the difference scheme under consideration. In this case the following message is printed·

No stability points

and the further computation is interrupted. This message does generally not imply that the difference scheme under consideration is unstable. One must re-specify the region (2.13.3) and repeat the computation using the program st2.ma. Several such attempts can lead to the detection of a piece of the stability region. Extending the region (2.13.3) on the basis of the information obtained from the previous attempts one can obtain the complete stability region (if it indeed has the finite size in the directions of the κ_1- and κ_2-axes).

It follows from (2.13.2) that the function $f(\kappa)$ undergoes a jump on the stability region boundary. In the discrete bilevel image $\{f_{j,k}\}$ the stability region boundaries will be determined as ordered sets of the *boundary cells*. Let us now give the definition of the boundary cell. Let (j, k) be the cell for which $f_{j,k} = 1$. Consider eight cells, which either have a common side with the (j, k) cell or have a common vertex with the (j, k) cell. Denote this set of cells by $P_8(j, k)$. We will consider the (j, k) cell to be the *boundary cell* if $f_{m,l} = 0$ for at least for one cell $(m, l) \in P_8(j, k)$ (see Fig. 2.33). The (j, k) cells are also termed the *pixels* in the theory of the digital pattern recognition (the word pixel was formed from the words *picture element*). The pixels of a contour can be traversed by a path, and it is always possible to choose a closed path for the traversal. In what follows, we describe the modified Pavlidis' algorithm, which enables one to trace the closed contours of boundaries as the closed sequences of boundary cells.

Step 1. *Tracing external contours.* The initial, or start pixel of each such contour is found by scanning the digital image $\{f_{j,k}\}$ along the horizontal rows of pixels from the left to the right and from the top to bottom. The boundary cell first (j, k) is taken as the initial cell for the subsequent contour tracing. The tracing algorithm can be described in terms of an observer who walks along pixels belonging to the set and selects the rightmost pixel available. The direction for the walk along the contour is determined from the requirement that the stable (j, k)

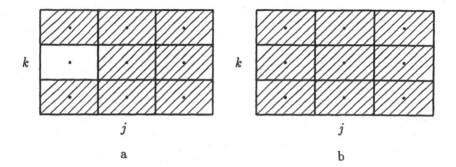

Figure 2.33: The cells in which $f_{j,k} = 1$ are hatched. (a) the (j, k) cell is the boundary cell; (b) the (j, k) cell is the interior cell.

cells lie to the right of the observer. Therefore, the external contours are traced in a clockwise manner.

Let us denote by A the starting cell of a contour. The next cell of the contour is one of its 8 neighbors, see Fig. 2.33. Therefore, it is convenient to assign a number S, $1 \leq S \leq 8$, to the search direction. The numbering of these eight directions is shown in Fig. 2.34. These directions will also be termed the *principal directions* in the following. Let us term the S-neighbor the cell, which belongs to the P_8 neighborhood of the (j, k) cell and which lies in the direction S from the (j, k) cell. For example, the 5-neighbor is the $(j - 1, k)$ cell. By using Fig. 2.34, it is easy to determine the indices (i1,j1) of the next cell along the contour based on the indices (i,j) of a given cell and the given search direction S, $1 \leq S \leq 8$:

```
S |   1       2       3       4       5       6       7       8

i1|  i + 1   i + 1    i     i - 1   i - 1   i - 1    i     i + 1

j1|  j       j + 1   j + 1  j + 1    j      j - 1   j - 1  j - 1
```

The function inds determines the indices i1 and j1 as functions of the direction S:

```
inds[is_, i_, j_]:=
(* -- This procedure determines the indices (i1,j1) of the
      next cell along the contour basing on the indices
```

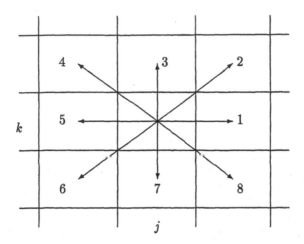

Figure 2.34: Principal directions and their numbering.

```
         (i,j) of a given cell and the given search direction
         "is", 1 <= is <= 8 ------------------------------- *)
( jb = 0; is2 = is; If[is2 > 8, is2 = is - 8];
 If[is2 == 1, i1 = i + 1; j1 = j;
 If[i1 < 1||i1 > ix || j1 < 1 || j1 > jy, jb = 1];
 Return[{i1, j1, jb}]];
   If[is2 == 2, i1 = i + 1; j1 = j + 1;
   If[i1 < 1||i1 > ix || j1 < 1 || j1 > jy, jb = 1];
   Return[{i1, j1, jb}]];
     If[is2 == 3, i1 = i; j1 = j + 1;
     If[i1 < 1||i1 > ix || j1 < 1 || j1 > jy, jb = 1];
     Return[{i1, j1, jb}]];
       If[is2 == 4, i1 = i - 1; j1 = j + 1;
       If[i1 < 1||i1 > ix || j1 < 1 || j1 > jy, jb = 1];
       Return[{i1, j1, jb}]];
         If[is2 == 5, i1 = i - 1; j1 = j;
         If[i1 < 1||i1 > ix || j1 < 1 || j1 > jy, jb = 1];
         Return[{i1, j1, jb}]];
           If[is2 == 6, i1 = i - 1; j1 = j - 1;
           If[i1 < 1||i1 > ix || j1 < 1 || j1 > jy, jb = 1];
```

```
     Return[{i1, j1, jb}]];
   If[is2 == 7, i1 = i; j1 = j - 1;
   If[i1 < 1||i1 > ix || j1 < 1 || j1 > jy, jb = 1];
   Return[{i1, j1, jb}]];
 If[is2 == 8, i1 = i + 1; j1 = j - 1;
 If[i1 < 1||i1 > ix || j1 < 1 || j1 > jy, jb = 1];
 Return[{i1, j1, jb}]] )
```

This function returns the values of i1 and j1. In addition, it returns an integer value jb: jb = 1, if the next cell (i1,j1) proves to be outside the given rectangle P given by the inequalities (2.13.3). If the (i1,j1) cell is still inside the region P, the value 0 is assigned to jb.

Following Reference 32 let us denote by R the subset of (j, k) cells, at which $f_{j,k} = 1$. The procedure tracer in our program st2.ma reproduces, with the aid of *Mathematica*, the Algorithm 7.1 from Reference 32, p. 143, therefore, we will only give the listing of tracer here:

```
tracer[nsg_, i0_, j0_]:=
(* ----------------- Tracing a contour --------------- *)
( i = i0;    j = j0;    is = 0;
fbin[[i, j]] = fbin[[i, j]] + 1;
is = is + 1;    jc[[is, nsg]] = 9;    iser = 7;    ifir = 1;
    xs[[is, nsg]] = xl + (i - 0.5)*hx;
    ys[[is, nsg]] = yb + (j - 0.5)*hy;

Label[m8]; icnt = 0; ifnd = 0;

Label[m13]; is1 = iser - 1; If[is1 == 0, is1 = 8];
           {i1, j1, jb} = inds[is1, i, j, ix, jy];
 If[jb == 1|| fbin[[i1, j1]] == 0, Goto[m9]];
           iser = iser - 2; If[iser <= 0, iser = iser + 8];

Label[m11]; is = is + 1; i=i1; j=j1;
          fbin[[i, j]] = fbin[[i, j]] + 1;
xs[[is, nsg]] = xl + (i - 0.5)*hx;
ys[[is, nsg]] = yb + (j - 0.5)*hy;
jc[[is, nsg]] = is1; ifnd = 1;
              If[i!= i0 || j != j0, Goto[m8], Goto[m7]]];

Label[m9]; {i1, j1, jb} = inds[iser, i, j, ix, jy];
           If[jb == 1||fbin[[i1, j1]] == 0, Goto[m10]];
```

```
is1 = iser; Goto[m11];

Label[m10]; is1 = iser + 1; If[is1 > 8, is1 = is1 - 8];
{i1, j1, jb} = inds[is1, i, j, ix, jy];
 If[jb == 1|| fbin[[i1, j1]] == 0, Goto[m12]];
       Goto[m11];
Label[m12]; iser = iser + 2; If[iser > 8, iser = iser - 8];
            ifir = 0;
If[ifnd == 1, Goto[m7]]; icnt = icnt + 1;
                              If[icnt < 3, Goto[m13]];
Label[m7]; kss[[nsg]] = is )
```

The function **tracer**, in particular, determines the length **kss[[k]]** of the kth external contour, $k \geq 1$, as the number of pixels belonging to the contour. The tracing of a contour is terminated as soon as the "observer" arrives at the starting cell of a contour.

The algorithm must be applied once for each hole of a region, in addition to one application for each external contour. When a pixel is marked as the current point of a contour, its value is incremented by 1. Then, at the end of the tracing, pixels of the contour will have values 2 or greater. In particular, the value will be one plus the number of times the pixel has been visited during the traversal. These values are used when the interior is searched for holes.

Step 2. *Detection of internal contours.* For this purpose the contour cells of external contours obtained at Step 1 are used as the initial cells for scanning inside the external contours. If we find a point located on a downward arc, we start a search to the right. Such pixels can be characterized easily by the requirement that the previous element of the chain code containing the sequential search directions S must have values 5 to 8 while the next element should be in the range 6 to 8. We include the value 5 for the former element in order to take care of inflection points. We have implemented, with the aid of *Mathematica*, Algorithm 7.2 from Reference 32, p. 146, for the tracing of both the external contours and the contours of unstable holes. Note that this algorithm takes into account the possible existence of contour pixels which are common to both the external contour and the contour of a hole. It is easy to see that the contours of holes are traced counterclockwise. Since there may be new contours inside the hole contours, hole contours must also be used as the start contours for the detection of further contours. This process is repeated recursively until no new contours are detected. Algorithm 7.2 from Reference 32 is efficient because elements of the picture not belonging to a contour need be scanned only once, and those belonging to a contour are scanned only twice.

In what follows we present the fragment of the program stab2, which implements the Algorithm 7.2:

```
(* ----------- Tracing the contours of holes ----------- *)
 k1 = 1; k2 = nsg; nsg1 = nsg; hx = (xr - xl)/ix;
                               hy = (yt - yb)/jy;
              hx1 = 1/hx; hy1 = 1/hy; k = 1;

Label[120];
      ig = kss[[k]]; istar = 1; ifin = ig;

Label[117]; {id, istar2} = remove[istar, ifin, k];
        istar = istar2;
          If[id == -1, Goto[13]];  If[id != 9, Goto[115]];
id0 = id; {id,istar2} = remove[istar, ifin, k];
                istar = istar2;

Label[115]; If[id0 < 5||id0 > 8, Goto [116]];
              If[id < 6||id > 8, Goto[116]];
x1 = xs[[istar, k]]; y1 = ys[[istar, k]];
   ilef = Floor[0.51 + (x1 - xl)*hx1];
                             j2 = Floor[0.51 + (y1 - yb)*hy1];
ilp = ilef + 1; irm = ix - 1;  i2 = ilp;

   Label[labilp];
    ia = fbin[[i2 - 1, j2]]; ib = fbin[[i2, j2]];
     ic = fbin[[i2 + 1, j2]]; If[ia != 0, Goto[137]];
   If[ib != 1||ib != 2||ic != 0, Goto[127]];
         i0 = i2; j0 = j2; nsg1 = nsg1 + 1;
 Print["The contour No. ", nsg1, " is detected."];
 tracer[nsg1, i0, j0, xl, yb]; Goto[116];

Label[127];
         If[ia == 0 && ib > 2 && id ==-1, Goto[116]];

Label[137];
         If[ia == 1 && ib == 2&&id == -1, Goto[116]];
        i2 = i2 + 1; If[i2 <= irm, Goto[labilp]]; Goto[13];

Label[116]; id0 = id; Goto[117];
```

```
Label[13];
          k = k + 1; If[k <= k2, Goto[120]];
If[nsg1 == nsg, Goto[121]];
nsg = k2; k1 = nsg + 1; k2 = nsg1; Goto[120];
Label[121]; nsg = nsg1; Print["Number of contours = ",nsg];
   Do[ isg = kss[[k]];
Print["The length of the contour No. ", k, " is", isg,
       " points"], {k, nsg}];
```

Step 3. *Refinement of the positions of contour points.* In the bilevel image $\{f_{j,k}\}$ the strong discontinuities of the function (2.13.2) are smeared over one cell of the rectangular grid in the (κ_1, κ_2) plane. Therefore, the absolute error in locating the positions of contours does not exceed $[(\Delta\kappa_1)^2 + (\Delta\kappa_2)^2]^{0.5}$. These errors can be easily reduced to a user-specified amount ε with the aid of the refinement procedure, which we describe below. This procedure implements a bisection process along a normal to the contour at each point (j, k) of a contour and it thus augments the above presented Pavlidis' algorithm.

Let us present the steps of our refinement procedure **bisec**.

Step 3.1. Let m be a number of a point $(\kappa_1^{(m)}, \kappa_2^{(m)})$ of the contour, $m \geq 1$. We join the points $(\kappa_1^{(m-1)}, \kappa_2^{(m-1)})$ and $(\kappa_1^{(m+1)}, \kappa_2^{(m+1)})$ by a straight line segment, see Fig. 2.19. Now let us construct the normal to this line segment, which passes through the $(\kappa_1^{(m)}, \kappa_2^{(m)})$ point. Using the *Mathematica* function ArcTan[x, y] $= \arctan(y/x)$ we find the angle α_m between the above normal and the positive direction of the κ_1 axis.

Step 3.2. The direction determined by the angle α_m is approximated by one of the eight principal directions shown in Fig. 2.34.

Step 3.3. Compute the coefficients a_1, b_1, a_2, b_2 in the parametric representation

$$\kappa_1 = a_1 t + b_1, \quad \kappa_2 = a_2 t + b_2 \tag{2.13.4}$$

of the line which approximates the direction determined by the angle α_m and which passes through the (κ_1^m, κ_2^m) point. We assume that $t = 0$ at the left end A and $t = 1$ at the right end B (see Fig. 2.35), where the points A and B lie on different sides of the contour and belong to the line (2.13.4).

Step 3.4. Since one of the points A or B lies outside the stability region, and the other point is inside the stability region, we can apply the bisection process in the interval $0 \leq t \leq 1$ for the purpose of determining the position of the discontinuity of the function $F(t) = f(a_1 t + b_1, a_2 t + b_2)$. We assume $t_l = 0$, $t_r = 1$ and then go to Step 3.5.

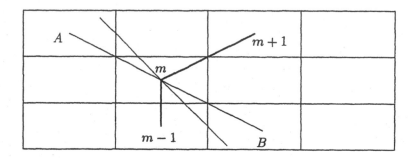

Figure 2.35: The refinement is performed along the approximate normal *AB*.

Step 3.5. Compute $t = 0.5(t_l + t_r)$. If $sgn(F(t)) = sgn(F(t_r))$, we assume $t_r = t$, otherwise $t_l = t$.

Step 3.6. Check the convergence criteria

$$|a_1(t_l - t_r)| < \varepsilon, \quad |a_2(t_l - t_r)| < \varepsilon \tag{2.13.5}$$

where ε is a given positive number. If one of the inequalities (2.13.5) is not satisfied, we return to Step 3.5.

Step 3.7. Compute the refined values of the coordinates of the m-th point by the formulas (2.13.4).

Step 3.8. Go to the next contour point by increasing the number m by 1, and return to Step 3.1.

In this way the positions of the points of all of the contours are refined. In what follows we present the listing of the function bisec, which implements the above eight steps of the refinement procedure.

```
bisec[x_, y_, ik_, k_, mk_, adir_List]:=
(* The refinement of the (x,y) coordinates of the ik th
   point of the k th contour; mk is the length of the
   kth contour --------------------------------------- *)
( i = Floor[1.0 + (x - xlt)*hx1];
                  j = Floor[1.0 + (y -ybt)*hy1];
i1 = ik + 1; i2 = ik - 1;
```

```
If[ik == mk, i1 = i1 - 1; i2 = i2 - 1];
   If[ik == 1, i2 = 1];
(* -- Determining the number k2 of the principal direction
   corresponding to the direction of a normal to the
   contour at (i, j) point. ------------------------ *)
  x1 = xxxx[[i1]]; y1 = ytab[[i1]];
               x2 = xxxx[[i2]]; y2 = ytab[[i2]];
     dx = x2 - x1; dy = y1 - y2;

anrm = atan3[dx, dy];   zm = 1000.0; il = -2;
Do[ zk = Abs[adir[[ja]] - anrm];
                      If[zk < zm, zm = zk;k2 = ja],
      {ja, 9}];
            If[k2 == 9, k2 = 1];
(* -- Determining the indices (il, jl) of a cell lying on a
   normal to the contour -------------------------- *)
 If[k2 == 1, il = i + 1; jl = j; ir = i - 1; jr = j];
 If[il > -2, Goto[refin]];

If[k2 == 2, il = i + 1; jl = j + 1; ir = i - 1; jr= j - 1];
   If[il > -2, Goto[refin]];
     If[k2 == 3, il = i; jl = j + 1; ir = i; jr = j - 1];
      If[il > -2, Goto[refin]];

If[k2 == 4, il = i - 1; jl = j + 1; ir = i + 1;jr = j - 1];
       If[il > -2, Goto[refin]];
     If[k2 == 5, il = i - 1; jl = j; ir = i + 1; jr = j];
       If[il > -2, Goto[refin]];
If[k2 == 6, il = i - 1; jl = j - 1; ir = i + 1;jr = j + 1];
       If[il > -2, Goto[refin]];
   If[k2 == 7, il = i; jl = j - 1; ir = i; jr = j + 1];
   If[il > -2, Goto[refin]];
If[k2 == 8, il = i + 1; jl = j - 1; ir = i - 1;jr = j + 1];

Label[refin];
    x1n = xlt + (il - 0.5)*hx; y1n = ybt + (jl - 0.5)*hy;
    x2n = xlt + (ir - 0.5)*hx; y2n = ybt + (jr - 0.5)*hy;
t1 = 0.0; tr = 1.0; a1 = x2n - x1n; a2 = y2n - y1n;
 isl = fbis[x1n, y1n, ss1, cc1, nspg];
    isr = fbis[x2n, y2n, ss1, cc1, nspg]; x3 = x; y3 = y;
    If[isl != isr, Goto[bisproc]];
                      If[isl== 0, Goto[recalc]];
If[il == 0||ir == 0, x3 = xl];
                 If[il == ix + 1||ir == ix + 1, x3 = xr];
```

```
If[jl == 0||jr == 0,y3 = yb];
                    If[jl == jy + 1||jr == jy + 1, y3 = yt];
Return[{x3, y3}];
Label[recalc]; ir = i; jr = j; x2n = x; y2n = y; isr = 1;
                    a1 = x2n - x1n; a2 = y2n - y1n;


Label[bisproc];
   t = N[(tl + tr)*0.5]; xc = a1*t + x1n; yc = a2*t + y1n;
          nc = fbis[xc, yc, ss1, cc1, nspg];
If[nc == isr, tr = t, tl = t];
   If[N[Abs[a1*(tl - tr)]] > eps, Goto[bisproc]];
             If[N[Abs[a2*(tl - tr)]] > eps, Goto[bisproc]];
   x3 = a1*t + x1n; y3 = a2*t + y1n;
   Return[{x3, y3}]    )
```

This function returns the list of the refined coordinates {x3, y3} based on the original "raw" coordinates x, y of a current point of a contour.

As a test of this modified Pavlidis' algorithm, let us consider the binary function $f(\kappa_1, \kappa_2)$ of the form

$$f(\kappa_1, \kappa_2) = \begin{cases} 0, & -.07 \leq \kappa_1 \leq .07, \ -.4 \leq \kappa_2 \leq -.3 \\ f_3(\kappa_1, \kappa_2) & \text{otherwise} \end{cases} \quad (2.13.6)$$

where

$$f_3(\kappa_1, \kappa_2) = \begin{cases} 0, & -0.05 \leq \kappa_1 \leq 0.05, \ 0.3 \leq \kappa_2 \leq 0.4 \\ f_2(\kappa_1, \kappa_2) & \text{otherwise} \end{cases}$$

$$f_2(\kappa_1, \kappa_2) = \begin{cases} 0, & -0.05 \leq \kappa_2 \leq 0.05 \\ f_1(\kappa_1, \kappa_2) & \text{otherwise} \end{cases}$$

$$f_1(\kappa_1, \kappa_2) = \begin{cases} 1, & 0.25 \leq (\kappa_1^2 + \kappa_2^2)^{0.5} \leq 0.5 \\ 0 & \text{otherwise} \end{cases}$$

Let us take $\kappa_{jl} = -0.6$, $\kappa_{jr} = 0.6$ in (2.10.3) and $J_1 = J_2 = 30$. In Fig. 2.36 we show the contours traced in the corresponding image $\{f_{j,k}\}$ with the aid of the Pavlidis' algorithm. We can see that the effects of the quantization of the distribution (2.13.6) over 900 cells of a mesh in the (κ_1, κ_2) plane lead to the distortions of the circular form of the contour segments. It is to be noted that there are, in this example, a number of contour pixels which are common to both the external contour and the contour of a hole. There are two holes in the example of Fig. 2.36.

To perform the corresponding computation, one must compile the file st2t.ma, see the listing in Appendix, and then click in the cell of input data

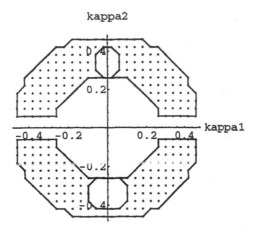

Figure 2.36: The result of execution of steps 1 and 2 of the modified Pavlidis' algorithm.

```
stab2[-0.6, 0.6, -0.6, 0.6, 30, 30, 1, 0.005]
```

In Fig. 2.37 we show the result of the application of the above procedure for the refinement of the positions of contour points. Choose $\varepsilon = 0.005$ in (2.13.5). It may be seen from Fig. 2.37 that the circular form of a number of contour segments has been restored. The dots are placed in Figs. 2.36 and 2.37 at points (j, k) at which $f_{j,k} = 1$, that is at stable points. This is necessary in order to easily distinguish between the stable and unstable subregions of the digital picture. The shape of the unstable holes in Fig. 2.37 proved to be distorted as a result of the execution of the refinement step (the holes here are rectangles). This distortion is explained by the relative proximity of the boundaries to the holes in this example (see also Fig. 2.36). The standard remedy is, in this case, the use of a finer mesh in the (κ_1, κ_2) plane. For example, the use of a 40 × 40 mesh will lead to a drastic improvement of the representation of the small holes in this computational example.

In Fig. 2.38 we show the result of the application of the above modified Pavlidis' algorithm for the determination of the stability region of the MacCormack scheme (2.12.3). In this computational example, the value of nspg was determined automatically as nspg = 12. This required about 80% of the overall computing time. Comparing Fig. 2.38, b with

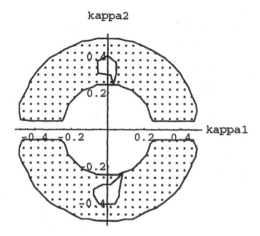

Figure 2.37: The refined contours.

Fig. 2.24 we can see that the algorithm of the present section produces the correct result.

To perform the corresponding computation load the file st2.ma, which is available on the attached diskette, and then click anywhere in the cell of input data

```
stab2[-1.2, 1.2, -1.2, 1.2, 30, 30, 2, 0.005]
```

We will present, in subsequent sections of this book, further examples of the applications of the program st2.ma, which is our main program for the stability analyses of the difference initial-value problems because of its universality.

Computer algebra is often considered as one of the branches of the artificial intelligence (AI) as is digital pattern recognition. We have shown that the combined use of algorithms from both of the above AI fields can successfully be used for the development of an universal algorithm for the automatic determination of the stability region for difference schemes.

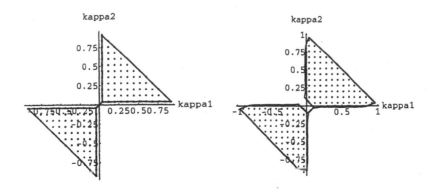

Figure 2.38: The stability region of the MacCormack scheme (2.12.3). (a) the result of execution of steps 1 and 2 of the modified Pavlidis' algorithm; (b) the refined contour, with $\varepsilon = 0.005$ for the bisection process.

2.14 CURVILINEAR SPATIAL GRIDS

2.14.1 A Simple Algebraic Grid Generation Method

Many problems of the mathematical physics involve a domain of spatial variables. The boundaries of these domains are often curvilinear. This gives rise to the problem of the difference approximation of spatial derivatives in the mesh cells in the neighborhood of curved boundaries. To illustrate this point, let us consider a problem of the flow of gas in a Laval nozzle, see Fig. 2.39.

For the numerical computation of a gas flow in this nozzle, let us introduce an uniform rectangular spatial grid in the (x, y) plane. We can see in Fig. 2.39 that there are, near the nozzle wall, cells which are only partially filled by gas. Let us call these cells the *partial cells* (they are dashed in Fig. 2.39). Assume that central differences are used in the interior cells for the approximation of the derivatives $\partial u/\partial x$ and $\partial u/\partial y$. In principle, it is possible to retain a second-order approximation in the partial cells by considering various special cases of the intersection of the boundary with the cells. But this generally leads to a complicated algorithm. In addition, the stability condition typically involves the area of the cell as a multiplier. Therefore, the partial cells, which have a small area in comparison with the area of a rectangular cell, lead to a substantial reduction of the maximum time step allowed by the stability of the difference scheme. These difficulties can be substantially alleviated by using the *curvilinear spatial grids* fitted to the boundaries of the spatial

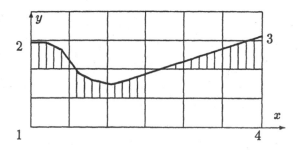

Figure 2.39: One half of a cross-section of the Laval nozzle

domain of interest. The points in these grids are usually constructed with the aid of the numerical grid generation computer programs. The area of the numerical grid generation is now a rapidly developing area.[33,34,35] Grid generation techniques may be subdivided into two big classes:

(i) algebraic techniques; [2,33,34,35]

(ii) techniques based on the numerical solution of certain PDEs.

There are two main types of grids: structured and unstructured. A typical unstructured grid is made up of triangles, while the main type of structured grid is called logically rectangular. The algebraic techniques use some kind of interpolation for the calculation of the grid nodes coordinates inside a spatial domain with given boundaries. In this section we describe a very simple algebraic method for the numerical grid generation in axisymmetric nozzles. Let $y = f_w(x)$ be the equation of the nozzle wall. Now introduce new coordinates ξ and η by the formulas

$$\xi = (i_x - 1)(x/x_{nz}), \quad \eta = (j_y - 1)[y/f_w(x)] \qquad (2.14.1)$$

where i_x and j_y represent the total number of grid nodes in the direction of the $x-$ and $y-$axes, respectively; x_{nz} represents the abscissa of the exit section of a nozzle. It is easy to see that the formulas (2.14.1) determine a one-to-one mapping

$$(x, y) \rightarrow (\xi, \eta)$$

which enables us to transform the region D of the physical space (see Fig. 2.40) into a rectangular region in the (ξ, η) plane. The ξ and η variables are called *curvilinear coordinates*. In the general case, it is assumed that

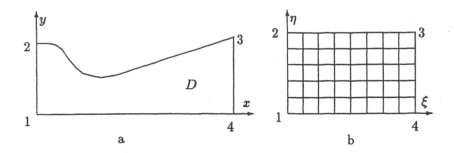

Figure 2.40: (a) the region D in the physical space; (b) the transformed region in the (ξ, η) plane

there exist the functions $x(\xi, \eta)$ and $y(\xi, \eta)$, which generate a one-to-one mapping

$$x = x(\xi, \eta), \quad y = y(\xi, \eta) \qquad (2.14.2)$$

from logical to physical space.

The generation of a boundary-conforming curvilinear grid is performed by the determination of the values of the curvilinear coordinates in the interior of a physical region from specified values (and/or slopes of the coordinate lines intersecting the boundary) on the boundary of the region.

One coordinate (ξ or η) will be constant on each segment of the physical boundary curve, while the other varies monotonically along the segment. For example, the line $\xi = 0$ (Fig. 2.40-b etc) corresponds to the boundary segment "12" in the physical plane (Fig. 2.40-a). The line $\eta = \eta_{max}$, where $\eta_{max} > 0$ is a given constant, corresponds to the nozzle wall in the physical plane (Fig. 2.40).

The equivalent problem in the transformed region is the determination of values of the physical (Cartesian) coordinates (x, y) in the interior of the transformed region from specified values of (x, y) and/or slopes on the boundary of this region. For example, in the case of the spatial region of Fig. 2.40-a, the spatial coordinates $x(\xi, \eta)$, $y(\xi, \eta)$ are specified on all four rectilinear segments of the rectangular region in the (ξ, η) plane. This is a more amenable problem for computation, since the boundary of the transformed region is comprised of horizontal and vertical segments in the (ξ, η) plane (Fig. 2.40-b)).

It follows from Eqs. (2.14.1) that the region in the (ξ, η) plane represents a rectangle of the length $(i_x - 1)$ in the direction of the η-axis. We can take, in principle, any stepsizes $\Delta \xi$ and $\Delta \eta$ for the uniform rectangular mesh in the (ξ, η) plane. The simplest choice is to take $\Delta\xi = \Delta\eta = 1$. Then we can specify the straight grid lines in the rectangle of the (ξ, η) plane by the simple equations

$$\xi_i = i - 1, \ i = 1, \ldots, i_x; \quad \eta_j = j - 1, \ j = 1, \ldots, j_y \qquad (2.14.3)$$

It is now easy to pass from the grid nodes (ξ_i, η_j) to their images in the (x, y) plane by using formulas (2.14.1):

$$x_i = \xi_i \cdot x_{nz}/(i_x - 1), \quad i = 1, \ldots, i_x$$
$$y_{ij} = f_w(x_i)\eta_j/(j_y - 1), \quad j = 1, \ldots, j_y \qquad (2.14.4)$$

The grid lines $x_i = const$ are simply the vertical straight line segments parallel with the y-axis. The lines $\eta_j = const$ have a form similar to the form of the nozzle wall.

2.14.2   The Program grid1.ma

To implement the simple algebraic method (2.14.3), (2.14.4), we have written a *Mathematica* program grid1.ma. We have taken a specific nozzle geometry from the paper.[36] The nozzle is made up of a subsonic part, a critical part, and a relatively short supersonic part. The nozzle of Reference 36 is shown in Fig. 2.41. The input parameters used in the program grid1.ma are described in Table 2.5.

Table 2.5. Parameters used in program grid1.ma

Parameter	Description
hc	the nozzle wall height h_c in the critical section of the nozzle, $h_c > 0$ (see Fig. 2.41)
cy0	the ordinate of the nozzle wall in the inlet section $x = 0$ related to h_c
cR0	the radius of the curvature of the nozzle wall in its subsonic part related to h_c
cR	the nozzle wall radius of the curvature in the critical section related to h_c
cx0	the abscissa of the critical nozzle section related to h_c
cx4	the nozzle length related to h_c, cx4 > cx0
thet1	the inclination angle ϑ_1 of the nozzle wall in the subsonic part of the nozzle; the angle is specified in radians
thet2	the inclination angle ϑ_2 of the nozzle wall in the supersonic part of the nozzle; the angle is specified in radians
ix	the number i_x of grid points on the ξ-axis, $i_x > 1$
jy	the number j_y of grid points on the η-axis, $j_y > 1$

The main function of the file grid1.ma is

$$\text{grid1}[hc, cy0, cR0, cR, cx0, cx4, \vartheta_1, \vartheta_2, i_x, j_y].$$

─────────────────── File grid1.ma ───────────────────

```
ClearAll[fw,grid1];
fw[x_] := ( f1 = If[0 <= x && x <= x1,
        y0 - R0 + Sqrt[(R0 - x)*(R0 + x)], 0];
 f2 = If[x1 <= x && x <= x2, y1 - ct1*(x - x1), f1];
 f3 = If[x2 <= x && x <= x3,
      hc1 + R - Sqrt[(R - x + x0)*(R + x - x0)], f2];
 f4 = If[x3 <= x && x <= x4, y3 + ct2*(x - x3), f3]; f4 )


grid1[hc_, cy0_, cR0_, cR_, cx0_, cx4_, thet1_, thet2_,
                                    ix_, jy_] :=
(
(* ------ Specification of the nozzle geometry ------- *)
R0 = cR0*hc; y0 = cy0*hc; x0 = cx0*hc; R = cR*hc;
    hc1 = hc;
x1 = R0*N[Sin[thet1]]; ct1 = N[Tan[thet1]];
                ct2 = N[Tan[thet2]];
x2 = x0 - R*N[Sin[thet1]]; x3 = x0 + R*N[Sin[thet2]];
x4 = cx4*hc; y1 = y0 - R0*(1 - N[Cos[thet1]]);
   y3 = hc + R*(1 - N[Cos[thet2]]);
  Print["The values of the nozzle geometry parameters"];
  Print["hc = ",hc, ", R0 = ", R0, ", R = ", R,
        ", x0 = ",x0];
  Print["thet1 = ", thet1, ", thet2 = ", thet2,
        ", nozzle length = ", x4];

(*  yw is the list for the ordinates of the nozzle wall  *)
    yw = {};  xorg = 0.0;
    Do[ xi = (i - 1)*x4/(ix - 1); yi = fw[xi];
    AppendTo[yw, yi], {i, ix}];
(* --- x, y are the lists for the abscissas x[[i]] and the
       ordinates y[[i, j]] of the nodes of the curvilinear
       spatial grid ------------------------------- *)
       x = Table[N[(i - 1)*x4/(ix - 1)], {i, ix}];
  y = Table[N[yw[[i]]*(j - 1)/(jy - 1)], {i, ix}, {j, jy}];

(* ------- The graphical output of the grid -------- *)
(* ------------ The lines eta = const ------------ *)
  mt = {}; xtab = {};
 Do[xi = (i - 1)*x4/(ix - 1); AppendTo[xtab, xi], {i, ix}];
   Do[ytab = {};
```

```
            Do[yi = y[[i,j]]; AppendTo[ytab, yi], {i, ix}];
   xytab = Table[{xtab[[i]], ytab[[i]]}, {i, ix}];
   mt1 = ListPlot[xytab,
     AxesLabel  -> {"x", "y"},
                      AxesOrigin -> {xorg, 0},
     PlotJoined -> True,
     PlotRange  -> {{0, x4}, {0, y0}},
                      DisplayFunction -> Identity];
              AppendTo[mt, mt1],
                            {j, jy}];
    mt3 = Show[mt, DisplayFunction -> $DisplayFunction];
(* ------------ The lines xi = const ------------------ *)
   Do[ xtab = {}; ytab = {};
      xi = (i - 1)*x4/(ix - 1); AppendTo[xtab, xi];
   AppendTo[ytab, 0.0];
      yi = yw[[i]]; AppendTo[xtab, xi]; AppendTo[ytab, yi];
   xytab = Table[{xtab[[j]], ytab[[j]]}, {j, 2}];
      mt1 = ListPlot[xytab,
                            PlotJoined -> True,
                            DisplayFunction -> Identity];
          AppendTo[mt, mt1],
                      {i, ix}];
   mt3 = Show[mt, DisplayFunction -> $DisplayFunction];
(* ----------- End of the procedure grid1 ------------ *)
)
```

The function fw[x_] enables us to obtain the value of the nozzle ordinate
for the given value of the abscissa x. We have used the polar coordinates
and other simple geometric formulas for finding the equations for the
different portions of the nozzle wall. In a more general case, the nozzle
wall may be specified in the form of a table yw of the ordinate values on
the nozzle wall.

The actual determination of the coordinates (x_i, y_{ij}) of the grid nodes
is made in the following two lines of the program grid1.ma:

```
   x = Table[N[(i - 1)*x4/(ix - 1)], {i, ix}];
 y = Table[N[yw[[i]]]*(j - 1)/(jy - 1)], {i, ix}, {j, jy}];
```

In Fig. 2.42 we show the curvilinear spatial grid obtained at $i_x = 61$,
$j_y = 21$ in (2.14.2), (2.14.3). The program grid1.ma is very fast. The
numerical generation of the 61×21 mesh required only a few seconds
of computer time. To perform this computation load the file grid1.ma
and then click in the cell of input data

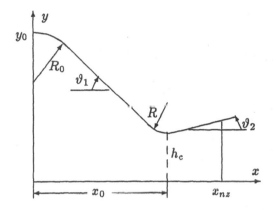

Figure 2.41: Axisymmetric nozzle ($R/h_c = 0.625$).

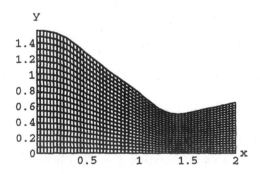

Figure 2.42: The curvilinear grid in the nozzle

```
grid1[0.5, 3.125, 1.0, 0.625, 2.805,4.005,Pi/4,Pi/12,61,21]
```

In Fig. 2.43 we show the curvilinear grid obtained for another nozzle from Reference 36. The cell of input data for this nozzle is:

```
grid1[0.5, 1.617, 1.617, 0.1, 2.0, 2.6, Pi/9, Pi/9, 61, 21]
```

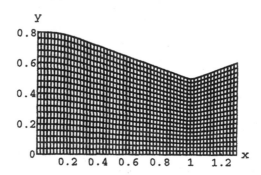

Figure 2.43: The curvilinear grid for the nozzle with $\vartheta_1 = \vartheta_2 = 20°$.

The contour of the nozzle in Fig. 2.43 has a small radius of curvature in the region of the nozzle throat. For this case it was recommended to make a stretching of the coordinate ξ in the vicinity of the throat.[36] That is, it is better to use a transformation $\xi = f(x)$ instead of $\xi = x$. The expression for the stretching function $f(x)$ may be found in Reference 36.

Exercise 2.29. Obtain a series of curvilinear grids in the nozzle of Fig. 2.41 at different values of the angles ϑ_1 and ϑ_2 with the aid of the program **grid1.ma**.

ANSWERS TO THE EXERCISES

2.2. $u''(x_j) = (u_{j+1} - 2u_j + u_{j-1})/(h^2)$

2.3. $u''(x_j) = (u_j - 2u_{j-1} + u_{j-2})/(h^2)$

2.4. $O(\tau) + O(h^2)$

2.5. Numerical computation using scheme (2.2.11) proves to be impossible, because this scheme is *unstable*.

2.6. $0 \leq |\kappa| \leq 1$

2.8.

$$\frac{u_j^{n+1} - u_j^n}{\tau} + a\frac{u_{j+1}^n - u_{j-1}^n}{2h} = |a|\frac{h}{2}\frac{u_{j+1}^n - 2u_j^n + u_{j-1}^n}{h^2}$$

2.9. $0 \leq |\kappa| \leq 1$

2.10. Scheme (2.2.11) is unstable.

2.11. $0 \leq |\kappa| \leq 1$

2.12.

$$\frac{u_j^{n+1} - u_j^n}{\tau} + a\frac{u_{j+1}^n - u_{j-1}^n}{2h} = \frac{a^2\tau}{2h^2}(u_{j+1}^n - 2u_j^n + u_{j-1}^n)$$
$$+ \frac{1}{2}(\varphi_j^{n+1} + \varphi_j^n) - \frac{a\tau}{4h}(\varphi_{j+1}^n - \varphi_{j-1}^n)$$

2.16. $0 \leq |\kappa| \leq 1$

2.21. $B_1 = I - (\tau/(4h))A(T_1 - T_{-1})$, $B_2 = (\tau/(4h))A(T_1 - T_{-1})$

2.24. $|\kappa_1| + |\kappa_2| \leq 1$, $\kappa_1 \cdot \kappa_2 \leq 0$.

REFERENCES

1. **Roache, P.J.,** *Computational Fluid Dynamics*, Hermosa, Albuquerque, New Mexico, 1976.
2. **Fletcher, C.A.J.,** *Computational Techniques for Fluid Dynamics*, Vols. I,II, Springer-Verlag, Berlin, 1988.
3. **Strikwerda, J.C.,** *Finite Difference Schemes and Partial Differential Equations*, Wadsworth & Brooks/Cole Advanced Books & Software, Pacific Grove, California, 1989.
4. **Konovalov, A.N.,** *Problems of Multiphase Fluid Filtration*, World Scientific, Singapore, 1994.
5. **Richtmyer, R.D. and Morton, K.W.,** *Difference Methods for Initial-value Problems*, Second Edition, Interscience Publishers, a division of John Wiley and Sons, New York, 1967.
6. **Courant, R., Friedrichs, K.O. and Lewy, H.,** Über die partiellen Differenzengleichungen der mathematischen Physik, *Mathematische Annalen* 100, 32, 1928.
7. **Godunov, S.K. and Ryabenkii, V.S.,** *Difference Schemes.. An Introduction to the Theory* (in Russian). Nauka, Moscow, 1977. [English transl.: Godunov, S.K. and Ryabenkii, V.S., *Difference Schemes. An Introduction to the Underlying Theory* (Studies in Mathematics and its Applications, 19), Elsevier Science Publishing Co., Inc., New York, 1987]
8. **Shokin, Yu.I.,** *Differential Approximation Method*, Nauka, Siberian Division, Novosibirsk, 1979 [English transl.: Shokin, Y.I.: *The Method of differential approximation* (Translated from the Russian by K.G. Roesner), Springer-Verlag, Berlin, 1983].
9. **Shokin, Yu.I. and Yanenko, N.N.,** *The Method of Differential Approximation. Application to Gas Dynamics* (in Russian), Nauka, Siberian Division, Novosibirsk, 1985.
10. **Hirt, C.W.,** Heuristic stability theory for finite-difference equations, *Journal of Computational Physics,* 2, 339, 1968.
11. **Lax, P. and Wendroff, B.,** Systems of conservation laws, *Communications on Pure and Applied Mathematics,* 13(2), 217, 1960.
12. **Noh, W.F. and Protter, M.H.,** Difference methods and the equations of hydrodynamics, *Journal of Mathematics and Mechanics,* 12(2), 149, 1963.
13. **Richtmyer, R.D.,** A Survey of Difference Methods for Nonsteady Fluid Dynamics, NCAR Techn. Note 63-2, Boulder, Colorado, 1963.

14. **Roždestvenskiĭ, B.L. and Janenko, N.N.**, *Systems of Quasilinear Equations and Their Applications to Gas Dynamics*. Second Edition (in Russian), Nauka, Moscow, 1978. [English transl.: *Systems of Quasilinear Equations and Their Applications to Gas Dynamics*, Translations of Mathematical Monographs, Vol.55 (American Mathematical Society, Providence, Rhode Island, 1983)].

15. **Oran, E.S. and Boris, J.P.**, *Numerical Simulation of Reactive Flow*, Elsevier, New York, 1987.

16. **MacCormack, R.W.**, *The effect of Viscosity in Hypervelocity Impact Cratering*, AIAA Paper 69-354, 1969.

17. **Godunov, S.K.**, Difference method for numerical computation of discontinuous solutions of equations of fluid dynamics, *Matem. Sbornik* (in Russian), 47, 271, 1959.

18. **Godunov, S.K., Zabrodin, A.V., Ivanov, M.Ya., Krayko, A.N. and Prokopov, G.P.**, *Numerical Solution of Multidimensional Gas Dynamics Problems* (in Russian), Nauka, Moscow, 1976.

19. **Harten, A., Hyman, J.M. and Lax, P.D.**, On finite-difference approximations and entropy conditions for shocks. *Communications on Pure and Applied Mathematics*, 29(3), 297, 1976.

20. **Karamyshev, V.B.**, *Monotone Schemes and Their Applications to Gas Dynamics* (in Russian), published by the Novosibirsk State University, Novosibirsk, 1994.

21. **Harten, A.**, High resolution schemes for hyperbolic conservation laws, *Journal of Computational Physics*, 49(3), 357, 1983.

22. **Chakravarthy, S.R. and Osher, S.**, A new class of high accuracy TVD schemes for hyperbolic conservation laws, *AIAA Paper*, No. 85-0363, 1985.

23. **Kolgan, V.P.**, Application of the principle of the minimal values of the derivative to the construction of finite difference schemes for the computation of discontinuous gas dynamics solutions, *Uchenye Zapiski TsAGI* (in Russian), 3(6), 68, 1972.

24. **Crank, J. and Nicolson, P.**, A practical method for numerical integration of solutions of partial differential equations of heat conduction type, *Proceedings of the Cambridge Philosophical Society*, 43(50), 1947.

25. **Marchuk, G.I.**, *Methods of Numerical Mathematics*. Third Edition (in Russian), Nauka, Moscow, 1989.

26. **Chu, C.K. and Sereny, A.**, Boundary conditions in finite difference fluid dynamics codes, *Journal of Computational Physics*, 15(4), 476, 1974.

27. **Samarskii, A.A. and Gulin, A.V.**, *Stability of Difference Schemes* (in Russian), Nauka, Moscow, 1973.

28. **Gustafsson, B., Kreiss, H.-O. and Sundström, A.**, Stability theory of difference approximations for mixed initial boundary value problems II. *Mathematics of Computation*, 26(119), 649, 1972.

29. **Godunov, S.K.**, *Difference Methods for the Solution of Gas Dynamics Equations. Lectures for the Students of the Novosibirsk State University* (in Russian), published by Novosibirsk State University, Novosibirsk, 1962.

30. **Gustafsson, B.**, The convergence rate for difference approximations to mixed initial boundary value problems, *Mathematics of Computation*, 29(130), 396, 1975.

31. **Peyret, R. and Taylor, T.D.**, *Computational Methods for Fluid Flow*, Springer-Verlag, New York, 1983.

32. **Pavlidis, T.**, *Algorithms for Graphics and Image Processing*, Springer-Verlag, Berlin, Heidelberg, 1982.

33. **Thompson, J.F., Warsi, Z.U.A. and Mastin, C.W.**, *Numerical Grid Generation-Foundations and Applications*, Elsevier Science Publishing Company, New York, 1985.

34. **Warsi, Z.U.A.**, *Fluid Dynamics. Theoretical and Computational Approaches*, CRC Press, Boca Raton, 1993.

35. **Knupp, P. and Steinberg, S.**, *Fundamentals of Grid Generation*, CRC Press, Boca Raton, 1994.

36. **Laval, P.**, Time-dependent calculation method for transonic nozzle flows, *Proceedings of the Second International Conference on Numerical Methods in Fluid Dynamics, September 15-19, 1970, University of California, Berkeley* / Ed. M. Holt. Lecture Notes in Physics, Vol. 8. Springer-Verlag, Berlin, 1971, p. 187.

Chapter 3

**FINITE DIFFERENCE METHODS
FOR PARABOLIC PDEs**

3.1 BASIC TYPES OF BOUNDARY CONDITIONS FOR PARABOLIC PDEs

In this chapter we want to discuss some difference methods for the numerical modeling of *diffusion processes*. Such processes take place, for example, in the propagation of the heat through a isotropic medium at rest,[1] in viscous fluid flows, in flows with chemical reactions,[2] in problems on the propagation of electromagnetic waves,[2] in problems of fluid flows through porous media,[3] etc.

Diffusion processes, in the absence of sources, are governed by the PDE of the form

$$\frac{\partial u}{\partial t} = \sum_{j=1}^{L} \frac{\partial}{\partial x_j}\left[D(u,\mathbf{x},t)\frac{\partial u}{\partial x_j}\right] \qquad (3.1.1)$$

where $\mathbf{x} = (x_1,\ldots,x_L)$ is the radius vector of a point in the Euclidean space R^L, $L \geq 1$, $x_1,\ldots,x_L$ are the spatial coordinates of a point, $D(u,\mathbf{x},t)$ is the diffusion coefficient; where $D > 0$. For heat diffusion the function u in Eq. (3.1.1) is the temperature of a medium, $D = k/(c\rho)$, where k is the heat conduction coefficient, c is the specific heat of the medium, ρ is the density of the medium, and $k/(c\rho) = const > 0$. Then Eq. (3.1.1) becomes the *heat equation*. In the case of one spatial variable x and the constant diffusion coefficient $D \equiv \nu = const > 0$ the heat equation has the form

$$\frac{\partial u}{\partial t} = \nu \frac{\partial^2 u}{\partial x^2} \qquad (3.1.2)$$

The solutions of (3.1.2) have the property that strong discontinuities are smoothed at $t > 0$.[4] This facilitates the application of finite-difference method for the approximate numerical solution of the diffusion equations.

The simplest problem for heat propagation is the initial-value problem for equation (3.1.2), that is the problem

$$
\begin{aligned}
\partial u/\partial t &= \nu \partial^2 u/\partial x^2, \qquad \nu = const > 0 \\
u(x,0) &= u_0(x), \qquad -\infty < x < \infty
\end{aligned}
\tag{3.1.3}
$$

where $u_0(x)$ is a given distribution of temperature at time $t = 0$.

Applied problems are usually solved in spatial domains of finite sizes. Let us assume that Eq. (3.1.2) is to be solved under certain initial and boundary conditions in a finite interval $a \leq x \leq b$, where $-\infty < a < b < \infty$. Then the following initial and boundary conditions can be formulated for Eq. (3.1.2):

$$
u(x,0) = u_0(x), \qquad a \leq x \leq b
\tag{3.1.4}
$$

$$
u(a,t) = \varphi_1(t), \quad u(b,t) = \varphi_2(t), \quad 0 \leq t \leq T
\tag{3.1.5}
$$

where $[0,T]$ is a given interval on the t-axis, and $\varphi_1(t)$ and $\varphi_2(t)$ are functions given on the boundary. Problem (3.1.2), (3.1.4), (3.1.5) is called the *first boundary-value problem*.[4]

Instead of the right boundary condition

$$
u(b,t) = \varphi_2(t), \quad 0 \leq t \leq T
$$

one can use a *Neumann* boundary condition

$$
\partial u(b,t)/\partial x = \varphi_3(t), \quad 0 \leq t \leq T
\tag{3.1.6}
$$

that is the value of the solution derivative is given at the right end of the spatial integration interval $a \leq x \leq b$. A more general form of the boundary conditions, which can be used either at the left end $x = a$ or at the right end $x = b$, is

$$
a_1(t)\frac{\partial u(x_0,t)}{\partial x} + a_2(t)u(x_0,t) = a_3(t) \quad 0 \leq t \leq T
\tag{3.1.7}
$$

where $a_1(t)$, $a_2(t)$ and $a_3(t)$ are the given functions, and $x_0 = a$ or b. The boundary condition (3.1.7) is called the mixed, or *Robin* boundary condition. The PDE (3.1.2) can be solved for different combinations of the boundary conditions at $x = a$ and $x = b$. The questions of the existence and uniqueness of solutions for the initial- and boundary-value problems are considered in the textbooks on the equations of the mathematical physics.[4]

In the case of more than one spatial variables in (3.1.1), that is when $L = 2$ or $L = 3$, the boundary conditions (3.1.5) generalize to the *Dirichlet* boundary condition

$$u(\mathbf{x}, t)\big|_{\Gamma} = \varphi_1(\mathbf{x}, t), \quad \mathbf{x} \in \Gamma,\ 0 \le t \le T$$

where Γ is the boundary of a spatial domain Ω, in which the solution of Eq. (3.1.1) is to be determined. The Robin boundary condition (3.1.6) generalizes to the form

$$\frac{\partial u}{\partial n} + \alpha u = \beta, \quad \mathbf{x} \in \Gamma, \quad 0 \le t \le T$$

where α and β are the given functions of $\mathbf{x}$ and t, and $\partial u / \partial n$ is the derivative normal to the boundary.

3.2 SIMPLE SCHEMES FOR THE ONE-DIMENSIONAL HEAT EQUATION

3.2.1 An Explicit Three-Point Scheme

Consider the question of the difference discretization of the initial-value problem (3.1.3). Let us construct the uniform mesh G_h in the (x, t) plane as in Fig. 2.1. Let us approximate the derivative $\partial u / \partial t$ in (3.1.2) by a forward difference in time. For the approximation of the spatial derivative $\partial^2 u / \partial x^2$ we can use the three-point difference quotient derived in Section 2.1. As a result we obtain the following explicit difference equation:

$$\frac{u_j^{n+1} - u_j^n}{\tau} = \nu \frac{u_{j+1}^n - 2u_j^n + u_{j-1}^n}{h^2} \tag{3.2.1}$$

$$j = 0, \pm 1, \pm 2, \ldots; \qquad n = 0, 1, \ldots, N$$
$$u_j^0 = u_0(jh), \quad j = 0, \pm 1, \ldots$$

Scheme (3.2.1) has the order of approximation $O(\tau) + O(h^2)$. It is easy to find the stability condition of the scheme by the Fourier method:

$$\frac{\nu \tau}{h^2} \le 1/2 \tag{3.2.2}$$

By using the *method of the separation of variables*[4] one can find the following exact solution of the heat equation (3.1.2):

$$u(x, t) = e^{-\nu \beta^2 t}(A_1 \cos \beta x + A_2 \sin \beta x) \tag{3.2.3}$$

where β, A_1 and A_2 are the constants and $\beta \neq 0$, $|A_1|+|A_2| \neq 0$. Let us formulate the following initial and boundary values of the type (3.1.4), (3.1.5):

$$u(x,0) = u_0(x) = A_1 \cos \beta x + A_2 \sin \beta x, \quad a \leq x \leq b \qquad (3.2.4)$$

$$u(a,t) = \varphi_1(t), \ u(b,t) = \varphi_2(t), \quad 0 \leq t \leq T \qquad (3.2.5)$$

where

$$\varphi_1(t) = e^{-\nu \beta^2 t}(A_1 \cos \beta a + A_2 \sin \beta a) \qquad (3.2.6)$$

$$\varphi_2(t) = e^{-\nu \beta^2 t}(A_1 \cos \beta b + A_2 \sin \beta b) \qquad (3.2.7)$$

Then the exact solution of the initial- and boundary-value problem (3.1.2), (3.2.4)-(3.2.7) should have the form (3.2.3). Thus, this problem can be taken as a test problem for checking the accuracy of difference schemes for Eq. (3.1.2).

We have implemented the difference scheme (3.2.1) augmented by the initial condition (3.2.4) and the boundary conditions (3.2.5)-(3.2.7) in the *Mathematica* program heat1.ma, which we have made by modifying our program advbac.ma from Section 2.2. The input parameters used in the program heat1.ma are described in Table 3.1.

Table 3.1. Parameters used in program heat1.ma

Parameter	Description
nu	the positive coefficient ν in the heat equation (3.1.2)
beta	the constant β in the exact solution (3.2.3)
A1	the constant A_1 in the exact solution (3.2.3)
A2	the constant A_2 in the exact solution (3.2.3), $|A_1| + |A_2| \neq 0$
a1	the abscissa a_1 of the left end of the interval on the x-axis
b1	the abscissa b_1 of the right end of the above interval, $a_1 < b_1$
M	the number M of the grid node coinciding with the right boundary $x = b_1$ of the spatial integration interval (see also (2.1.5))
cap	the Courant number $\kappa = \nu \tau / (h^2)$, where τ is the time step of the difference scheme, and h is the step of the uniform grid in the interval $a_1 \leq x \leq b_1$
numbt	the number of the time steps to be made by the difference scheme, numbt > 1
Npic	the number of the pictures of the difference solution graphs for $Npic$ moments of time, $Npic > 1$

The main function of the program heat1.ma is:

$$\text{heat1}\,[\nu, \beta, A_1, A_2, a_1, b_1, M, \kappa, numbt, Npic]\,.$$

To program the scheme (3.2.1) we have replaced the corresponding lines in the package **advbac.ma** by the lines

```
(* -- Do-loop over grid nodes on the x-axis --------- *)
     fxr = uj[[1]] - uj[[0]];
        Do[ fxl = fxr;
(* -- The explicit scheme (3.2.1) for the heat
     equation (3.1.2) --------------------------- *)
     fxr = uj[[j + 1]] - uj[[j]];
       up[[j]] = uj[[j]] + cap*(fxr - fxl),
         {j, 1, M2}];
(*  The boundary condition on the left boundary x = a1 *)
     up[[0]] = uexact[a1, t, nu, beta, A1, A2];
(*  The boundary condition on the right boundary x = b1 *)
     up[[M]] = uexact[b1, t, nu, beta, A1, A2];
```

In the above fragment, **M2 = M - 1**.

The initial condition (3.2.4) was programmed in the form of the function **uinit[...]**, which has the form

```
uinit[x_, beta_, A1_, A2_] :=
     (xb = beta*x; A1*Cos[xb] + A2*Sin[xb] )
```

The complete listing of the *Mathematica* package **heat1.ma** implementing the explicit scheme (3.2.1) with the conditions (3.2.4)-(3.2.6) is available on the attached diskette, see also Appendix. In Fig. 3.1 we present the computational results which were obtained by using scheme (3.2.1) at $\kappa = \nu\tau/(h^2) = 0.5$. It can be seen that the computation remains stable as the time t increases.

In Fig. 3.2 we show the numerical results obtained by using the same scheme (3.2.1) in the case $\kappa = \nu\tau/(h^2) = 0.6$. It can be seen that the amplitude of the numerical solution errors increases with time, so that at $t = 54\tau$ (see the graph in the right lower corner of Fig. 3.2) the value $max_x|u(x,t)|$ exceeds the exact amplitude of the numerical solution by a factor of about 2.13. Which means that the computation is unstable. This result confirms the validity of the stability condition (3.2.2).

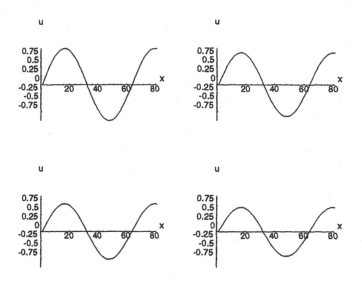

Figure 3.1: The results of the computation by scheme (3.2.1). $\kappa = 0.5$.

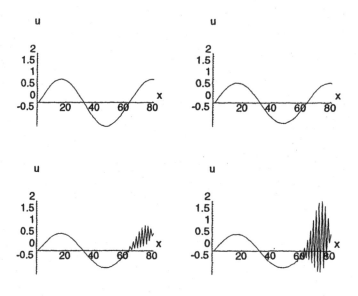

Figure 3.2: The results of the computation by scheme (3.2.1). $\kappa = 0.6$.

The numerical solutions presented in Figures 3.1 and 3.2 were obtained at $\nu = 2$, $\beta = 4$, $A_1 = 0$, $A_2 = 1$ in (3.2.3); the number of the mesh cells on the x-axis is $M = 80$ and $a = 0$, $b = 2$ in (3.2.4)-(3.2.7). It is easy to transform the analytic formula (3.2.3) to the form

$$u(x,t) = e^{-\nu \beta^2 t} \sqrt{A_1^2 + A_2^2} \sin(\beta x + \alpha) \qquad (3.2.8)$$

where

$$\sin \alpha = A_1 / \sqrt{A_1^2 + A_2^2}, \quad \cos \alpha = A_2 / \sqrt{A_1^2 + A_2^2}$$

It follows from formula (3.2.8) that the solution $u(x,t)$ is a spatially periodic function, and the period is

$$P = 2\pi/\beta \qquad (3.2.9)$$

It is clear from the geometric considerations that the halved period of the harmonic oscillator (3.2.8) can be represented in the numerical solution, if at least two mesh cells are within the halved period $P/2 = \pi/\beta$. We thus arrive at the following inequality:

$$2h \leq \pi/\beta \qquad (3.2.10)$$

If, at a given value of β, the mesh is so crude that $h > \pi/(2\beta)$, then the individual harmonics cannot be represented adequately by the numerical solution. For example, at $h = 2/80 = 1/40$ we obtain from (3.2.10) that the frequency β should satisfy the inequality $\beta \leq 40\pi/2 \approx 60$. The inequality (3.2.10) does not depend on the choice of a specific difference scheme, and it gives a *resolution limit* for any difference method applied to the approximate solution of the problem (3.1.2), (3.2.4)-(3.2.7).

It follows from Eq. (3.2.8) that the amplitude of the harmonic oscillations decays faster with time as the frequency β increases. This effect can be seen in Fig. 3.3 obtained at $\beta = 8$.

In Fig. 3.4 we show the numerical solution obtained at $h = 1/40$, $\beta = 50$, that is near the resolution limit of a difference method. We can see that the quality of the numerical solution is still satisfactory. In Fig. 3.5 we show the numerical results obtained at $h = 1/40$, $\beta = 70$, that is when the inequality (3.2.10) is violated. We can see that at $t = 0$ the amplitudes of the peaks are different, because the mesh on the x-axis is too crude for this initial condition. In order to obtain a satisfactory numerical solution in this case one must take a finer mesh on the x-axis, whose step h satisfies the inequality (3.2.10).

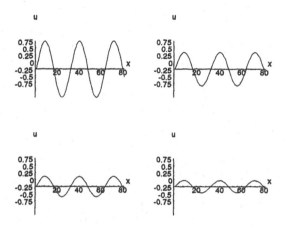

Figure 3.3: The results of the computation by scheme (3.2.1). $\beta = 8$, $\kappa = 0.5$.

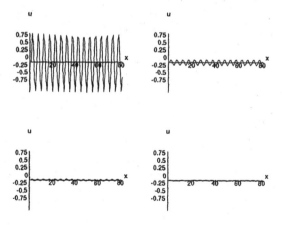

Figure 3.4: The results of the computation by scheme (3.2.1). The graphs (from the left to the right and from top to bottom) correspond to the moments of time $t = 0$, $t = \tau$, $t = 2\tau$ and $t = 3\tau$, respectively and $\beta = 50$, $\kappa = 0.5$.

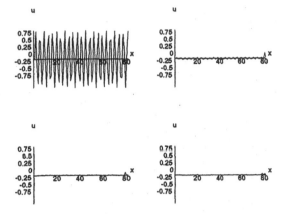

Figure 3.5: The results of the computation by scheme (3.2.1). The graphs (from the left to the right and from top to bottom) correspond to the moments of time $t = 0$, $t = \tau$, $t = 2\tau$ and $t = 3\tau$, respectively. $\beta = 70$, $\kappa = 0.5$.

3.2.2 An Implicit Scheme

It may be seen from the formula (3.2.2) that in the cases when $\nu = O(1)$ the stability imposes a severe restriction on the time step τ: $\tau = O(h^2)$. This problem can be alleviated by using an implicit difference approximations instead of the explicit scheme (3.1.8). Let us consider the following one-parametric family of difference schemes for equation (3.1.2) [1,2]:

$$\frac{u_j^{n+1} - u_j^n}{\tau} = \nu \frac{\theta(u_{j+1}^{n+1} - 2u_j^{n+1} + u_{j-1}^{n+1}) + (1 - \theta)(u_{j+1}^n - 2u_j^n + u_{j-1}^n)}{h^2}$$

(3.2.11)

where θ is a constant nonnegative nondimensional weight parameter. The scheme (3.1.10) coincides with the explicit scheme (3.1.8) in the case $\theta = 0$. In the case $\theta > 0$ the stencil of scheme (3.1.10) has the same six-point form as in Fig. 2.12. Using the Fourier method for the stability investigation of scheme (3.1.10), we can easily obtain the following stability condition (see also[1]):

$$\frac{2\nu\tau}{h^2} \begin{cases} \leq 1/(1 - 2\theta), & 0 < \theta < 1/2 \\ \text{no limitations}, & 1/2 \leq \theta \leq 1 \end{cases}$$

(3.2.12)

Since the difference scheme (3.1.10) is a scalar two-level difference scheme with constant coefficients, the condition (3.1.11) is both necessary and sufficient for stability.

Note that when (3.2.11) is implicit, $\theta > 0$, and if we solve for u_j^{n+1} then the solution depends on u_{j+1}^{n+1}, u_{j-1}^{n+1} in contrast to the case of explicit schemes. For the three-point implicit scheme like scheme (3.2.11) the special variant of the Gaussian elimination method described above in Section 2.9 proves to be very efficient.

3.2.3   Nonlinear Diffusion

Let us now consider the case when the diffusion coefficient D in (3.1.1) is a function of u: $D = D(u)$. Then we have in the case of one spatial variable, a PDE of the form

$$\frac{\partial u}{\partial t} = \frac{\partial}{\partial x}\left[D(u)\frac{\partial u}{\partial x}\right] \tag{3.2.13}$$

If the time step τ is sufficiently small, then the variations of the values u_j^n remain small when going from the time level n to the time level $n+1$, and the nonlinear diffusion coefficient $D(u)$ changes only a small amount, so the nonlinear terms may be computed in terms of the known grid variables u^n:

$$\frac{u_j^{n+1} - u_j^n}{\tau} = \theta\frac{D(u_{j+1/2}^n)(u_{j+1}^{n+1} - u_j^{n+1}) - D(u_{j-1/2}^n)(u_j^{n+1} - u_{j-1}^{n+1})}{h^2}$$
$$+ (1-\theta)\frac{D(u_{j+1/2}^n)(u_{j+1}^n - u_j^n) - D(u_{j-1/2}^n)(u_j^n - u_{j-1}^n)}{h^2}$$

where $u_{j\pm1/2}^n = (1/2)(u_j^n + u_{j\pm1}^n)$. However, in cases of large time steps this technique for the treatment of the nonlinear diffusion may lead to an instability, and the numerical solution may change substantially at a time step.

A possible way for the solution of this problem is the use of a completely implicit difference scheme. Let us introduce the function $\Phi(u)$ by the formula

$$\Phi(u) = \int D(u)du$$

Then

$$D(u) = \Phi'(u) \tag{3.2.14}$$

and we can rewrite the equation (3.1.12) as follows:

$$\frac{\partial u}{\partial t} = \frac{\partial^2\Phi(u)}{\partial x^2} \tag{3.2.15}$$

The implicit scheme approximating (3.1.14) may be written as

$$\frac{u_j^{n+1} - u_j^n}{\tau} = \frac{\Phi(u_{j+1}^{n+1}) - 2\Phi(u_j^{n+1}) + \Phi(u_{j-1}^{n+1})}{h^2} \qquad (3.2.16)$$

so that all the grid values in the right hand side are computed at the $(n+1)$th time level. The efficient variant of the Gaussian elimination method, which was described above in Section 2.6, is difficult to use directly because of the presence of the nonlinear terms in the right hand side of (3.1.15). Let us assume that the value u_j^{n+1} is computed by iteration, so that

$$u_j^{n+1} = u_j^{(m)} + \Delta u_j^{(m)}, \quad m = 0, 1, 2, \dots$$

where $\Delta u_j^{(m)}$ are the corrections of the previous iteration $u_j^{(m)}$; $u_j^{(0)} = u_j^n$. Assuming that the corrections $\Delta u_j^{(m)}$ are small for all j, we now linearize the difference equation (3.1.15) by using the Taylor series expansion of $\Phi(u_j^{n+1})$:

$$\Phi(u_j^{n+1}) = \Phi(u_j^{(m)} + \Delta u_j^{(m)}) =$$
$$\Phi(u_j^{(m)}) + (d\Phi(u_j^{(m)})/du)\Delta u_j^{(m)} + O((\Delta u_j^{(m)})^2)$$
$$= \Phi(u_j^{(m)}) + D(u_j^{(m)})\Delta u_j^{(m)} + O((\Delta u_j^{(m)})^2) \qquad (3.2.17)$$

We have taken into account the formula (3.1.13) here. Substituting the right hand side of (3.1.16) instead of $\Phi(u^{n+1})$ into (3.1.15) and neglecting the terms of the orders $O((\Delta u_j^{(m)})^k)$, $k = 2, 3, \dots$, we obtain the following difference equation:

$$\frac{\Delta u_j^{(m)}}{\tau} - \frac{1}{h^2}\left[D(u_{j+1}^{(m)})\Delta u_{j+1}^{(m)} - 2D(u_j^{(m)})\Delta u_j^{(m)} + D(u_{j-1}^{(m)})\Delta u_{j-1}^{(m)}\right]$$
$$= \frac{u_j^n - u_j^{(m)}}{\tau} + \frac{\Phi(u_{j+1}^{(m)}) - 2\Phi(u_j^{(m)}) + \Phi(u_{j-1}^{(m)})}{h^2} \qquad (3.2.18)$$

Usually 2-5 iterations in m are sufficient to obtain the desired accuracy of the numerical solution.

Exercise 3.1. Obtain the stability condition (3.1.9) of scheme (3.1.8) with the aid of the Fourier method.

Hint. You can use the program st1.ma from Section 2.3 for obtaining the expression for the amplification factor of scheme (3.1.8). For this purpose re-specify the body of the function sch in accordance with the left hand side of difference equation (3.1.8). Then re-specify the formula for the nondimensional variable c1 by replacing the line 1 = 1 /. dt

-> c1*h/a; by the line 1 = 1 /.   dt -> c1*h^ 2/nu;, where nu= ν, the diffusion coefficient.

Exercise 3.2. Modify the program **heat1.ma** in such a way that it shows the exact solution (3.2.3) along with the numerical solution on the same picture.

Exercise 3.3. The temporal evolution of the numerical solution of the problem (3.1.2), (3.2.4)-(3.2.7) can be shown especially clearly if the surface $u = u(x,t)$ is plotted. Modify the program **heat1.ma** so that it shows also the surface $u = u(x,t)$ for $0 \leq t \leq t_N$.

Hint. Use the two-dimensional table **ut** and the program **mc2d.ma** from Section 2.12.

Exercise 3.4. Obtain the stability condition (3.2.12) of scheme (3.2.11) with the aid of the Fourier method.

Exercise 3.5. Rewrite the difference equation (3.2.18) in the form

$$a_j \Delta u_{j-1}^{(m)} + b_j \Delta u_j^{(m)} + c_j \Delta u_{j+1}^{(m)} = f_j$$

and find the limitations on $D(u)$, h, τ at which the condition (2.6.16) of the diagonal dominance is satisfied.

Exercise 3.6. Consider the heat equation

$$\frac{\partial u}{\partial t} = \nu \left(\frac{\partial^2 u}{\partial x^2} + \frac{\partial^2 u}{\partial y^2} \right) \tag{3.2.19}$$

in the case of two spatial variables x, y, where $\nu = const > 0$. Let us approximate Eq. (3.2.19) by the explicit difference scheme with forward difference in time and central differences in the spatial variables similar to scheme (3.2.1). Find the necessary and sufficient stability condition of this scheme with the aid of the Fourier method.

Exercise 3.7. Consider the heat equation

$$\frac{\partial u}{\partial t} = \nu \left(\frac{\partial^2 u}{\partial x^2} + \frac{\partial^2 u}{\partial y^2} + \frac{\partial^2 u}{\partial z^2} \right) \tag{3.2.20}$$

in the case of three spatial variables x, y, z, where $\nu = const > 0$. Let us approximate Eq. (3.2.20) by the explicit difference scheme with forward difference in time and central differences in the spatial variables similar to scheme (3.2.1). Find the necessary and sufficient stability condition of this scheme with the aid of the Fourier method. Compare this stability condition with that from the foregoing exercise in the particular case of equal steps of the uniform spatial computing mesh in different coordinate directions.

3.3 DIFFERENCE SCHEMES FOR ADVECTION-DIFFUSION EQUATION

3.3.1 An Explicit Scheme

If the substance is transported at the nonzero velocity $\mathbf{v}$, the equation (3.1.1) becomes insufficient for the mathematical description of the diffusion processes. To take into account the transport or convection of the substance we must consider the following PDE:[1,2]

$$\frac{\partial u}{\partial t} + \sum_{j=1}^{L} v_j \frac{\partial u}{\partial x_j} = \sum_{j=1}^{L} \frac{\partial}{\partial x_j} \left[D(u, \mathbf{x}, t) \frac{\partial u}{\partial x_j} \right] \qquad (3.3.1)$$

where $v_1, \ldots, v_L$ are the components of the velocity vector $\mathbf{v}$ along the axes $x_1, \ldots, x_L$, respectively.

Let us consider the particular case when $L = 1$, $x_1 \equiv x$, $v_1 = a = const$, $D \equiv \nu = const > 0$. Then the equation (3.3.1) reverts to

$$\frac{\partial u}{\partial t} + a \frac{\partial u}{\partial x} = \nu \frac{\partial^2 u}{\partial x^2} \qquad (3.3.2)$$

The equation (3.3.2) is often termed the *advection-diffusion equation*.

Let us now approximate the time derivative in (3.3.2) by a forward difference and the derivatives $\partial u/\partial x$ and $\partial^2 u/\partial x^2$ by the central differences at the nth time level. Then we obtain the following explicit difference scheme:

$$\frac{u_j^{n+1} - u_j^n}{\tau} + a \frac{u_{j+1}^n - u_{j-1}^n}{2h} = \nu \frac{u_{j+1}^n - 2u_j^n + u_{j-1}^n}{h^2} \qquad (3.3.3)$$

This scheme has the order of approximation $O(\tau) + O(h^2)$. The stability region for this scheme can easily be obtained with the aid of the program st2.ma, which was presented in Section 2.13. Let us introduce the following nondimensional variables:

$$\kappa_1 = a\tau/h, \ \ \kappa_2 = \nu\tau/(h^2) \qquad (3.3.4)$$

In Fig. 3.6 we show the stability region obtained by the program st2.ma available on the attached diskette. It may be seen from this figure that the stability region is bounded by the parabola $\kappa_2 = \frac{1}{2}\kappa_1^2$ from below. The line $\kappa_2 = 0.5$ is the upper boundary of the stability region. Therefore, we may write the stability condition for the scheme (3.3.3) as follows:

$$\kappa_1^2 \leq 2\kappa_2 \leq 1 \qquad (3.3.5)$$

The same result was obtained in Reference 5 with the aid of a careful analytical investigation of the behaviour of the amplification factor of

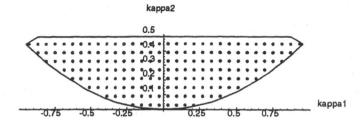

Figure 3.6: The stability region is dotted. The 40×10 mesh in the (κ_1, κ_2) plane.

scheme (3.3.3). At $\kappa_2 = 0$ (the absence of diffusion) the scheme (3.2.3) is unstable, as this was shown in Section 2.3 (see Exercise 2.8). This instability follows also from Fig. 3.6. At positive ν the inequality $2\kappa_2 \leq 1$ proves to be very restrictive for the time step. Note that this limitation is caused by the central difference approximation of the diffusion term in the scheme (3.3.3).

Let us rewrite the difference equation (3.3.3) as follows:

$$u_j^{n+1} = (1 - 2\kappa_2)u_j^n + (\kappa_2 - \frac{1}{2}\kappa_1)u_{j+1}^n + (\kappa_2 + \frac{1}{2}\kappa_1)u_{j-1}^n \quad (3.3.6)$$

The PDE (3.3.2) is known to possess the property[6]

$$\sup_x |u(x, t)| \leq \sup_x |u(x, t')| \quad \text{if } t > t'$$

This inequality means that the maximum value of $|u(x, t)|$ should not increase as t increases. It is therefore reasonable to require that the scheme (3.3.3) possesses a similar property. Let us assume that the condition

$$\frac{1}{2}|\kappa_1| \leq \kappa_2 \quad (3.3.7)$$

is satisfied along with the stability condition (3.3.5). Then we have from Eq. (3.3.6) that

$$|u_j^{n+1}| \leq (1 - 2\kappa_2)|u_j^n| + (\kappa_2 - \frac{1}{2}\kappa_1)|u_{j+1}^n| +$$
$$(\kappa_2 + \frac{1}{2}\kappa_1)|u_{j-1}^n| \leq |u_j^n|$$

thus

$$\max_j |u_j^{n+1}| \leq \max_j |u_j^n| \quad (3.3.8)$$

If the inequality (3.3.7) is not satisfied, the inequality (3.3.8) can be violated in general. Let us take, for example, the initial data $u(x,0)$ as follows[6]

$$u_j^0 = 1, \ j \le 0; \quad u_j^0 = -1, \ j > 0$$

Then it is easy to calculate with the aid of Eq. (3.3.6) that

$$u_0^1 = 1 + (\kappa_1 - 2\kappa_2) > 1$$

The inequality (3.3.7) can be rewritten in the form

$$h \le \frac{2\nu}{a} \tag{3.3.9}$$

Let us introduce the quantity

$$P_h = \frac{ah}{\nu} \tag{3.3.10}$$

This is a nondimensional parameter, which is termed the *cell Péclet number*, or the *grid Péclet number* in the theory of heat flow. In the theory of fluid flow, the quantity ah/ν corresponds to the *Reynolds number*. Thus the parameter P_h can also be called the *cell Reynolds number*. Condition (3.3.9) is a condition on the mesh spacing h that must be satisfied in order for the solution to the scheme to behave qualitatively like that of the parabolic differential equation. The condition $|P_h| \le 2$ does not follow from the stability condition (3.3.5). The inequalities (3.3.5) were indeed obtained from the stability consideration in the L_2 norm, see also Section 2.10. On the other hand, for the difference schemes of the odd approximation orders one can also consider the stability in the C norm.[1] If we define the grid analog of the C norm as

$$\| u^n \|_C = \max_j |u_j^n|$$

then the inequality (3.3.8) can be rewritten as

$$\| u^{n+1} \|_C \le \| u^n \|_C$$

The inequality (3.3.8) can thus be considered as the stability definition of scheme (3.3.3) in the C norm. From the above consideration it follows that the stability of scheme (3.3.3) in the C norm takes place if the inequalities

$$0 \le \kappa_2 \le 0.5, \quad \kappa_2 \ge \frac{1}{2}|\kappa_1|$$

are satisfied. We show the corresponding stability region in Fig. 3.7.

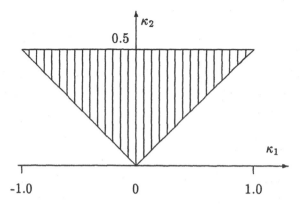

Figure 3.7: The stability region is dashed.

It can be seen from Figs. 3.6 and 3.7 that the stability region in the C norm is somewhat smaller than the stability region in the L_2 norm. The reason for this is that the functions obtained as the series in the spatially periodic functions ($\sin kx$, $\cos kx$) form a class of functions which is narrower than the set of arbitrary grid functions, see Reference 7 for further details.

3.3.2   Allen's Scheme

Many explicit and implicit difference schemes have been proposed for the numerical integration of equation (3.3.2) with the purpose of relaxing the stability conditions (3.3.5). Consider the two-step scheme proposed by J.Allen:[7,8,5]

$$\tilde{u}_j^{n+1} = u_j^n - \frac{\kappa_1}{2}(u_{j+1}^n - u_{j-1}^n) + \kappa_2(u_{j+1}^n + u_{j-1}^n - 2\tilde{u}_j^{n+1}) \quad (3.3.11)$$

$$u_j^{n+1} = u_j^n - \frac{\kappa_1}{2}(\tilde{u}_{j+1}^{n+1} - \tilde{u}_{j-1}^{n+1}) + \kappa_2(\tilde{u}_{j+1}^{n+1} + \tilde{u}_{j-1}^{n+1} - 2u_j^{n+1}) \quad (3.3.12)$$

It may be seen from the comparison of the difference equations (3.3.3) and (3.3.11) that the term $-2u_j^n$ in the approximation of the diffusion term of scheme (3.3.3) has been replaced by the term $-2\tilde{u}_j^{n+1}$. As a result of this the difference equation (3.3.11) appears to be implicit. But since it involves only one value $\tilde{u}_j^{n+1}$ from the new time level $n+1$, it can be explicitly resolved with respect to $\tilde{u}_j^{n+1}$:

$$\tilde{u}_j^{n+1} = [u_j^n - \frac{\kappa_1}{2}(u_{j+1}^n - u_{j-1}^n) + \kappa_2(u_{j+1}^n + u_{j-1}^n)]/(1 + 2\kappa_2) \quad (3.3.13)$$

The second equation (3.3.12) can be resolved in a similar way with respect to u_j^{n+1}. The scheme (3.3.11), (3.3.12) has the order of approximation $O(\tau) + O(h^2)$.

For the Fourier stability analysis we must at first eliminate the values $\tilde{u}^{n+1}$ from Eq. (3.3.12) by using Eq. (3.3.11). It follows from the formula (3.3.13) that the expression for $\tilde{u}^{n+1}$ is rational. We have faced the difficulties when we tried to substitute directly the fraction (3.3.13) into Eq. (3.3.12) in the sense that the *Mathematica* system produced an incorrect result. Thus it appears that it is desirable to avoid the algebraic manipulation with rational expressions within the framework of the *Mathematica* system.

It is possible, under certain assumptions, to eliminate the intermediate values in any implicit difference two-stage scheme

$$\tilde{S}_0 \tilde{u}^{n+1} = \tilde{S}_1 u^n \qquad (3.3.14)$$

$$S_0 u^{n+1} = S_{11} u^n + S_{12} \tilde{u}^{n+1} \qquad (3.3.15)$$

Let us indeed assume that the difference operator $\tilde{S}_0$ is nonsingular. Then we can multiply the both sides of Eq. (3.3.15) by $\tilde{S}_0$ from the left:

$$\tilde{S}_0 S_0 u^{n+1} = \tilde{S}_0 S_{11} u^n + \tilde{S}_0 S_{12} \tilde{u}^{n+1} \qquad (3.3.16)$$

Let us further assume that the operator $\tilde{S}_0$ commutes with the operator S_{12}, that is $\tilde{S}_0 S_{12} = S_{12} \tilde{S}_0$. Then we have from Eq. (3.3.16) with regard for Eq. (3.3.14) that

$$\begin{aligned} \tilde{S}_0 S_0 u^{n+1} - \tilde{S}_0 S_{11} u^n - S_{12} \tilde{S}_0 \tilde{u}^{n+1} = \\ \tilde{S}_0 S_0 u^{n+1} - \tilde{S}_0 S_{11} u^n - S_{12} \tilde{S}_1 \tilde{u}^n = 0 \end{aligned} \qquad (3.3.17)$$

The difference equation (3.3.17) thus involves only the grid values u^{n+1} and u^n and, in addition, it is fraction-free. Therefore, it can be used for the input in our program st2.ma. We have programmed the left-hand side of Eq. (3.3.17) in our package st2.ma as the function

```
(* Allen's scheme for the 1D advection-diffusion equation*)
  Allen =
( ut1 = u[j, n + 1]*(1 + 2*cp2)^2;

  vp = u[j+1,n] - (cp1/2)*(u[j+2,n] - u[j,n]) +
       cp2*(u[j+2,n] + u[j,n]));

  vm = u[j-1,n] - (cp1/2)*(u[j,n] - u[j-2,n]) +
```

```
        cp2*(u[j,n] + u[j-2,n]));

sc = ut1 - u[j,n]*(1 + 2*cp2) + (cp1/2)*(vp - vm) -
     cp2*(vp + vm); sc )
```

In the case of Allen's scheme $(3.3.11)$-$(3.3.12)$ the operators $\tilde{S}_0$ and S_0 are simply the constant multipliers: $\tilde{S}_0 = S_0 = 1 + 2\kappa_2$. We have introduced the nondimensional quantities cp1= κ_1 and cp2= κ_2 already at this stage to avoid possible errors in the result produced by the *Mathematica* system, because the expressions $(3.3.4)$ for κ_1 and κ_2 also represent the rational expressions. In what follows we present the left-hand side of Eq. $(3.3.17)$ produced by the *Mathematica* system:

```
-((1 + 2*cp2)*u[j, n]) + (1 + 2*cp2)^2*u[j, 1 + n] +
(cp1*(-u[-1 + j, n] + (cp1*(-u[-2 + j, n] + u[j, n]))/2 -
cp2*(u[-2 + j, n] + u[j, n]) + u[1 + j, n] -
(cp1*(-u[j, n] + u[2 + j, n]))/2 + cp2*(u[j, n] +
u[2 + j, n])))/2 - cp2*(u[-1 + j, n] - (cp1*(-u[-2 + j, n]
+ u[j, n]))/2 + cp2*(u[-2 + j, n] + u[j, n]) + u[1 + j, n]
- (cp1*(-u[j, n] + u[2 + j, n]))/2 + cp2*(u[j, n] +
u[2 + j, n]))
```

The following expression for the modulus $|\lambda|^2$ of the amplification factor of Allen's scheme was obtained:

$$|\lambda|^2 =$$

```
(4 - 4*cp1^2 + 4*ca^2*cp1^2 + cp1^4 - 2*ca^2*cp1^4 +
    ca^4*cp1^4 + 16*cp2 + 16*ca*cp2 - 8*cp1^2*cp2 -
    8*ca*cp1^2*cp2 + 8*ca^2*cp1^2*cp2 + 8*ca^3*cp1^2*cp2 +
    32*cp2^2 + 32*ca*cp2^2 + 32*ca^2*cp2^2 - 8*cp1^2*cp2^2
    + 8*ca^4*cp1^2*cp2^2 + 32*cp2^3 + 32*ca*cp2^3 +
    32*ca^2*cp2^3 + 32*ca^3*cp2^3 + 16*cp2^4 +
    32*ca^2*cp2^4 + 16*ca^4*cp2^4 + 2*cp1^4*sa^2 -
    2*ca^2*cp1^4*sa^2 - 8*cp1^2*cp2*sa^2 +
    24*ca*cp1^2*cp2*sa^2 - 16*cp2^2*sa^2 +
    48*ca^2*cp1^2*cp2^2*sa^2 - 32*cp2^3*sa^2 -
    32*ca*cp2^3*sa^2 - 32*cp2^4*sa^2 - 32*ca^2*cp2^4*sa^2 +
    cp1^4*sa^4 + 8*cp1^2*cp2^2*sa^4 + 16*cp2^4*sa^4)/
    (4*(1 + 2*cp2)^4)
```

Here the variables ca$=\cos\xi_1$, sa$=\sin\xi_1$. It may be seen that the obtained expression for $|\lambda|^2$ has a bulky form. Let us apply the Pythagor's identity $\cos^2\xi_1 + \sin^2\xi_1 = 1$ for the reduction of the length of the expression for $|\lambda|^2$. For this purpose we can use the transformation rule

```
r = r /. ca^2 -> 1 - sa^2;
```

Here the variable r denotes $|\lambda|^2$ in our program st2.ma. The application of the above transformation rule has enabled us to reduce substantially the length of the expression for $|\lambda|^2$:

$$|\lambda|^2 =$$

```
(1 + 4*cp2 + 4*ca*cp2 + 12*cp2^2 + 8*ca*cp2^2 +
    4*ca^2*cp2^2 + 16*cp2^3 + 16*ca*cp2^3 + 16*cp2^4 -
    cp1^2*sa^2 - 4*cp1^2*cp2*sa^2 + 4*ca*cp1^2*cp2*sa^2 -
    8*cp2^2*sa^2 - 8*cp1^2*cp2^2*sa^2 +
    16*ca^2*cp1^2*cp2^2*sa^2 - 16*cp2^3*sa^2 -
    16*ca*cp2^3*sa^2 - 32*cp2^4*sa^2 + cp1^4*sa^4 +
    8*cp1^2*cp2^2*sa^4 + 16*cp2^4*sa^4)/
    (1 + 2*cp2)^4
```

Sometimes also the transformation rule

```
r = r /. ca^4 -> 1 - 2*sa^2 + sa^4;
```

may prove to be useful. In the case of the Allen's scheme the application of this additional rule has not led to the further reduction of the length of the expression for $|\lambda|^2$.

It is easy to see from the above expression for $|\lambda|^2$ that in the particular case of the advection equation, that is at $\nu = 0$, the formula for $|\lambda|^2$ takes a simple form

$$|\lambda|^2 = 1 - \kappa_1^2 \sin^2\xi_1(1 - \kappa_1^2 \sin^2\xi_1) \tag{3.3.18}$$

The Allen's scheme (3.3.11)-(3.3.12) at $\nu = 0$ coincides with the two-stage Runge-Kutta scheme (3.4.4) for the advection equation, which is discussed in detail in the next Section. It is shown therein on the basis of the expression (3.3.18) that in the case $\nu = 0$ the scheme (3.3.11)-(3.3.12) is stable at $|\kappa_1| \le 1$.

It is interesting to study the question on the size of the stability region of Allen's scheme (3.3.11)-(3.3.12) in the positive direction of the κ_2-axis. Let us elucidate the following question: Is there a point of intersection of the stability region boundary with the κ_2-axis? It is obvious that $\kappa_1 = 0$ on the κ_2-axis. Substituting the value cp1 = 0 into the above expression for $|\lambda|^2$, we can easily obtain the formula

$$\begin{aligned}
|\lambda|^2 = (1 + 4\kappa_2 + 4\kappa_2 \cos\xi_1 + 12\kappa_2^2 + 8\kappa_2^2 \cos\xi_1 + \\
4\kappa_2^2 \cos^2\xi_1 + 16\kappa_2^3 + 16\kappa_2^3 \cos\xi_1 + 16\kappa_2^4 - \\
8\kappa_2^2 \sin^2\xi_1 - 16\kappa_2^3 \sin^2\xi_1 - 16\kappa_2^3 \cos\xi_1 \sin^2\xi_1 - \\
32\kappa_2^4 \sin^2\xi_1 + 16\kappa_2^4 \sin^4\xi_1)/(1 + 2\kappa_2)^4
\end{aligned} \qquad (3.3.19)$$

It is difficult to study the behavior of this function at finite values of κ_2. But it is easy to study the asymptotics of this function as $\kappa_2 \to \infty$:

$$|\lambda|^2 \approx \frac{16\kappa_2^4 - 32\kappa_2^4 \sin^2\xi_1 + 16\kappa_2^4 \sin^4\xi_1}{16\kappa_2^4} = \cos^4\xi_1 \le 1$$

This relation implies that the stability region of scheme (3.3.11)-(3.3.12) has an infinite extension along the κ_2-axis. In Fig. 3.8,b we present the stability region of Allen's scheme (3.3.11)-(3.3.12), which was obtained with the aid of the program st2.ma. We have found a part of the stability

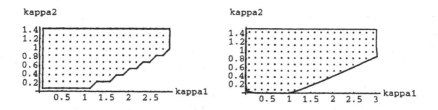

Figure 3.8: The stability region is dotted. (a) the result of the execution of Steps 1 and 2 of the Pavlidis' algorithm; (b) the refined contour of the stability region boundary. The mesh of 20×10 cells in the (κ_1, κ_2) plane.

region in the right half-plane $\kappa_1 \ge 0$ in connection with the fact that the stability region of Allen's scheme is symmetric with respect to κ_2-axis. We have taken the values $\kappa_{1l} = 0$, $\kappa_{1r} = 3$, $\kappa_{2l} = 0$, $\kappa_{2r} = 1.5$ for the rectangular domain (2.13.3). It may be seen from Fig. 3.8b that the stability region boundary intersects the κ_1-axis at $\kappa_1 = 1$, which agrees with the above presented theoretical result $|\kappa_1| \le 1$ for the stability in the absence of the diffusion. It may also be seen that at sufficiently large κ_2 the stability region fills the overall interval $0 \le \kappa_1 \le 3$. The interested reader can take larger values of κ_{1r} and κ_{2r} to study the further behaviour of the stability region.

The big size of the stability region of Allen's scheme in the direction of the κ_2-axis is explained, of course, by the partially implicit treatment of the diffusion term in this scheme.

3.4 RUNGE-KUTTA METHODS

It follows from the foregoing sections of Chapters 2 and 3 that the explicit difference schemes impose substantial limitations on the time step τ. At the same time, there have been developed a number of techniques for constructing the explicit schemes with extended stability intervals.

The idea of the application of the Runge-Kutta methods for the construction of difference schemes approximating the partial differential equations was proposed by V.V.Rusanov[10], see also References 11 to 15. Let us explain the basic idea of these methods at the example of Eq. (2.1.1). In accordance with the *method of lines* for the solution of PDEs let us replace the operator of the spatial differentiation $a\partial u/\partial x$ by a difference operator approximating the operator $a\partial u/\partial x$. For this purpose one can use, for example, the central difference:

$$Pu^n = a[u(x+h,t) - u(x-h,t)]/(2h) \qquad (3.4.1)$$

After that one considers instead of the original PDE (2.1.1) the ordinary differential equation

$$\frac{du}{dt} + Pu = 0 \qquad (3.4.2)$$

where the quantities $x+h$, $x-h$ and h are the parameters. Now one can approximate Eq. (3.4.2) by some Runge-Kutta scheme to obtain a difference scheme approximating the original PDE (2.1.1).

V.V. Rusanov[10,11] used the Runge-Kutta methods for the construction of third-order difference schemes for the PDEs of hyperbolic type. It was later proposed in Reference 16 to use the Runge-Kutta methods for the construction of explicit difference schemes possessing the extended stability intervals. The structure of the Runge-Kutta schemes proposed in References 16,17, and 18 is generally simpler than the structure of Rusanov's schemes.[10,11] Therefore, we at first describe the Jameson's schemes[16−18] and then explain the difference between the Jameson's schemes and the Rusanov's schemes. It should be noted that the Jameson's schemes have gained relatively wide acceptance for the numerical solution of applied problems. A review of the relevant works may be found in Reference 19.

In what follows we describe a number of the specific Runge-Kutta schemes, which approximate the scalar PDE of the hyperbolic or parabolic type

$$\frac{\partial u}{\partial t} + Lu = 0 \tag{3.4.3}$$

where the differential operator Lu involves only the partial derivatives with respect to spatial variables. Let us assume that the difference operator Pu in Eq. (3.4.2) approximates the operator Lu.

The simplest one-stage scheme

$$u^{n+1} = u^n - \tau P u^n$$

for Eq. (2.1.1) was shown in Chapter 2 to be unstable. Let us now consider the following two-stage Runge-Kutta scheme for Eq. (2.1.1):

$$\begin{aligned}
u^{(0)} &= u^n \\
u^{(1)} &= u^{(0)} - \alpha \tau P u^{(0)} \\
u^{(2)} &= u^{(0)} - \alpha \tau P u^{(1)} \\
u^{n+1} &= u^{(2)}
\end{aligned} \tag{3.4.4}$$

Let us choose the intermediate parameter α in (3.4.4) from the requirement that this scheme has the weakest possible limitation for the Courant number $\kappa = a\tau/h$. For this purpose let us at first eliminate the intermediate value $u^{(1)}$:

$$\begin{aligned}
u^{n+1} = u^n - \tau P u^{(1)} &= u^n - \tau P(u^n - \alpha \tau P u^n) \\
&= [I - \tau P + \alpha(\tau P)^2]u^n
\end{aligned} \tag{3.4.5}$$

where I is the identity operator. Let $z = \mathcal{F}(-\tau P)$ be the Fourier symbol of the difference operator $-\tau P$, that is $\mathcal{F}(-\tau P)$ is the result of the Fourier transform of the operator $-\tau P$. Then it is easy to obtain that $z = -i\kappa \sin \xi$, $\xi = kh$, where k is the wave number. Consequently the amplification factor g of scheme (3.4.5) has the form

$$g = 1 + z + \alpha z^2 = 1 - i\kappa \sin \xi - \alpha \kappa^2 \sin^2 \xi \tag{3.4.6}$$

We obtain from (3.4.6) that

$$\begin{aligned}
|g|^2 &= 1 + \kappa^2 \sin^2 \xi(1 - 2\alpha) + \alpha^2 \kappa^4 \sin^4 \xi \\
&= 1 - \kappa^2 \sin^2 \xi[(2\alpha - 1) - \alpha^2 \kappa^2 \sin^2 \xi]
\end{aligned}$$

If $|g|$ is not to exceed 1.0 for small values of ξ, then α should not be less than 0.5. If $|g|^2 < 1.0$ for $\sin \xi = 1$, then it will be less than 1 for all other values of ξ. Therefore, we must have

$$\kappa^2 \leq (2\alpha - 1)/(\alpha^2)$$

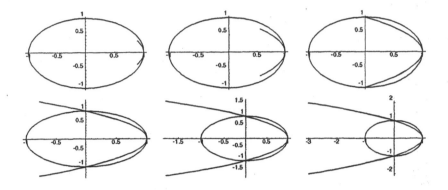

Figure 3.9: The curves $(Re\,g, Im\,g)$ for $\kappa = \kappa_j = 2j/6$, $j = 1, \ldots, 6$.

The largest value of the right hand side of the expression is 1 for $\alpha = 1$. The most stable algorithm of the class (3.4.6) is thus described by the formula

$$g = 1 + z + \alpha z^2 \qquad (3.4.7)$$

which is stable for $\kappa^2 \leq 1$. In Fig. 3.9, which is similar to Fig. 2.7, we show the curves $(Re\,g, Im\,g)$ corresponding to the optimal two-stage Runge-Kutta scheme (3.4.4) with $\alpha = 1$.

The generalization of the above procedure for the case of higher degrees of z in the expression for the amplification factor $g(\kappa, \xi)$ leads to the polynomial

$$Q(z) = 1 + z + \beta_2 z^2 + \cdots + \beta_m z^m \qquad (3.4.8)$$

such that $|Q(z)| \leq 1.0$ over some interval of the imaginary axis centered at the origin. Let $I_M(z)$ denote a polynomial of degree m in z, whose stability region includes the largest interval of the imaginary axis, together with the coordinate origin. The leading terms of this polynomial are $1 + z$, otherwise it will not represent the time advancement operator. The polynomial $I_2(z) = 1 + z + z^2$ was found for (3.4.7) above. Similarly, $I_3(z)$ can be determined by direct methods, as was shown in Reference 20. An explicit formula for $I_m(z)$ was given in[21], which is valid for any positive integer m:

$$I_m(z) = (-i)^{m-1}[T_{m-1}(i\zeta) + i(1 + \zeta^2)U_{m-2}(i\zeta)] \qquad (3.4.9)$$

where $\zeta = z/(m-1)$, $T_m(\cdot)$ is the Chebyshev polynomial of the first kind, $T_m(\cos(\zeta_1)) = \cos(m\zeta_1)$, and $U_m(\cdot)$ is a Chebyshev polynomial of the second kind, $U_m(\cos(\zeta_1)) = \sin[(m + 1)\zeta_1]/\sin(\zeta_1)$. Van der Houwen[20] found the coefficients of the polynomials $I_m(z)$ for $m = 1, 2, 3, 5, 7, 9$. Pike et al.[21] have presented tables for the coefficients of the first eleven polynomials $I_m(z)$ (that is for $m = 1, \ldots, 11$). The required polynomials $Q(z)$ may be written in the form

$$Q(z) = 1 + \gamma_1 z[1 + \gamma_2 z(1 + \gamma_3 z(1 + \gamma_4 z(1 + \cdots + \gamma_m z)))] \qquad (3.4.10)$$

for their convenient use within the context of Jameson schemes. The coefficients $\gamma_1, \ldots, \gamma_m$ are expressed in terms of the coefficients of the polynomial $I_m(z)$. In particular, we always have $\gamma_1 = 1$. Note that γ_2 is always equal to $1/2$ for odd m thus producing second-order accuracy of the numerical method in time. Let us present the values of the coefficients $\gamma_1, \ldots, \gamma_m$ for the cases $m = 2, 3, 4, 5$ following:[21]

$$\left.\begin{array}{l} m = 2: \ \gamma_1 = 1, \ \gamma_2 = 1 \\ m = 3: \ \gamma_1 = 1, \ \gamma_2 = \frac{1}{2}, \ \gamma_3 = \frac{1}{2} \\ m = 4: \ \gamma_1 = 1, \ \gamma_2 = \frac{5}{9}, \ \gamma_3 = \frac{4}{15}, \ \gamma_4 = \frac{1}{3} \\ m = 5: \ \gamma_1 = 1, \ \gamma_2 = \frac{1}{2}, \ \gamma_3 = \frac{3}{8}, \ \gamma_4 = \frac{1}{6}, \ \gamma_5 = \frac{1}{4} \end{array}\right\} \qquad (3.4.11)$$

Now let us proceed from formula (3.4.10) to the operator form of the difference scheme. Taking into account Eq. (3.4.2) we can write the difference scheme leading to the polynomial (3.4.10) in the operator form as

$$u^{n+1} = \Lambda u^n \qquad (3.4.12)$$

where

$$\Lambda = I + \gamma_1(-\tau P)(I + \gamma_2(-\tau P)(I + \cdots + \gamma_m(-\tau P))) \qquad (3.4.13)$$

The difference scheme (3.4.12)-(3.4.13) can be implemented as the following m-stage process:

$$\left.\begin{array}{l} u^{(0)} = u^n \\ u^{(1)} = u^{(0)} - \gamma_m \tau P u^{(0)} \\ u^{(2)} = u^{(0)} - \gamma_{m-1} \tau P u^{(1)} \\ \ \vdots \\ u^{(m)} = u^{(0)} - \gamma_1 \tau P u^{(m-1)} \\ u^{n+1} = u^{(m)} \end{array}\right\} \qquad (3.4.14)$$

A more general construction than (3.4.14) was also considered in Reference 14-16:

$$\left.\begin{array}{l} u^{(0)} = u^n \\ u^{(1)} = u^{(0)} - \alpha_1 \tau R^{(0)} \\ \ \vdots \\ u^{(m-1)} = u^{(0)} - \alpha_{m-1} \tau R^{(m-2)} \\ u^{(m)} = u^{(0)} - \tau R^{(m-1)} \\ u^{n+1} = u^{(m)} \end{array}\right\} \qquad (3.4.15)$$

where

$$R^{(q)} = \sum_{r=0}^{q} \{\beta_{qr} P u^{(r)} - \gamma_{qr} D u^{(r)}\}, \quad q = 0, \ldots, m-1 \qquad (3.4.16)$$

The symbols $Du^{(r)}$ in (3.4.16) are the operators of artificial dissipation whereas the β_{qr} must satisfy the *consistency condition*

$$\sum_{r=0}^{q} \beta_{qr} = 1 \qquad (3.4.17)$$

A simpler procedure is to recalculate the residual $R^{(q)}$ at each stage using the most recently updated value of u. Then (3.4.16) becomes

$$R^{(q)} = P u^{(q)} - D u^{(q)}$$

The coefficients $\alpha_1, \ldots, \alpha_{m-1}$ in (3.4.15) are expressed in terms of the coefficients $\gamma_1, \ldots, \gamma_m$ in (3.4.14) with the aid of the obvious formula

$$\alpha_j = \gamma_{m-j+1}; \quad j = 1, \ldots, m-1$$

As an example of one of the schemes of the class (3.4.15), (3.4.16) for the equation (3.4.2) we present the classic fourth-order Runge-Kutta scheme (see also Reference 18):

$$\left.\begin{aligned}
u^{(0)} &= u^n \\
u^{(1)} &= u^{(0)} - \tfrac{\tau}{2} P u^{(0)} \\
u^{(2)} &= u^{(0)} - \tfrac{\tau}{2} P u^{(1)} \\
u^{(3)} &= u^{(0)} - \tau P u^{(2)} \\
u^{(4)} &= u^{(0)} - \\
&\quad \tfrac{\tau}{6}(P u^{(0)} + 2 P u^{(1)} + 2 P u^{(2)} + P u^{(3)}) \\
u^{n+1} &= u^{(4)}
\end{aligned}\right\} \qquad (3.4.18)$$

In order to better understand the difference between Jameson's and Rusanov's schemes we present the latter[10] as applied to the quasi-linear equation

$$\partial u / \partial t + \partial f(u) / \partial x = 0 \qquad (3.4.19)$$

This scheme may be written in the form

$$\left.\begin{aligned}
u_i^{[0]} &= u_i^n \\
u_i^{[1]} &= \tfrac{1}{2}(u_i^{[0]} + u_{i+1}^{[0]}) - \tfrac{1}{3}\tfrac{\tau}{h}(f_{i+1}^{[0]} - f_i^{[0]}) \\
u_i^{[2]} &= u_i^{[0]} - \tfrac{2}{3}\tfrac{\tau}{h}(f_i^{[1]} - f_{i-1}^{[1]}) \\
u_i^{[3]} &= u_i^{[0]} - \tfrac{\omega}{24}\delta^4 u_i^{[0]} - \\
&\quad \tfrac{\tau}{4h}[(I - \tfrac{2}{3}\delta^2)\mu \delta f_i^{[0]}] - \tfrac{3}{4}\tfrac{\tau}{h}\mu \delta f_i^{[2]} \\
u_i^{n+1} &= u_i^{[3]}
\end{aligned}\right\} \qquad (3.4.20)$$

Here the following difference operators are used:

$$f_i \equiv f(u_i), \ \mu f_i = \frac{1}{2}(f_{i+1/2} + f_{i-1/2}), \quad \delta f_i = f_{i+1/2} - f_{i-1/2}$$

$$f_{i+1/2} = f(u_{i+1/2}), \ u_{i+1/2} = \frac{1}{2}(u_i + u_{i+1}), \ \text{etc.}$$

The term $\frac{\omega \delta^4}{24} u_i^{[0]}$ is introduced to induce stability into the scheme. The scheme (3.4.20) approximates the original PDE (3.4.19) with the error $O(h^3) + O(\tau^3)$. Comparing Rusanov's scheme (3.4.20) with the Jameson schemes one can easily see the main difference: *distinct* difference operators for the approximation of the term $\partial f(u)/\partial x$ in Eq. (3.4.19) are used in (3.4.20) at different stages.

As an example of the application of the Runge-Kutta scheme (3.4.18) let us consider the two-dimensional advection equation (2.12.3). The spatial differencing operator Pu is now expressed by the formula

$$Pu = A(u_{j+1,k}^n - u_{j-1,k}^n)/(2h_1) + B(u_{j,k+1}^n - u_{j,k-1}^n)/(2h_2) \quad (3.4.21)$$

Denote by $\mathcal{F}(-\tau P)$ the result of the Fourier transform of the operator τP. Then it is easy to find that

$$\mathcal{F}(-\tau P) = -i(\kappa_1 \sin \xi_1 + \kappa_2 \sin \xi_2)$$

where the nondimensional quantities κ_1 and κ_2 are defined by formulas (2.12.4). Let $z = -\mathcal{F}(\tau P)$. Then the amplification factor $g(\kappa, \xi)$ of the difference scheme under consideration is expressed by the formula

$$g(\kappa, \xi) = 1 + z + \frac{1}{2!}z^2 + \frac{1}{3!}z^3 + \frac{1}{4!}z^4 \quad (3.4.22)$$

From the requirement $|g(\kappa, \xi)| \leq 1$ we easily obtain the inequality

$$|z|^2 = (\kappa_1 \sin \xi_1 + \kappa_2 \sin \xi_2)^2 \leq 8 \quad (3.4.23)$$

Let us at first consider the first quadrant of the (κ_1, κ_2) plane, where $\kappa_1 \geq 0$, $\kappa_2 \geq 0$. It is clear from (3.4.23) that the largest limitation for the step τ will occur at $\xi_1 = \xi_2 = \pi/2$, which leads to the inequality $\kappa_1 + \kappa_2 \leq 2\sqrt{2}$. Now take the second quadrant, in which $\kappa_1 \leq 0$, $\kappa_2 \geq 0$. It is clear that in this case the most restrictive limitation for τ is obtained at $\xi_1 = -\pi/2$, $\xi_2 = \pi/2$, which leads to the inequality $-\kappa_1 + \kappa_2 \leq 2\sqrt{2}$. Considering the remaining two quadrants of the (κ_1, κ_2) plane in a similar way, we arrive at a conclusion that the stability region of the Runge-Kutta scheme (3.4.18) for Eq. (2.12.2) is determined by the formula

$$|\kappa_1| + |\kappa_2| \leq 2\sqrt{2} \quad (3.4.24)$$

Comparing the stability condition (2.12.16) of the MacCormack scheme (2.12.3) with the stability condition (3.4.24) of the Runge-Kutta scheme (3.4.18), (3.4.21) we can see that the stability condition of the Runge-Kutta scheme is far less restrictive. However, scheme (3.4.18) requires 4 evaluations of the operator Pu for obtaining the value u^{n+1}. Schmidt et al.[17] have introduced a quantitative characteristic of the efficiency of a difference scheme, which will be denoted by ef:

$$ef = C/N$$

where C is the Courant number of the scheme; in the case of scheme (3.4.18) $C = 2\sqrt{2}$ in accordance with (3.4.24). N is the needed number of evaluations of the values of difference operator Pu for obtaining the value u^{n+1}. $ef = 1/2$ in the case of the MacCormack scheme (2.12.3); in the case of the Runge-Kutta scheme (3.4.18) $ef = 2\sqrt{2}/4 \approx 0.707$. Thus the Runge-Kutta scheme (3.4.18), (3.4.21) is more efficient than the MacCormack scheme (2.12.3).

It should be noted that the same polynomial (3.4.10) can be obtained in infinitely many ways. For example, the scheme of the class (3.4.15) with

$$R^{(q)} = Pu^{(q)}, \quad \alpha_1 = \frac{1}{4}, \; \alpha_2 = \frac{1}{3}, \; \alpha_3 = \frac{1}{2}, \quad m = 4 \qquad (3.4.25)$$

was considered in Reference 17 and 18. In the case of the difference operator Pu of the form (3.4.21) this scheme also leads to the amplification factor (3.4.22). Therefore, the stability condition of this Runge-Kutta scheme is also given by formula (3.4.24). However, scheme (3.4.15), (3.4.25) is more advantageous, in terms of the needed computer memory, than the classical Runge-Kutta scheme (3.4.18).[21]

In the case of the advection-diffusion equation (3.3.1) ($L = 2$) the operator Pu may be formed with the aid of the central differences as

$$Pu_{j,k}^n = v_1 \frac{u_{j+1,k}^n - u_{j-1,k}^n}{2h_1} + v_2 \frac{u_{j,k+1}^n - u_{j,k-1}^n}{2h_2}$$
$$- \frac{D_{j+1/2,k}^n(u_{j+1,k}^n - u_{j,k}^n) - D_{j-1/2,k}^n(u_{j,k}^n - u_{j,k-1}^n)}{h_1^2}$$
$$- \frac{D_{j,k+1/2}^n(u_{j,k+1}^n - u_{j,k}^n) - D_{j,k-1/2}^n(u_{j,k}^n - u_{j,k-1}^n)}{h_2^2}$$

One has to solve again the problem on finding those optimal values of the coefficients $\alpha_1, \ldots, \alpha_{m-1}$ in (3.4.15), which ensure the weakest limitation for the time step. In particular, it was shown in Reference 19 that the three-stage Runge-Kutta scheme of the form[22,23,24]

$$u^{(0)} = u^n$$
$$u^{(1)} = u^{(0)} - \tau P u^{(0)}$$
$$u^{(2)} = \frac{3}{4}u^{(0)} + \frac{1}{4}u^{(1)} - \frac{\tau}{4}P u^{(1)}$$
$$u^{(3)} = \frac{1}{3}u^{(0)} + \frac{2}{3}u^{(2)} - \frac{2}{3}\tau P u^{(2)}$$
$$u^{n+1} = u^{(3)}$$

for Eq. (3.3.1) with $L = 2$ is the most efficient for the solution of non-stationary problems with viscosity, in which there are subdomains of flow stagnation (that is the subdomains, in which $|v_1| \approx 0$, $|v_2| \approx 0$).

The extension of the above Runge-Kutta schemes for the case of the PDE systems may be performed in a straightforward way: one can replace the scalar u in (3.4.3), (3.4.2) and (3.4.15) by the vector **u** of dependent variables.

3.5   FINITE VOLUME METHOD

3.5.1   Curvilinear Grids of Quadrilateral Cells

In this Section we present one more method for obtaining the difference discretizations of PDEs in cases of curvilinear spatial grids. We have already presented earlier in Section 2.14 a method for difference discretization on such grids, which was based on a transformation of the spatial region in the plane of physical coordinates (x, y) to a rectangular region in the plane of certain new variables (ξ, η), which are called the curvilinear coordinates.

In contrast to the method of Section 2.14, the finite volume method enables one to obtain the difference approximations to PDEs in the Cartesian coordinates x, y without the necessity of introducing the curvilinear coordinates.

Another advantage of the finite volume method is that it is equally well applicable in cases of various spatial irregular grids: (i) the curvilinear grids of quadrilateral cells; (ii) the triangular grids; (iii) the grids of pentagonal or hexagonal cells; (iv) the grids of Dirichlet cells, etc.

In the method of finite volume the PDE under consideration is at first presented in an integral form. The spatial domain D is subdivided into the elementary finite (small) volumes within which the integration of the equation is performed.

Let us describe the finite volume method at the example of the equation

$$\frac{\partial u}{\partial t} + \frac{\partial F(u)}{\partial x} + \frac{\partial G(u)}{\partial y} = \frac{\partial}{\partial y}\left[a(u)\frac{\partial u}{\partial x}\right] + \frac{\partial}{\partial y}\left[b(u)\frac{\partial u}{\partial y}\right] \qquad (3.5.1)$$

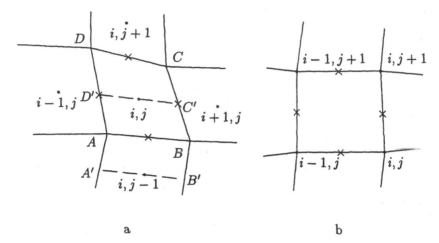

Figure 3.10: (a) cell-centered scheme; (b) nodal scheme. • flow property u, × flux property $\mathbf{H} \cdot \mathbf{n} ds$.

where $F(u)$, $G(u)$, $a(u)$ and $b(u)$ are the given functions. Let us introduce the notations

$$\bar{F}(u) = F(u) - a(u)\frac{\partial u}{\partial x}, \quad \bar{G}(u) = G(u) - b(u)\frac{\partial u}{\partial y} \qquad (3.5.2)$$

Then we can rewrite Eq. (3.5.1) as follows:

$$\frac{\partial u}{\partial t} + \frac{\partial \bar{F}(u)}{\partial x} + \frac{\partial \bar{G}(u)}{\partial y} = 0 \qquad (3.5.3)$$

Let us at first consider the derivation of the difference approximations to the first-order derivatives $\partial F(u)/\partial x$ and $\partial G(u)/\partial y$. The form of the specific difference formulas for the approximation of the first-order derivatives depends on the disposition of (x, y) points to which the values of flow parameters are assigned. In the case of the cell-centered scheme (see Fig. 3.10a) the flow parameters are assigned to certain points inside the cells. In the case of nodal scheme the values of flow parameters are computed at the points of intersection of the grid lines belonging to two families of lines (see Fig. 3.10b).

In the fluid dynamics simulations it is convenient to use the cell-centered scheme. Then the change of the flow parameters at the cell center in the passage from the time level n to the time level $n + 1$ is due to the fluxes across the boundaries of the (i, j) cell. Let us at first derive the difference formula for the approximation of the first-order derivatives at the cell center (i, j). Let us take a two-dimensional finite volume $ABCD$, see Fig. 3.10a. Denote by $V_{i,j}$ the region bounded by the contour $ABCDA$. Let us integrate Eq. (3.5.3) over the volume $V_{i,j}$:

$$\iint_{V_{i,j}} \left(\frac{\partial u}{\partial t} + \frac{\partial \bar{F}(u)}{\partial x} + \frac{\partial \bar{G}(u)}{\partial y} \right) dx dy = 0. \qquad (3.5.4)$$

We can apply the Green's theorem to the second and third items of the integrand in (3.5.4):

$$\iint_{V_{i,j}} \left(\frac{\partial \bar{F}(u)}{\partial x} + \frac{\partial \bar{G}(u)}{\partial y} \right) dx dy = \oint_{ABCDA} \mathbf{H} \cdot \mathbf{n} ds \qquad (3.5.5)$$

where $\mathbf{H} = (\bar{F}, \bar{G})$. The Green's formula (3.5.5) may be rewritten in the case of the Cartesian coordinates x, y as follows:

$$\iint_{V_{i,j}} \left(\frac{\partial \bar{F}(u)}{\partial x} + \frac{\partial \bar{G}(u)}{\partial y} \right) dx dy = \oint_{ABCDA} (\bar{F} dy - \bar{G} dx) \qquad (3.5.6)$$

where the contour ABCDA is traced in a counter-clockwise manner.

The term "finite volume method" has appeared for the first time in the paper MacCormack's and Paullay's. Although it should be noted that M.L. Wilkins[26] had used the Green's formula (3.5.5) in his numerical method for elasticity problems long before the paper.[25]

Taking into account Eq. (3.5.6) we can rewrite Eq. (3.5.4) as follows:

$$\frac{\partial}{\partial t} \iint_{V_{i,j}} u \, dx dy + \oint_{ABCDA} (\bar{F} dy - \bar{G} dx) = 0 \qquad (3.5.7)$$

By using the mean-value theorem we can approximate the integral over the region $V_{i,j}$ in (3.5.7) as follows:

$$\frac{\partial}{\partial t} \iint_{V_{i,j}} u \, dx dy \approx A_{i,j} \cdot \partial u_{i,j} / \partial t \qquad (3.5.8)$$

where $A_{i,j}$ is the area of the quadrangle $ABCD$ shown in Fig. 3.10a, and $\partial u_{i,j}/\partial t$ is some approximation of the time derivative $\partial u/\partial t$.

The mean-value theorem can also be used for the approximation of the integral over the contour $ABCDA$ in (3.5.7):

$$\oint_{ABCDA} (\bar{F} dy - \bar{G} dx) = \sum_{ABCDA} (\bar{F} \Delta y - \bar{G} \Delta x) \qquad (3.5.9)$$

where

$$\Delta x_{AB} = x_B - x_A, \quad \Delta y_{AB} = y_B - y_A$$

$$\bar{F}_{AB} = \frac{1}{2}(\bar{F}_{i,j-1} + \bar{F}_{i,j}), \quad \bar{G}_{AB} = \frac{1}{2}(\bar{G}_{i,j-1} + \bar{G}_{i,j}) \qquad (3.5.10)$$

The expressions for Δx_{BC}, etc., are similar to Eqs. (3.5.10) and will not be written here for brevity.

Substituting the approximations (3.5.8), (3.5.9) into Eq. (3.5.7) we obtain the following difference equation for the approximation of Eq. (3.5.3) in the absence of the second derivative terms:

$$A_{ij}\frac{du_{ij}}{dt} + \frac{1}{2}(F_{i,j-1} + F_{i,j})\Delta y_{AB} - \frac{1}{2}(G_{i,j-1} + G_{i,j})\Delta x_{AB}$$

$$+ \frac{1}{2}(F_{i,j} + F_{i+1,j})\Delta y_{BC} - \frac{1}{2}(G_{i,j} + G_{i+1,j})\Delta x_{BC}$$

$$+ \frac{1}{2}(F_{i,j} + F_{i,j+1})\Delta y_{CD} - \frac{1}{2}(G_{i,j} + G_{i,j+1})\Delta x_{CD}$$

$$+ \frac{1}{2}(F_{i,j} + F_{i-1,j})\Delta y_{DA} - \frac{1}{2}(G_{i,j} + G_{i-1,j})\Delta x_{DA} = 0 \quad (3.5.11)$$

Let us consider the particular case of an uniform rectangular grid whose lines are parallel with the $x-$ and $y-$axes. Then $\Delta x_{AB} = \Delta x$, $\Delta y_{AB} = 0$, etc., and Eq. (3.5.11) takes a simpler form

$$\Delta x \Delta y \frac{du_{ij}}{dt} - \frac{1}{2}(G_{i,j-1} + G_{i,j})\Delta x + \frac{1}{2}(F_{i,j} + F_{i+1,j})\Delta y$$

$$+ \frac{1}{2}(G_{i,j} + G_{i,j+1})\Delta x - \frac{1}{2}(F_{i-1,j} + F_{i,j})\Delta y = 0$$

or

$$\frac{d}{dt}u_{i,j} + \frac{F_{i+1,j} - F_{i-1,j}}{2\Delta x} + \frac{G_{i,j+1} - G_{i,j-1}}{2\Delta y} = 0$$

We can see that the finite volume approximations of the spatial derivatives revert to the central differences.

Let us now consider the question on the derivation of the approximations of the approximations to the second-order derivatives within the context of the finite volume method. Similarly to Eq. (3.5.6) we can write the formula

$$\iint_{V_{i,j}} \left[\frac{\partial}{\partial x}\left(-a(u)\frac{\partial u}{\partial x}\right) + \frac{\partial}{\partial y}\left(-b(u)\frac{\partial u}{\partial y}\right)\right] dx\,dy$$

$$= \oint_{ABCDA} \left(\left[-a(u)\frac{\partial u}{\partial x}\right] dy - \left[-b(u)\frac{\partial u}{\partial y}\right] dx \right) \quad (3.5.12)$$

Let us now approximate the integral in the right hand side of Eq. (3.5.12) similarly to Eq. (3.4.11):

$$\oint_{ABCDA} \left(\left[-a(u)\frac{\partial u}{\partial x}\right] dy + \left[b(u)\frac{\partial u}{\partial y}\right] dx \right) \approx$$

$$-\left[a(u)\frac{\partial u}{\partial x}\right]_{i,j-1/2}\Delta y_{AB} + \left[b(u)\frac{\partial u}{\partial y}\right]_{i,j-1/2}\Delta x_{AB} -$$

$$\left[a(u)\frac{\partial u}{\partial x}\right]_{i+1/2,j}\Delta y_{BC} + \left[b(u)\frac{\partial u}{\partial y}\right]_{i+1/2,j}\Delta x_{BC} -$$

$$\left[a(u)\frac{\partial u}{\partial x}\right]_{i,j+1/2}\Delta y_{CD} + \left[b(u)\frac{\partial u}{\partial y}\right]_{i,j+1/2}\Delta x_{CD} -$$

$$\left[a(u)\frac{\partial u}{\partial x}\right]_{i-1/2,j}\Delta y_{DA} + \left[b(u)\frac{\partial u}{\partial y}\right]_{i-1/2,j}\Delta x_{DA} \qquad (3.5.13)$$

For the approximation of the derivative $(\partial u/\partial x)_{i,j-1/2}$ we can use the following formula (see also References 27,28):

$$\left[a(u)\frac{\partial u}{\partial x}\right]_{i,j-1/2} = \left(\frac{1}{S_{A'B'C'D'}}\right)\iint a(u)\frac{\partial u}{\partial x}dxdy$$

$$\approx \left(\frac{1}{S_{A'B'C'D'}}\right)a(u_{i,j-1/2})\int_{A'B'C'D'A'} udy \qquad (3.5.14)$$

$$\left[b(u)\frac{\partial u}{\partial y}\right]_{i,j-1/2} = \left(\frac{1}{S_{A'B'C'D'}}\right)\iint b(u)\frac{\partial u}{\partial y}dxdy$$

$$\approx -\left(\frac{1}{S_{A'B'C'D'}}\right)b(u_{i,j-1/2})\int_{A'B'C'D'A'} udx \qquad (3.5.15)$$

We can approximate the integrals over the contour in Eqs. (3.5.14) and (3.5.15) by the formulas

$$\int_{A'B'C'D'A'} udy \approx u_{i,j-1}\Delta y_{A'B'} + u_B\Delta y_{B'C'} + u_{i,j}\Delta y_{C'D'} +$$

$$u_A\Delta y_{D'A'}$$

$$\int_{A'B'C'D'A'} udx \approx u_{i,j-1}\Delta x_{A'B'} + u_B\Delta x_{B'C'} + u_{i,j}\Delta x_{C'D'} +$$

$$u_A\Delta x_{D'A'}$$

The values u_A and u_B entering these formulas can be computed as the mean values obtained on the basis of the four nodal values of u around the point A or B:

$$u_A = \frac{1}{4}(u_{i,j} + u_{i-1,j} + u_{i-1,j-1} + u_{i,j-1})$$

$$u_B = \frac{1}{4}(u_{i,j} + u_{i+1,j} + u_{i+1,j-1} + u_{i,j-1})$$

If the grid distortion is not very large, then

$$\Delta y_{A'B'} \approx -\Delta y_{C'D'} \approx \Delta y_{AB}, \ \Delta y_{B'C'} \approx -\Delta y_{D'A'} \approx \Delta y_{j-1,j}$$

and

$$S_{A'B'C'D'} \approx \Delta x_{AB}\Delta y_{j-1,j} - \Delta y_{AB}\Delta x_{j-1,j}$$

Then we can write the following approximate formulas:

$$\left[a(u)\frac{\partial u}{\partial x}\right]_{i,j-1/2} \approx a(u_{i,j-1/2}) \cdot \frac{\Delta y_{AB}(u_{i,j-1}-u_{i,j})+\Delta y_{j-1,j}(u_B-u_A)}{S_{A'B'C'D'}}$$

$$\left[b(u)\frac{\partial u}{\partial y}\right]_{i,j-1/2} \approx b(u_{i,j-1/2})\times$$
$$\frac{-[\Delta x_{AB}(u_{i,j-1}-u_{i,j})+\Delta x_{j-1,j}(u_B-u_A)]}{S_{A'B'C'D'}}$$

(3.5.16)

The approximations for the terms

$$\left[a(u)\frac{\partial u}{\partial x}\right]_{i+1/2,j}, \quad \left[b(u)\frac{\partial u}{\partial y}\right]_{i+1/2,j}$$

etc., which enter the formula (3.5.13), can be obtained similarly to the approximations (3.5.16). We will not present them here for brevity.

3.5.2 Triangular Grids

Let us now consider the case of a triangular spatial grid. Such grids are widely used for the numerical solution of various applied problems in the field of elasticity, fluid dynamics, etc.

The spatial domain is divided into a large number of small triangular cells. Let us write the integral form of Eq. (3.5.3) similarly to Eq. (3.5.7):

$$\frac{\partial}{\partial t} \iint_{\Omega_i} u\,dx\,dy + \oint_{\Gamma_i} (\bar{F}dy - \bar{G}dx) = 0 \qquad (3.5.17)$$

where i denotes the number of the point at which the flow parameters are computed, and Ω_i is a control volume containing the ith point; Γ_i is the boundary of the Ω_i.

We can again distinguish between the cell-centered scheme (Fig. 3.11a) and the nodal scheme (Fig. 3.11b) for the computation of the contour integral in (3.5.17). Let us at first consider the cell-centered scheme. In this case the control volumes are taken as the triangular cells of the mesh. In Fig. 3.11 such control volumes are shown by thick lines. The flow variables are stored at the center of each cell and assumed to represent an averaged value over the entire control volume. In order to evaluate the boundary integral in Eq. (3.5.17), estimates of the flow variables along the edges of the triangular cells are needed. These are taken as the average of the values in both cells on either side of that edge.[29] Let us mark the centers of the cells having common side with the cell Ω_i by 1, 2 and 3 (see Fig. 3.11,a). Then we approximate the contour integral in (3.5.17) in accordance with the above evaluation scheme as

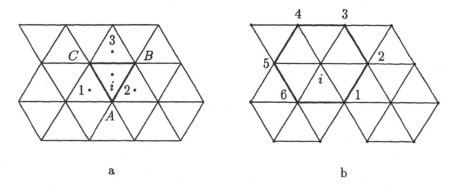

a   b

Figure 3.11: (a) cell-centered scheme; (b) nodal scheme.

follows:

$$\oint_{\Gamma_i} (\bar{F}dy - \bar{G}dx) \approx \frac{1}{2}(\bar{F}_i + \bar{F}_2)\Delta y_{AB} - \frac{1}{2}(\bar{G}_i + \bar{G}_2)\Delta x_{AB}$$

$$+ \frac{1}{2}(\bar{F}_i + \bar{F}_3)\Delta y_{BC} - \frac{1}{2}(\bar{G}_i + \bar{G}_3)\Delta x_{BC} +$$

$$\frac{1}{2}(\bar{F}_i + \bar{F}_1)\Delta y_{CA} - \frac{1}{2}(\bar{G}_i + \bar{G}_1)\Delta x_{CA} \quad (3.5.18)$$

where $\Delta x_{AB} = x_B - x_A$, $\Delta y_{AB} = y_B - y_A$, etc. According to Reference 29, the cell-centered scheme is equivalent to central differencing on a Cartesian grid and is second-order accurate for a smooth mesh.

For the approximation of the integral over the volume Ω_i in Eq. (3.5.17) we can again use the mean-value theorem:

$$\frac{\partial}{\partial t} \iint_{\Omega_i} u\, dx\, dy \approx A_i \cdot \frac{du_i}{dt} \quad (3.5.19)$$

where A_i is the area of the control volume Ω_i, and du_i/dt is a difference approximationof the time derivative du/dt at point i. The area A_i can be computed with the aid of the Green's formula:

$$A_i = \iint_{\Omega_i} dx\, dy = \oint_{\Gamma_i} x\, dy = - \oint_{\Gamma_i} y\, dx \quad (3.5.20)$$

so that

$$A_i = \frac{1}{2}[(x_A + x_B)\Delta y_{AB} + (x_B + x_C)\Delta y_{BC} + (x_C + x_A)\Delta y_{CA}] \quad (3.5.21)$$

Substituting the approximations (3.5.18), (3.5.19) and (3.5.21) into Eq. (3.5.17), we obtain the difference equation

$$A_i \frac{du_i}{dt} + \frac{1}{2}(\bar{F}_i + \bar{F}_2)\Delta y_{AB} - \frac{1}{2}(\bar{G}_i + \bar{G}_2)\Delta x_{AB}$$
$$+ \frac{1}{2}(\bar{F}_i + \bar{F}_3)\Delta y_{BC} - \frac{1}{2}(\bar{G}_i + \bar{G}_3)\Delta x_{BC} +$$
$$\frac{1}{2}(\bar{F}_i + \bar{F}_1)\Delta y_{CA} - \frac{1}{2}(\bar{G}_i + \bar{G}_1)\Delta x_{CA} = 0 \qquad (3.5.22)$$

The time integration of Eq. (3.5.22) can be performed with the aid of a Runge-Kutta scheme as in References 28-31.

Let us now consider the formulas for the approximation of the contour integral in Eq. (3.5.17) in the case of the nodal scheme. The flow variables are stored at the vertices of the triangles. The control volume for a particular node i is taken to be the union of all triangles with a vertex at i, as shown in Fig. 3.10,b. The contour integral in Eq. (3.5.17) is approximated by the trapezoidal integration rule:

$$\oint_{\Gamma_i} (\bar{F} dy - \bar{G} dx)$$
$$\approx \frac{1}{2} \sum_{k=1}^{6} [(\bar{F}_k + \bar{F}_{k+1})\Delta y_{k,k+1} - (\bar{G}_k + \bar{G}_{k+1})\Delta x_{k,k+1}]$$
$$\frac{\partial}{\partial t} \iint_{\Omega_i} u dx dy \approx A_i \cdot \frac{du_i}{dt}$$
$$A_i = \frac{1}{2} \sum_{k=1}^{6} (x_k + x_{k+1})\Delta y_{k,k+1}$$

where

$$\Delta x_{k,k+1} = x_{k+1} - x_k, \quad \Delta y_{k,k+1} = y_{k+1} - y_k, \quad k = 1, \ldots, 6$$
$$x_7 \equiv x_1, \quad y_7 \equiv y_1, \quad \bar{F}_7 \equiv \bar{F}_1, \quad \bar{G}_7 \equiv \bar{G}_1$$

Further details concerning the approximation of the second derivative terms can be found in Reference 30 and 31 for the cell-centered scheme and in Reference 31 and 32 for the nodal scheme.

The finite volume method can be applied also for the discretization of the elliptic PDEs, see, for example, Reference 28.

We will consider in Section 4.6 the order of approximation of the differencing formulas obtained by the finite volume method on different grids.

3.6   THE ADI METHOD

As it was shown in Section 3.2 (see also Exercise 3.6), the explicit difference schemes for the one- and two-dimensional heat equation impose a severe restriction on the time step τ dictated by the stability: $\tau = O(h^2)$. In this connection the idea of using the *implicit* difference schemes for the numerical solution of parabolic PDEs was intensively investigated.

Let us consider the two-dimensional heat equation

$$\frac{\partial u}{\partial t} = \nu \left(\frac{\partial^2 u}{\partial x^2} + \frac{\partial^2 u}{\partial y^2} \right), \quad \nu = const > 0 \qquad (3.6.1)$$

in a square region $D = \{0 < x < 1, \ 0 < y < 1\}$ of the (x, y) plane. We will solve Eq. (3.6.1) at a given initial condition

$$u(x, y, 0) = u_0(x, y), \quad (x, y) \in D \qquad (3.6.2)$$

and the boundary conditions

$$u(x, y, t) = \varphi(x, y, t), \quad (x, y) \in \Gamma, \ t \in [0, T] \qquad (3.6.3)$$

where Γ is the boundary of D, and $[0, T]$ is a given interval on the t-axis, $0 < t < \infty$.

Let us now introduce in the square D an uniform rectangular mesh G_h with the steps h_1 along the x-axis and the step h_2 along the y-axis, so that the coordinate lines of the mesh are given by the equations

$$x = x_j = jh_1, \quad j = 0, 1, \ldots, M_1 \qquad (3.6.4)$$
$$y = y_k = kh_2, \quad k = 0, 1, \ldots, M_2 \qquad (3.6.4)$$

Let us approximate Eq. (3.6.1) by the following implicit difference scheme: ̄

$$\frac{u_{j,k}^{n+1} - u_{j,k}^n}{\tau} = \nu \left(\frac{u_{j+1,k}^{n+1} - 2u_{jk}^{n+1} + u_{j-1,k}^{n+1}}{h_1^2} + \right.$$

$$\left. \frac{u_{j,k+1}^{n+1} - 2u_{jk}^{n+1} + u_{j,k-1}^{n+1}}{h_2^2} \right) \qquad (3.6.5)$$

$$j = 1, \ldots, M_1 - 1; \quad k = 1, \ldots, M_2 - 1$$

The stencil of scheme (3.6.5) has the form shown in Fig. 3.12. We can see that five unknown grid values u^{n+1} enter the scheme (3.6.5) at the $(n + 1)$th time level.

The method of the *matrix factorization* can be used for the numerical solution of the algebraic system (3.6.5), see, e.g.,References 33 and 34.

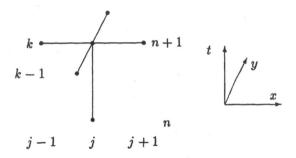

Figure 3.12: The stencil of scheme (3.6.5).

This method can be regarded as an extension of the scalar factorization (see Section 2.9) for the case of a system of vector equations. Below in Section 4.4 we show how these vector equations can be formulated on the basis of a scheme similar to Eq. (3.6.5). It easy to show with the aid of the Fourier method that the difference scheme (3.6.5) is absolutely stable. It is important to note that the implementation of the matrix factorization on a computer requires $O(h^{-4})$ arithmetic operations in the case of a square mesh with the steps $h_1 = h_2 = h$. Thus, although the scheme (3.6.5) is absolutely stable, its realization is related to considerable computer expenses.

The *alternating direction implicit* (ADI) method proposed in 1955 simultaneously by D.W. Peacemen, H.H. Rachford and J. Douglas[35,36] made a significant contribution to the development of efficient difference methods for the parabolic equations. In order to present the ADI scheme in a more compact form let us introduce the difference operators

$$\Lambda_1 u_{jk} = (u_{j+1,k} - 2u_{jk} + u_{j-1,k})/(h_1^2)$$
$$\Lambda_2 u_{jk} = (u_{j,k+1} - 2u_{jk} + u_{j,k-1})/(h_2^2)$$
(3.6.6)

Then we can write the ADI scheme as follows.[29]

$$\frac{u^{n+1/2} - u^n}{\tau/2} = \Lambda_1 u^{n+\frac{1}{2}} + \Lambda_2 u^n$$
(3.6.7)

$$\frac{u^{n+1} - u^{n+1/2}}{\tau/2} = \Lambda_1 u^{n+\frac{1}{2}} + \Lambda_2 u^{n+1}$$
(3.6.8)

At the first half-step of the length $\tau/2$ the operator $L_1 = \nu\frac{\partial^2}{\partial x^2}$ is approximated implicitly, and the operator $L_2 = \nu\frac{\partial^2}{\partial y^2}$ is approximated explicitly.

At the second half-step the operator L_1 is approximated explicitly, and the operator L_2 is approximated implicitly. Let us show that the ADI scheme (3.6.7), (3.6.8) has the second order of approximation both in space and time. For this purpose let us write the scheme (3.6.7), (3.6.8) in the form

$$A_1 u^{n+\frac{1}{2}} - B_1 u^n = 0 \qquad (3.6.9)$$

$$A_2 u^{n+1} - B_2 u^{n+\frac{1}{2}} = 0 \qquad (3.6.10)$$

where

$$\begin{aligned} A_1 &= I - \tfrac{\tau}{2}\Lambda_1, \quad A_2 = I - \tfrac{\tau}{2}\Lambda_2 \\ B_1 &= I + \tfrac{\tau}{2}\Lambda_2, \quad B_2 = I + \tfrac{\tau}{2}\Lambda_1 \end{aligned} \qquad (3.6.11)$$

I is the identity operator, i.e. $Iu^n \equiv u^n$. Let us multiply the both sides of Eq. (3.6.9) from the left by the operator B_2, and then multiply the both sides of Eq. (3.6.10) by A_1 from the left by A_1 and add the both sides of the obtained equations. As a result we arrive at the equation

$$A_1 A_2 u^{n+1} - B_1 B_2 u^n + (B_2 A_1 - A_1 B_2) u^{n+1/2} = 0$$

Assuming that the operators Λ_1 and Λ_2 commute we obtain the scheme

$$A_1 A_2 u^{n+1} - B_1 B_2 u^n = 0 \qquad (3.6.12)$$

Substituting the expressions (3.6.11) into Eq. (3.6.12) we obtain after some algebra the following difference scheme in "integral steps":

$$\frac{u^{n+1} - u^n}{\tau} = \frac{\Lambda_1 + \Lambda_2}{2}(u^n + u^{n+1}) - \frac{1}{4}\tau \Lambda_1 \Lambda_2 (u^{n+1} - u^n) \qquad (3.6.13)$$

It is easy to show that Eq. (3.6.13) approximates the original heat equation (3.6.1) with the second order of approximation at point

$$\left(jh_1, kh_2, (n + \frac{1}{2})\tau\right)$$

Let us now prove the unconditional stability of scheme (3.6.7), (3.6.8). Let us set

$$u_{jk}^n = \eta_n e^{i(jm_1 h_1 + km_2 h_2)} \quad u_{jk}^{n+\frac{1}{2}} = \eta_{n+\frac{1}{2}} e^{i(jm_1 h_1 + km_2 h_2)} \qquad (3.6.14)$$

where m_1 and m_2 are the real wavenumbers. Substituting Eqs. (3.6.14) into Eqs. (3.6.7) and (3.6.8) we obtain that

$$\rho_1 = \frac{\eta_{n+1/2}}{\eta_n} = \frac{1 - \frac{1}{2}a_2}{1 + \frac{1}{2}a_1}$$

$$\rho_2 = \frac{\eta_{n+1}}{\eta_n + 1/2} = \frac{1 - \frac{1}{2}a_1}{1 + \frac{1}{2}a_2}$$

$$\rho = \frac{(1 - \frac{1}{2}a_1)(1 - \frac{1}{2}a_2)}{(1 + \frac{1}{2}a_1)(1 + \frac{1}{2}a_2)} = \rho_1 \rho_2 \qquad (3.6.15)$$

where ρ is the amplification factor of scheme (3.6.7), (3.6.8), and

$$a_s = 4r_s \sin^2 \frac{m_s h_s}{2}, \quad r_s = \frac{\nu\tau}{h_s^2}, \quad s = 1, 2 \qquad (3.6.16)$$

It follows from Eq. (3.6.15) that $|\rho| \le 1$ at any τ.

Let us now perform the estimation of the computational complexity of the computer implementation of the ADI scheme (3.6.7), (3.6.8). It is easy to see that the solution of the system of difference equations (3.6.7) can be found by a scalar sweep (see Section 2.9) along each line $y = y_k = kh_1$, $k = 1, \ldots, M_2 - 1$. It is not difficult to show that the solution of a tridiagonal system along the M_1 grid nodes requires $O(M_1)$ arithmetic operations, see also Reference 29. Therefore, the overall number of the arithmetic operations needed for solving system (3.6.7) is the quantity of the order $M_2 \cdot O(M_1) = O(M_1 M_2)$.

The system of difference equations (3.6.8) is solved by inverting the tridiagonal matrices (see Section 2.9) along the lines $x = x_j = jh_1$. Therefore, it requires $O(M_1 M_2)$ operations for its solution. Thus, the ADI scheme requires $O(M_1 M_2)$ arithmetic operations for its computer implementation. If, in particular, $M_1 = M_2 = M$, then we obtain the estimate of the computational complexity of the form $O(M^2)$. And the matrix factorization method requires $O(M^4)$ operations, as this was mentioned above. That is in the case of using the ADI scheme (3.6.7), (3.6.8) we have a gain of two orders of magnitude of the quantity M in comparison with the implicit scheme (3.6.5).

The ADI scheme (3.6.7), (3.6.8) requires the specification of the intermediate values on the boundary. Let us consider the case of the Dirichlet boundary condition (3.6.3). Then values of $u^{n+\frac{1}{2}}$ on the boundaries can be obtained by using the two steps (3.6.7), (3.6.8) of the ADI scheme. For step (3.6.7), $u^{n+1/2}$ is needed at $x = 0$ and $x = 1$. Let us rewrite Eqs. (3.6.7) and (3.6.8) in the form

$$u^{n+1/2} = u^n + \frac{\tau}{2}(\Lambda_1 u^{n+1/2} + \Lambda_2 u^n) \qquad (3.6.17)$$

$$u^{n+1} = u^{n+1/2} + \frac{\tau}{2}(\Lambda_1 u^{n+1/2} + \Lambda_2 u^{n+1}) \qquad (3.6.18)$$

By subtracting the both sides of Eq. (3.6.18) from the both sides of Eq. (3.6.17), we have (see also Reference 6)

$$u^{n+\frac{1}{2}} = \frac{1}{2}(\varphi^{n+1} + \varphi^n) + \frac{\tau}{4}(\Lambda_2\varphi^n - \Lambda_2\varphi^{n+1})$$

on the boundaries $x = 0$ and $x = 1$. Thus $u^{n+\frac{1}{2}}$ is determined where needed.

The boundary condition

$$u_{l,m}^{n+\frac{1}{2}} = \varphi_{l,m}^{n+\frac{1}{2}} = \varphi(x_l, y_m, t_{n+1/2})$$

is very easy to implement but is only first-order accurate.[6]

Let us now consider a more general PDE of the form

$$u_t = b_{11}u_{xx} + 2b_{12}u_{xy} + b_{22}u_{yy}$$

This equation is parabolic if

$$b_{11}, b_{22} > 0 \quad \text{and} \quad b_{12}^2 < b_{11}b_{22}$$

It was shown in Reference 6 how to modify the basic ADI method to include the approximation of the mixed derivative terms.

Exercise 3.8. Show with the aid of the Fourier method that the scheme (3.6.5) is absolutely stable.

3.7 APPROXIMATE FACTORIZATION SCHEME

In this Section we will consider the question of the application of efficient implicit difference schemes for the numerical solution of the two-dimensional advection-diffusion equation (cf. (3.3.1))

$$\frac{\partial u}{\partial t} + A\frac{\partial u}{\partial x} + B\frac{\partial u}{\partial y} = \nu\left(\frac{\partial^2 u}{\partial x^2} + \frac{\partial^2 u}{\partial y^2}\right) \tag{3.7.1}$$

where A and B are the components of the advection velocity vector along the x- and y-axis, respectively; $\nu = const > 0$.

Let us assume that Eq. (3.7.1) is to be solved numerically in a square region $D = \{0 < x, y < 1\}$. We can introduce an uniform rectangular mesh G_h in D by formulas (3.6.4) and the central differencing operators[28]

$$L_x u_{jk}^n = (u_{j+1,k}^n - u_{j-1,k}^n)/(2h_1)$$
$$L_y u_{jk}^n = (u_{j,k+1}^n - u_{j,k-1}^n)/(2h_2)$$
$$L_{xx} u_{jk}^n = \frac{u_{j-1,k}^n - 2u_{jk}^n + u_{j+1,k}^n}{h_1^2} \tag{3.7.2}$$
$$L_{yy} u_{jk}^n = \frac{u_{j,k-1}^n - 2u_{jk}^n + u_{j,k+1}^n}{h_2^2}$$

Let us now define the quantity

$$\Delta u_{j,k}^n = u_{j,k}^{n+1} - u_{j,k}^n \tag{3.7.3}$$

which can be thought of as a correction to advance the solution to a new time level $(n+1)$. Then we can write a two-level implicit scheme

$$\frac{\Delta u_{j,k}^{n+1}}{\tau} + A(1-\beta)L_x u_{j,k}^n +$$
$$B(1-\beta)L_y u_{j,k}^n + A\beta L_x u_{j,k}^{n+1} + B\beta L_y u_{j,k}^{n+1} =$$
$$(1-\beta)(\nu L_{xx} u_{j,k}^n + \nu L_{yy} u_{j,k}^n) + \beta(\nu L_{xx} u_{j,k}^{n+1} + \nu L_{yy} u_{j,k}^{n+1}) \tag{3.7.4}$$

Here β is a user-specified weight parameter, $0 \le \beta \le 1$. We now observe that

$$A(1-\beta)L_x u_{j,k}^n + A\beta L_x u_{j,k}^{n+1} = AL_x u_{j,k}^n + A\beta L_x(u_{j,k}^{n+1} - u_{j,k}^n) =$$
$$AL_x u_{j,k}^n + A\beta L_x \Delta u_{j,k}^{n+1}$$

where the quantity $\Delta u_{j,k}^{n+1}$ is determined by formula (3.7.3). Regrouping the other terms in Eq. (3.7.4) we can transform it to the form

$$\frac{\Delta u_{j,k}^{n+1}}{\tau} + AL_x u_{j,k}^n + BL_y u_{j,k}^n + \beta(AL_x \Delta u_{j,k}^{n+1} + BL_y \Delta u_{j,k}^{n+1}) =$$
$$\nu(L_{xx} u_{j,k}^n + L_{yy} u_{j,k}^n) + \beta\nu(L_{xx}\Delta u_{j,k}^{n+1} + L_{yy}\Delta u_{j,k}^{n+1}) \tag{3.7.5}$$

We now rewrite Eq. (3.7.5) in a more compact form as follows:

$$(1 + \tau\beta\Lambda_1 + \tau\beta\Lambda_2)\Delta u_{j,k}^{n+1} = -\tau(\Lambda_1 + \Lambda_2)u_{j,k}^n \tag{3.7.6}$$

where

$$\Lambda_1 = AL_x - \nu L_{xx}, \quad \Lambda_2 = BK_y - \nu L_{yy}$$

To reduce the numerical solution of the algebraic system (3.7.6) to scalar sweeps we replace the difference operator on the left hand side of Eq. (3.7.6) with the aid of the *approximate factorization technique*[33] as follows:

$$(1 + \tau\beta\Lambda_1)(1 + \tau\beta\Lambda_2)\Delta u_{j,k}^{n+1} = -\tau(\Lambda_1 + \Lambda_2)u_{j,k}^n \tag{3.7.7}$$

Since

$$(1 + \tau\beta\Lambda_1)(1 + \tau\beta\Lambda_2) = (1 + \tau\beta\Lambda_1 + \tau\beta\Lambda_2) + \tau^2\beta^2\Lambda_1\Lambda_2$$

we see that Eq. (3.7.7) approximates Eq. (3.7.6) with the accuracy $O(\tau^2)$.

The approximate factorization (AF) scheme (3.7.7) can be implemented in the form of a two-stage algorithm. The first stage reduces to the solution of the following system of equations, which is valid along each grid line parallel with the x-axis:

$$(1 + \tau\beta\Lambda_1)\Delta u_{j,k}^* = -\tau(\Lambda_1 + \Lambda_2)u_{j,k}^n \qquad (3.7.8)$$

Taking into account the definitions (3.7.2) of the differencing operators L_x, L_y, L_{xx}, L_{yy} it is easy to see that the equations (3.7.8) represent a tridiagonal system, which can effectively be solved by the variant of the Gaussian elimination algorithm described in Section 2.9.

At the second stage the system

$$(1 + \tau\beta\Lambda_2)\Delta u_{j,k}^{n+1} = \Delta u_{j,k}^* \qquad (3.7.9)$$

is solved along each grid line parallel with the y-axis. The algorithm (3.7.8), (3.7.9) was proposed in Reference 37.

According to[28] the two-stage algorithm (3.7.8), (3.7.9) in the particular case $A = B = 0$ (i.e. there is no convection) is unconditionally stable at $\beta \geq 0.5$ and has the approximation error of the form $O(\tau^2) + O(h_1^2) + O(h_2^2)$ if $\beta = 0.5$.

The AF schemes can be extended in a similar way to approximate the systems of PDEs, for example, the Navier-Stokes equations of fluid dynamics. In this case, the AF schemes can be used also for the numerical solution of the stationary problems by the pseudo-unsteady method. The convergence of the pseudo-unsteady method is faster, if the difference scheme used allows large time steps. In the case of the AF schemes, the errors of the approximate factorization (of the order $O(\tau^2)$) increase at larger time steps. This causes a significant deterioration of the convergence speed of the pseudo-unsteady iteration process (see examples in References 38 and 39). Therefore, the implicit unfactored schemes now gain a more widespread acceptance (see the review of the relevant works in References 40 and 41). Although these schemes require the iterations at each time level, they enable one to perform the computations at very large Courant numbers, at which the AF schemes become unstable or inefficient.

The efficient computation of a convective transport by finite difference method represents a difficult problem. One way for the solution of this problem was discussed above in Chapter 2 (see Section 2.7 on the TVD schemes). However, the TVD schemes have a variable stencil, therefore, their implementation in the form of implicit difference schemes causes difficulties. One of the ways of using the TVD schemes within the context of the AF schemes is to use the explicit approximations of the convective terms by the TVD differencing operators, and the viscous terms are approximated implicitly as in the above AF scheme (3.7.8),

(3.7.9). A family of two- and three-level AF schemes of such a type was presented in Reference 42.

Exercise 3.9. Investigate the stability of the AF scheme (3.7.8), (3.7.9) with the aid of the Fourier method.

Exercise 3.10. Find the condition(s) at which the tridiagonal matrices of systems (3.7.7) and (3.7.8) satisfy the condition of the diagonal dominance.

Hint. Consider separately the cases $A > 0$, $A < 0$, $B > 0$, $B < 0$.

3.8 DISPERSION

We have already studied such important properties of the difference schemes as the approximation and stability. Both a low approximation order and the instability are the sources of the numerical solution errors.

Another source of the errors is determined by the *phase errors* of a difference scheme. In this section we will show how the *Mathematica* system can be used for the investigation of the phase errors of a given difference scheme approximating a hyperbolic or parabolic PDE. If there are for the numerical solution of the same applied problem several difference schemes having approximately the same stability limits and the same approximation order, then the preference can be given to a difference scheme, which has the least phase error.

3.8.1 The Case of One Spatial Variable

Let us again consider the Cauchy problem (2.1.1), (2.1.2). The solution of this problem can be found in the form of the Fourier series

$$u(x,t) = \sum_{k=-\infty}^{\infty} a_k e^{i(kx-\omega t)} \tag{3.8.1}$$

where ω is the wave frequency, k is the real wavenumber, $k = 2\pi/\lambda$, λ is the wave length. The speed at which the wave phase $kx - \omega t$ propagates in the space is called the *phase speed* and is equal to $v = \omega/k$. The dependence of the phase speed of the wave propagation on the wave length λ is called the *wave dispersion*.

Let us now take an individual Fourier component

$$u_k(x,t) = e^{i(kx-\omega t)} \tag{3.8.2}$$

Let us now find such value of ω in (3.8.2) at which the function (3.8.2) becomes the solution of Eq. (2.1.1). For this purpose let us substitute the right hand side of formula (3.8.2) into Eq. (2.1.1) instead of $u(x, t)$:

$$u_k(x, t)i(ak - \omega) = 0$$

We obtain from here that $\omega = ak$. Therefore, the Fourier component $u_k(x, t)$ of the form

$$u_k(x, t) = e^{ik(x-at)} \tag{3.8.3}$$

satisfies Eq. (2.1.1). We easily find from (3.8.3) that

$$u_k(x, t + \tau) = e^{-ika\tau} u_k(x, t) \tag{3.8.4}$$

It is easy to show that the expression

$$g_0(k, \tau) = e^{-ika\tau} \tag{3.8.5}$$

is the Fourier symbol of the step operator for Eq. (2.1.1) (see also Reference 43). Eq. (3.8.4) shows that during the time τ the Fourier component $u_k(x, t)$ changes by a factor $\exp(-ika\tau)$. We find from Eq. (3.8.5) that

$$\Phi = arg\, g_0 = \arctan \frac{Im\, g_0}{Re\, g_0} = -ka\tau \tag{3.8.6}$$

The quantity Φ represents the exact phase shift in the solution of Eq. (2.1.1) during the time τ.

Let us now substitute the Fourier component (3.8.3) into the two-level difference scheme

$$u_j^{n+1} = S(\tau, h)u_j^n \tag{3.8.7}$$

where $S(\tau, h)$ is the step operator of the difference scheme. As a result we obtain an amplification factor $g(\kappa, \xi)$ of a difference scheme, where $\xi = kh$ and $\kappa = a\tau/h$. Therefore, the phase shift in the difference solution during the time τ is equal to

$$\Phi_h = \arctan \frac{Im\, g}{Re\, g} \tag{3.8.8}$$

The quantity

$$E(\kappa, \xi) = \Phi_h - \Phi \tag{3.8.9}$$

is called the *phase error*, or the *dispersion* of the difference scheme (3.8.7).

Example 1. Let us consider the Lax-Wendroff scheme (2.4.3). In this case we have that

$$g(\kappa, \xi) = 1 - i\kappa \sin \xi - 2\kappa^2 \sin^2 \frac{\xi}{2}$$

Therefore,

$$E(\kappa, \xi) = \arctan \frac{-\kappa \sin \xi}{1 - 2\kappa^2 \sin^2 \frac{\xi}{2}} + \kappa\xi \tag{3.8.10}$$

This formula does not give too much insight into the behavior of $E(\kappa, \xi)$. In this connection let us at first study the behavior of $E(\kappa, \xi)$ for small values of $|\xi|$. Such values of ξ correspond to the long-wavelength disturbances, i.e. when $\xi = kh = 2\pi h/\lambda \ll 1$. Using the Taylor series expansions around the point $\xi = 0$ one can obtain that[6]

$$E(\kappa, \xi) = \frac{1}{6}\kappa \xi^3 (1 - \kappa^2) + O(\xi^5) \qquad (3.8.11)$$

We see from here that if $|\kappa|$ is close to 1, then the dispersion will be less.

In the exact continuous PDE solution, all components are advected at the phase speed $v = \omega/k = ak/k = a$. In the difference solution the phase speed differs for each Fourier component. Since

$$\Phi_h = \Phi + E = \kappa\xi\left[1 - \frac{\xi^2}{6}(1 - \kappa^2)\right] + O(\xi^5) \qquad (3.8.12)$$

we see that in the difference solution each Fourier component is carried along slower than in the exact solution $\Phi = -\kappa\xi$, because the factor $1 - (\xi^2/6)(1 - \kappa^2) < 1$ is present in the expression (3.8.12) for Φ_h. Thus, different Fourier components will spread apart, or disperse, as the numerical solution proceeds, and the phenomenon is frequently referred to as dispersion error. (For an early study of dispersion error, see Reference 44).

We now present a symbolic algorithm for the investigation of the dispersion errors of scalar difference schemes approximating the scalar PDE of the form (2.5.1). The basic steps of the algorithm are as follows:

Step 1. The input data are input into the *Mathematica* program. These are the following data:
a) the difference equation under consideration;
b) the original partial differential equation;
c) the order **kt** of the Taylor expansions around the point $\xi = 0$.

Step 2. Substitution of the Fourier component (3.8.2) into the original PDE (2.5.1) for the purpose of determining the expression for the frequency ω.

Step 3. Determination of Φ by the formula

$$\Phi = arg\, g_0 = \arctan \frac{Im\, g_0}{Re\, g_0} = -\omega\tau$$

Step 4. Symbolic computation of the amplification factor $g(\kappa, \xi)$ of the difference scheme under consideration.

Step 5. Symbolic computation of the expressions (3.8.8) and (3.8.9).

Step 6. Expansion of the expression (3.8.9) for $E(\kappa, \xi)$ into the truncated Taylor series at point $\xi = 0$.

We have implemented Steps 1-6 in the *Mathematica* program disp1.ma (this file is available on the attached disk). The symbolic computation of the amplification factor $g(\kappa, \xi)$ of the difference scheme (3.8.7) is performed in the same way as in our *Mathematica* program st2.ma. The built-in *Mathematica* function ArcTan[z] gives (in radians) the arc tangent $tan^{-1}(z)$ of the complex number z. For real z, the results are always in the range $-\pi/2$ to $\pi/2$.

For the Taylor expansion of the phase error (3.8.9) at point $\xi = 0$ we have used the *Mathematica* function

$$\text{Series}[f, \{x, x_0, n\}]$$

which generates a power series expansion for f about the point $x = x_0$ to order $(x - x_0)^n$. In our case n is equal to the user-specified value kt.

With the aid of the program disp1.ma we have considered a number of difference schemes. As the first example we have taken the Lax-Wendroff scheme (2.4.3) for Eq. (2.1.1). We have obtained at kt=5 the following expression for the Taylor expansion of the phase error $E(\kappa, \xi)$:

```
                  3                       2      4     5
        cp    cp     3      cp (1 + 5 cp  - 6 cp ) xi               6
 E =  (-- - ---) xi  -   ----------------------------------  + O[xi]
        6     6                           120
```

Here cp= κ, xi= ξ. Comparing this expression with the formula (3.8.11) we can see that the first term on the right hand side has the same form in both cases. In addition, the *Mathematica* system has produced in a few seconds also the next term of the Taylor series expansion, which is denoted by $O(\xi^5)$ in Eq. (3.8.11).

The *Mathematica* function

$$\text{Plot3D}[f, \{x, xmin, xmax\}, \{y, ymin, ymax\}]$$

generates a three-dimensional plot of f as a function of x and y. We have used this function to generate a surface $E = E(\kappa, \xi)$ in the region $-1 \le \kappa \le 1$, $0 \le \xi \le 2\pi$, see Fig. 3.13-a. It can be seen from Fig. 3.13-a that $E(\kappa, \xi) = 0$ at $|\kappa| = 1$. This agrees with Eq. (3.8.10): at $\kappa = 1$ $E(\kappa, \xi) = \arctan(-\tan \xi) + \xi = -\xi + \xi = 0$. Fig. 3.13-b shows in more detail the surface $E = E(\kappa, \xi)$ in the subregion $-0.5 \le \kappa \le 0.5$ of the stability region of the Lax-Wendroff scheme.

As the second example let us take the two-stage Jameson's scheme (3.4.4) with $\alpha = 1$. We have implemented the analysis of the dispersion of this scheme in the same notebook disp1.ma.

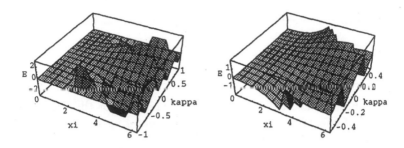

Figure 3.13: The phase error $E(\kappa, \xi)$ corresponding to the Lax-Wendroff scheme (2.4.3). (a) $-1 \leq \kappa \leq 1$; (b) $-0.5 \leq \kappa \leq 0.5$.

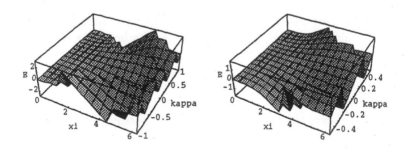

Figure 3.14: The phase error $E(\kappa, \xi)$ corresponding to the two-stage Jameson's scheme (3.4.4) with $\alpha = 1$. (a) $-1 \leq \kappa \leq 1$; (b) $-0.5 \leq \kappa \leq 0.5$.

The obtained symbolic expression for the phase error $E(\kappa, \xi)$ has in the case of the Jameson's scheme under consideration the form

```
                    2 cp Sin[xi]
E = -ArcTan[-----------------------] + ArcTan[Tan[cp xi]]
                   2     2
               2 - cp + cp  Cos[2 xi]
```

We have obtained at kt=5 the following expression for the Taylor series expansion of the phase error $E(\kappa, \xi)$ corresponding to the Jameson's scheme (3.4.4) with $\alpha = 1$:

```
             3                      3    5
      cp   2 cp     3      -cp    cp    cp     5          6
E = (-- - -----) xi   + (--- + --- - ---) xi  + O[xi]
      6     3            120     3     5
```

It can be seen from here that $E \neq 0$ at $|\kappa| = 1$ in contrast to the above considered Lax-Wendroff scheme (2.4.3), see also Fig. 3.14-a. The amplitudes of $E(\kappa, \xi)$ are of the same order of magnitude in the region $[-1 \leq \kappa \leq 1] \times [0 \leq \xi \leq 2\pi]$. The Jameson's scheme (3.4.4) has at $\alpha = 1$ the order of approximation $O(\tau) + O(h^2)$, and the Lax-Wendroff scheme is second-order accurate both in space and time. The both schemes have the same stability region $|\kappa| \leq 1$. However, the behavior of the phase error $E(\kappa, \xi)$ in the case of the Lax-Wendroff scheme is better near the boundaries $\kappa = \pm 1$ than in the case of the two-stage Jameson's scheme (3.4.4) with $\alpha = 1$. Therefore, the Lax-Wendroff scheme (2.4.3) is more preferable than the scheme (3.4.4) with $\alpha = 1$.

We now discuss the peculiarities of the dispersion analysis of difference schemes for parabolic PDEs. The dispersion of difference schemes approximating the one-dimensional advection-diffusion equation (3.3.2) was studied extensively by C.A.J. Fletcher.[28]

Let us substitute the Fourier component (3.8.2) into Eq. (3.3.2) in order to determine the form of the expression for ω. We easily find that

$$\omega = ak - i\nu k^2 \qquad (3.8.13)$$

Thus, the Fourier component of the form

$$u_k(x,t) = e^{-\nu k^2 t} e^{i(kx - akt)} \qquad (3.8.14)$$

satisfies the advection-diffusion equation (3.3.2). It follows from Eq. (3.8.13) that the decay of the wave amplitude is determined by the diffusive term νu_{xx} in Eq. (3.3.2), and the propagation speed is constant. Since $k = 2\pi/\lambda$, the short-wavelength disturbances decay faster with time than the long-wavelength disturbances.

The six-step algorithm presented above is applicable also to the difference schemes approximating the parabolic PDEs.

Example 1. As the first example let us consider the explicit scheme (3.3.3) for Eq. (3.3.2). For the dispersion analysis of schemes approximating the parabolic PDEs we have modified the above discussed program disp1.ma. The program for parabolic PDEs has the name disp1ad.ma and is available on the attached disk. By using this program we have obtained the following expression for the phase error (3.8.9):

```
                        cp1 Sin[xi]
E= -ArcTan[--------------------------] + ArcTan[Tan[cp1 xi]]
            1 - 2 cp2 + 2 cp2 Cos[xi]
```

The Taylor series expansion of this error around the point $\xi = 0$ was obtained on a computer in the form

```
                2               3              3     5
    cp1 (1 + 2 cp1  - 6 cp2) xi       -cp1   cp1   cp1
E = -------------------------------- + (---- - ---- - ---- +
                  6                      120    6      5

    cp1 cp2    3          2    5          6
>  -------- + cp1  cp2 - cp1  cp2 ) xi  + O[xi]
       4
```

Example 2. The Crank-Nicolson scheme for Eq. (3.3.2) has the form (see also Reference 28)

$$\frac{u_j^{n+1} - u_j^n}{\tau} + \frac{1}{2}a\left(\frac{u_{j+1}^n - u_{j-1}^n}{2h} + \frac{u_{j+1}^{n+1} - u_{j-1}^{n+1}}{2h}\right)$$

$$= \frac{1}{2}\nu\left(\frac{u_{j-1}^n - 2u_j^n + u_{j+1}^n}{h^2} + \frac{u_{j-1}^{n+1} - 2u_j^{n+1} + u_{j+1}^{n+1}}{h^2}\right) \qquad (3.8.15)$$

According to Reference 28 this difference scheme is absolutely stable. But in order to ensure the absence of the numerical solution oscillations in space one has to satisfy the inequality $P_h \leq 2$, where P_h is the cell Péclet number (3.3.10).[28] We have obtained by our program displadc.ma the following expression for the amplification factor $g(\kappa_1, \kappa_2, \xi)$:

```
  2 - 2 cp2 + 2 cp2 Cos[xi] - I cp1 Sin[xi]
{-------------------------------------------}
  2 + 2 cp2 - 2 cp2 Cos[xi] + I cp1 Sin[xi]
```

Here cp1= $\kappa_1 = a\tau/h$, cp2= $\nu\tau/(h^2)$. The phase error was computed as

```
                                    2        2
E = -ArcTan[4 cp1 Sin[xi]/(4 - 6 cp2  + 8 cp2  Cos[xi] -

      2           2       2
  2 cp2  Cos[2 xi] - cp1  Sin[xi] )] + ArcTan[Tan[cp1 xi]]
```

The Taylor series expansion of this error around the point $\xi = 0$ was obtained in the form

```
        3                       2        4       2    5
 cp1   cp1    3    cp1 (2 + 10 cp1  + 3 cp1  + 60 cp2 ) xi
(--- + ----) xi  - ---------------------------------------
  6     12                          240


        6
 + O[xi]
```

Comparing this expression with a similar expression obtained above for the explicit scheme (3.3.3) we can see that at small $|\xi|$ and at $\kappa_2 < \kappa_1^2/4$ the dispersion error E is smaller in the case of the Crank-Nicolson scheme (3.8.15). The plots of the surfaces $E = E(\kappa_1, \kappa_2, \xi)$ can give the further information about the behavior of the dispersion. In the case under consideration the function E depends on three variables κ_1, κ_2 and ξ. In this connection there arises the difficulty of the visualization of the hypersurface $E = E(\kappa_1, \kappa_2, \xi)$ in the four-dimensional Euclidean space of $(\kappa_1, \kappa_2, \xi, E)$ points. In Fig. 3.15 we show only the three-dimensional sections in the plane $\kappa_2 = 0.5$ of these surfaces corresponding to the explicit scheme (3.3.3) (Fig.3.15-a) and the Crank-Nicolson scheme (3.8.15) (Fig. 3.15-b). According to the stability condition (3.3.5) of scheme (3.3.3) the (κ_1, κ_2) points with $\kappa_2 = 0.5$ and $|\kappa_1| \le 1$ belong to the stability region of this scheme. We can see that the error E vanishes at $\kappa_2 = 0.5$ and $|\kappa_1| = 1$ in the case of the explicit scheme (3.3.3). This desirable property does unfortunately not take place in the case of the Crank-Nicolson scheme (3.8.15) (see Fig. 3.15-b).

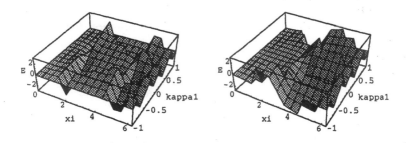

Figure 3.15: The phase error $E = E(\kappa_1, \kappa_2, \xi)$ in the section $\kappa_2 = 0.5$. See text for details.

3.8.2 The Case of Two Spatial Variables

The above presented dispersion analysis can be extended for the case of two spatial variables x, y and the time variable t. An extensive analysis of the phase errors of 16 difference schemes approximating the systems of hyperbolic PDEs was performed by E. Turkel.[45] In what follows we give an outline of the procedure for the analysis of the phase errors of difference schemes approximating the hyperbolic or parabolic PDE of the form

$$\frac{\partial u}{\partial t} = Lu \tag{3.8.16}$$

where L is a linear differential operator. We substitute the Fourier component of the form

$$u_{k_1 k_2}(x, y, t) = e^{i(k_1 x + k_2 y - \omega t)} \tag{3.8.17}$$

into the PDE (3.8.16) in order to determine the form of the expression for thefrequency ω:

$$\omega = f(k_1, k_2)$$

After that we find from the equation

$$u_{k_1 k_2}(x, y, t + \tau) = e^{-if(k_1, k_2)\tau} u_{k_1, k_2}(x, y, t)$$

the expression for the Fourier symbol of the step operator corresponding to the original PDE (3.8.16):

$$g_0(k_1, k_2, \tau) = e^{-if(k_1, k_2)\tau}$$

Let us now take the difference scheme

$$u_{jk}^{n+1} = S(\tau, h_1, h_2)u_{jk}^n \tag{3.8.18}$$

which approximates the PDE (3.8.16). Substituting the Fourier component (3.8.17) into Eq. (3.8.18) we obtain the expression for the amplification factor $g(\kappa, \xi)$, $M \geq 1$ and $\xi = (\xi_1, \xi_2)$, $\xi_1 = k_1 h_1, \xi_2 = k_2 h_2$. The quantity

$$E(\kappa, \xi) = \arctan \frac{Im\, g}{Re\, g} - \arctan \frac{Im\, g_0}{Re\, g_0}$$

represents the *phase error* of the difference scheme (3.8.18).

Note that the phase error for a given difference method can be reduced significantly by a corresponding modification of the method. The examples of such modifications are presented in Reference 46. It should be noted that the reduction of the phase error is often achieved at the cost of a smaller time step allowed by the stability.

The main purpose of this Section was to show how the *Mathematica* system can be applied for the analysis of the dispersion of difference schemes. The presented examples show that the advanced computer algebra systems like the *Mathematica* system prove to be a very efficient tool in investigating such an important property as dispersion.

Exercise 3.11. Compare the phase error of the first-order difference scheme (2.1.9) with the phase error of the Lax-Wendroff scheme (2.4.3).

Hint. Replace the input Lax-Wendroff scheme in the file disp1.ma by the scheme (2.1.9) and then obtain the surface $E = E(\kappa, \xi)$.

Exercise 3.12. Compare the phase error of the Crank-Nicolson scheme (2.9.1) with the phase error of the Lax-Wendroff scheme (2.4.3).

Exercise 3.13. Apply the four optimal Jameson schemes (3.4.11)-(3.4.14) for the approximation of the advection equation (2.1.1). Compare the phase errors of these schemes by using the program disp1.ma.

Exercise 3.14. Apply the optimal two-stage Jameson's scheme (3.4.4) ($\alpha = 1$) for the approximation of the advection-diffusion equation (3.3.2). Compare the phase error $E(\kappa_1, \kappa_2, \xi)$ of this scheme with that of the Crank-Nicolson scheme (3.8.15).

Exercise 3.15. Write a *Mathematica* program for the analysis of the phase error of the MacCormack scheme (2.12.3) for the two-dimensional advection equation (2.12.2). Plot the surfaces $E = E(\kappa_1, \kappa_2, \xi_1, \xi_2)$ for the following fixed values of the Courant numbers κ_1 and κ_2: (a) $\kappa_1 = 0.9$, $\kappa_2 = 0.05$; (b) $\kappa_1 = 0.05$, $\kappa_2 = 0.9$; (c) $\kappa_1 = 0.45$, $\kappa_2 = 0.45$. Compare the corresponding phase errors.

ANSWERS TO THE EXERCISES

3.3. $\frac{\tau}{h^2} D(u) > 0 \quad \forall u$

3.4. $\kappa_1 + \kappa_2 \leq 1/2,$ where $\kappa_j = \nu\tau/(h_j^2),\; j = 1, 2$

3.5. $\kappa_1 + \kappa_2 + \kappa_3 \leq 1/2,$ where $\kappa_j = \nu\tau/(h_j^2),\; j = 1, 2, 3$

3.10.

$$2\frac{\nu\tau}{h_1^2} < |\beta\frac{\tau A}{h_1}| \leq 1 + 2\frac{\nu\tau}{h_1^2},\; 2\frac{\nu\tau}{h_2^2} < |\beta\frac{\tau B}{h_2}| \leq 1 + 2\frac{\nu\tau}{h_2^2}$$

REFERENCES

1. **Richtmyer, R.D. and Morton, K.W.,** *Difference Methods for Initial-value problems,* Second Edition. Interscience Publishers, a division of John Wiley and Sons, New York, 1967.
2. **Oran, E.S. and Boris, J.P.,** *Numerical Simulation of Reactive Flow,* Elsevier, New York, 1987.
3. **Bear, J., and Bachmat, Y.,** *Introduction to Modeling of Transport Phenomena in Porous Media,* Kluwer Academic Publishers, Dordrecht, Netherlands, 1990.
4. **Tikhonov, A.N. and Samarskii, A.A.,** *Equations of the Mathematical Physics,* Fifth Edition (in Russian), Nauka, Moscow, 1977.
5. **Roache, P.J.,** *Computational Fluid Dynamics,* Hermosa, Albuquerque, New Mexico, 1976.
6. **Strikwerda, J.C.,** *Finite Difference Schemes and Partial Differential Equations,* Wadsworth & Brooks/Cole Advanced Books & Software, Pacific Grove, California, 1989.
7. **Scobelev, B.Yu. and Vorozhtsov, E.V.,** *On the Stability Investigation of Difference Schemes in Different Banach Spaces.* Preprint No. 94-10 of the Institute of Theoretical and Applied Mechanics of the Siberian Division of the Russian Academy of Sciences, Novosibirsk, 1994 (in Russian).
8. **Allen, J.S.,** *Numerical solutions of the compressible Navier-Stokes equations for the laminar near wake in supersonic flow,* Ph.D. Dissertation, Princeton University, Princeton, New Jersey, 1968.
9. **Allen, J.S. and Cheng, S.I.,** Numerical solutions of the compressible Navier-Stokes equations for the laminar near wake, *Physics of Fluids* 13, 37, 1970.
10. **Rusanov, V.V.,** Difference schemes of the third order accuracy for the shock-capturing computations of discontinuous solutions, *Doklady AN SSSR* (in Russian) 180, 1303, 1968.
11. **Rusanov, V.V.,** On difference schemes of third order accuracy for nonlinear hyperbolic systems, *Journal of Computational Physics* 5, 507, 1970.
12. **Burstein, S. and Mirin, A.,** Third order difference methods for hyperbolic equations, *Journal of Computational Physics* 5, 547, 1970.

13. **Balakin, V.B.,** On methods of the Runge-Kutta type for gas dynamics equations, *Zhurn. Vychisl. Matem. i Matem. Fiziki* (in Russian) 10, 1512, 1970.

14. **Warming, R.F., Kutler, P., and Lomax, H.,** Second and third order noncentered difference schemes for nonlinear hyperbolic equations. *AIAA Journal* 11, 189, 1973.

15. **Roždestvenskiĭ, B.L. and Janenko, N.N.,** *Systems of Quasilinear Equations and Their Applications to Gas Dynamics.* Second Edition (in Russian). Nauka, Moscow, 1978. [English transl.: *Systems of Quasilinear Equations and Their Applications to Gas Dynamics,* Translations of Mathematical Monographs, Vol.55 (American Mathematical Society , Providence, Rhode Island, 1983)].

16. **Jameson, A., Schmidt, W., and Turkel, E.,** *Numerical solution of the Euler equations by finite volume methods using Runge-Kutta time stepping schemes,* AIAA Paper, 1981, No. 1259.

17. **Jameson, A., and Schmidt, W.,** Some recent developments in numerical methods for transonic flows, *Computer Methods in Applied Mechanics and Engineering* 51, 467, 1985.

18. **Schmidt, W., and Jameson, A.,** Euler solvers as an analysis tool for aircraft aerodynamics, *Advances in Computational Transonics,* Vol. 4, "Recent Advances in Numerical Methods in Fluids" / Ed. W. Habashi. Pineridge Press, Swansea, 1985, p. 371.

19. **Ganzha, V.G., and Vorozhtsov, E.V.,** On the stability of Jameson schemes, *Bridging Mind and Model: Papers in Applied Mathematics* / Ed. P.J. Costa, St. Thomas Technology Press, St. Paul, MN, 1993, p. 237.

20. **Van der Houwen, P.J.,** *Construction of Integration Formulas for Initial Value Problems,* North-Holland, Amsterdam, 1977.

21. **Pike, J., and Roe, P.L.,** Accelerated convergence of Jameson's finite-volume Euler scheme using van der Houwen integrators, *Computers and Fluids* 13, 223, 1985.

22. **Shu, C.-W. and Osher, S.,** Efficient implementation of essentially non-oscillatory shock-capturing schemes, *Journal of Computational Physics* 77, 439, 1988.

23. **Shu, C.-W. and Osher, S.,** Efficient implementation of essentially non-oscillatory shock-capturing schemes II, *Journal of Computational Physics* 83, 32, 1989.

24. **Mulder, W., Osher, S. and Sethian, J.A.,** Computing interface motion in compressible gas dynamics, *Journal of Computational Physics* 100, 209, 1992.

25. **MacCormack, R.W. and Paullay, A.J.**, *Computational efficiency achieved by time splitting of finite difference operators*, AIAA Paper, 1972, No. 72-154.
26. **Wilkins, M.L.**, Calculation of elastic-plastic flow. In: *Methods in Computational Physics, Vol. 3* / Eds. B. Alder, S. Fernbach and M. Rotenberg, Academic Press, New York, 1964, p. 211-263.
27. **Peyret, R. and Taylor, T.D.**, *Computational Methods for Fluid Flow*, Springer-Verlag, New York, 1983.
28. **Fletcher, C.A.J.**, *Computational Techniques for Fluid Dynamics 1. Fundamental and General Techniques*, Springer-Verlag, Berlin, 1988.
29. **Mavriplis, D.J.**, Multigrid solution of the two-dimensional Euler equations on unstructured triangular meshes, *AIAA Journal* 26, 824, 1988.
30. **Jameson, A. and Mavriplis, D.J.**, Finite volume solution of the two-dimensional Euler equations on a regular triangular mesh, *AIAA Journal* 24, 611, 1986.
31. **Mavriplis, D. and Jameson, A.**, *Multigrid solution of the two-dimensional Euler equations on unstructured triangular meshes*, AIAA Paper, Jan. 1987, No. 87-0353.
32. **Jameson, A., Baker, T.J. and Weatherill, N.P.**, *Calculation of inviscid transonic flow over a complete aircraft*, AIAA Paper, 1986, No. 86-0103.
33. **Yanenko, N.N.**, *The Method of Fractional Steps: The Solution of Problems of Mathematical Physics in Several Variables*, Springer-Verlag, New York, 1971.
34. **Samarskii, A.A. and Gulin, A.V.**, *Numerical Methods* (in Russian), Nauka, Moscow, 1989.
35. **Peacemean, D.W. and Rachford, H.H., jr.**, The numerical solution of parabolic and elliptic differential equations, *Journal of the Society of Industrial and Applied Mathematics* 3, 28, 1955.
36. **Douglas J., jr.**, On the numerical integration of $\frac{\partial^2 u}{\partial x^2} + \frac{\partial^2 u}{\partial y^2} = \frac{\partial u}{\partial t}$ by implicit methods, *Journal of the Society of Industrial and Applied Mathematics* 3, 42, 1955.
37. **Douglas, J. and Gunn, J.E.**, A general formulation of alternating direction implicit methods, Part I, parabolic and hyperbolic problems, *Numerische Mathematik* 6, 428, 1964.

38. **Karamyshev, V.B.**, On an unfactored implicit scheme for solving gas dynamics problems, *Modeling in Mechanics* (in Russian) 6(23), No. 3, p. 60, 1992.

39. **Karamyshev, V.B.**, The predictor-corrector scheme for solving nonstationary problems of gas dynamics. 2. Some computational examples, *Modeling in Mechanics* (in Russian) 5(22), No. 3, p. 48,1991.

40. **MacCormack, R.W.**, On efficient numerical methods for solving the Navier-Stokes equations in three dimensions, *The Second Japan-Soviet Union Joint Symposium on Computational Fluid Dynamics, August 27-31, 1990. Proceedings, Vol. II* / Eds. Y. Yoshizawa and K. Oshima, Published by the Japan Society of Computational Fluid Dynamics, Tsukuba,1991, p. 1.

41. **Karamyshev, V.B.**, *Monotonous Schemes and Their Applications to Gas Dynamics* (in Russian), published by the Novosibirsk State University, Novosibirsk, 1994.

42. **Salary, K., Knupp, P., Roache, P. and Steinberg, S.**, TVD applied to radionuclide transport in fractured porous media. Proceedings of the IX International Conference on Computational Methods in Water Resources, Denver, Colorado, 9-12 June 1992, p. 141.

43. **Shokin, Yu.I. and Yanenko, N.N.**, *The Method of Differential Approximation. Application to Gas Dynamics* (in Russian), Nauka, Siberian Division, Novosibirsk, 1985.

44. **Stone, H.L. and Brian, P.L.T.**, Numerical solution of convective transport problems, *American Institute of Chemical Engineers Journal* 9, 681, 1963.

45. **Turkel, E.**, Phase error and stability of second order methods for hyperbolic problems I, *Journal of Computational Physics* 15, 226, 1974.

46. **Gottlieb, D. and Turkel, E.**, Phase error and stability of second order methods for hyperbolic problems II, *Journal of Computational Physics* 15, 251, 1974.

Chapter 4

NUMERICAL METHODS FOR ELLIPTIC PDEs

4.1 BOUNDARY-VALUE PROBLEMS FOR ELLIPTIC PDEs

The mathematical physics problems whose solutions depend only on the spatial variables and do not depend on the time t are usually called *stationary* problems. Otherwise these problems are called *nonstationary*, or *time-dependent*, or *unsteady* problems.

Let us consider, for example, a stationary thermal field in the plane of spatial variables x, y. Then the partial derivative $\partial u / \partial t$ in Eq. (3.1.18) is equal to zero, and we arrive at the equation

$$\Delta u = 0 \qquad (4.1.1)$$

where the operator Δ is termed the *Laplacian* operator,

$$\Delta = \frac{\partial^2}{\partial x^2} + \frac{\partial^2}{\partial y^2}$$

Equation (4.1.1) is called the *Laplace's* equation. This is the most widespread equation of the *elliptic* type.

In the presence of the heat sources we obtain the equation

$$\Delta u = -f(x, y) \qquad (4.1.2)$$

where $f = F/k$, F is the density of the heat sources, and k is the thermal conductivity coefficient; $f(x, y)$ is usually a given function of its arguments. The inhomogeneous Laplace's equation is often called the *Poisson* equation.

We now consider the second example of the application of the elliptic PDEs. Let a stationary flow of an incompressible fluid take place in some spatial region D with the boundary Γ, and let $\mathbf{v}(x, y, z)$ be the fluid velocity. If the flow is irrotational, then there exists such a scalar function $\varphi(x, y, z)$ that

$$\mathbf{v} = -grad\varphi \qquad (4.1.3)$$

where the function φ is called the *velocity potential*. If there are no sinks or sources, then the equation

$$div\mathbf{v} = 0 \qquad (4.1.4)$$

holds for an incompressible fluid. Substituting expression (4.1.3) into Eq. (4.1.4), we obtain the equation

$$div\ grad\ \varphi = 0$$

or

$$\Delta\varphi = 0$$

where the Laplacian operator in the case of three spatial variables x, y and z reads

$$\Delta = \frac{\partial^2}{\partial x^2} + \frac{\partial^2}{\partial y^2} + \frac{\partial^2}{\partial z^2}$$

The third example of the application of the Laplace's and Poisson equations can be found in the numerical generation of curvilinear spatial grids,[2] see also Section 4.5 below.

Further examples of applied problems leading to elliptic PDEs can be found in the field of electrodynamics,[1,3] elasticity theory,[3] filtration theory.[4]

Let us now formulate a number of typical boundary-value problems, which are considered for elliptic PDEs. Let D be a region in the (x, y) plane in which the solution of a PDE is to be found, and let Γ be the boundary of the region D. Then the boundary-value problems for equation (4.1.2) read:

find the function $u(x, y)$ satisfying inside D the equation (4.1.2) and the boundary condition of one of the following forms:

I. $u = f_1(x, y)$, $(x, y) \in \Gamma$ (the first boundary-value problem),

II. $\frac{\partial u}{\partial n} = f_2(x, y)$, $(x, y) \in \Gamma$ (the second boundary-value problem),

III. $\frac{\partial u}{\partial n} + f_4(u - f_3) = 0$, $(x, y) \in \Gamma$ (the third boundary-value problem),

where f_1, f_2, f_3 and f_4 are the given functions, and $\partial u/\partial n$ is a derivative along the outer normal to the boundary Γ. The first boundary-value problem for the Laplace's equation is often called the *Dirichlet* problem, and the second boundary-value problem is called the *Neumann* problem.

An important observation concerning Eq. (4.1.2) with the Neumann condition

$$\frac{\partial u}{\partial n} = f_2(x, y), \ (x, y) \in \Gamma \qquad (4.1.5)$$

is that for a solution to exist, the data must satisfy the constraint

$$\iint_D f\,dx\,dy = -\int_\Gamma f_2 d\Gamma \qquad (4.1.6)$$

This relationship is called an *integrability condition* and is easily proved by the divergence theorem as follows:

$$\iint_D f\,dx\,dy = -\iint_D \Delta u \; dx\,dy = -\int_\Gamma \mathbf{n} \cdot grad \; u \; d\Gamma =$$

$$-\int_\Gamma \frac{\partial u}{\partial n} d\Gamma = -\int_\Gamma f_2 d\Gamma$$

The vector $\mathbf{n}$ is the outer unit normal vector to the boundary Γ. The integrability condition (4.1.6) has the physical interpretation that the heat sources in the region must balance with the heat flux on the boundary for a steady temperature to exist (see also Reference 1).

We now briefly mention the *maximum principle* for second-order elliptic PDEs. This principle is used, for example, in the numerical grid generation (see also Section 4.5). The next two theorems are expressions of maximum principles.[1,5]

Theorem 4.1. *Let L be a second-order elliptic operator defined by*

$$Lu = au_{xx} + 2bu_{xy} + cu_{yy}$$

that is, the coefficients a and c are positive and b satisfies $b^2 < ac$. If a function u satisfies $Lu \geq 0$ in a bounded domain D, then u has its maximum value on the boundary of D.

Theorem 4.2. *If the elliptic equation*

$$au_{xx} + 2bu_{xy} + cu_{yy} + d_1 u_x + d_2 u_y + eu = 0$$

holds in a domain D, with a and c positive and e nonpositive, then the solution $u(x,y)$ cannot have a positive local maximum or a negative local minimum in the interior of D.

4.2 A SIMPLE ELLIPTIC SOLVER

Let us take a square region D in the x, y plane, that is

$$D = \{(x,y)|\; 0 \leq x, y \leq 1\} \qquad (4.2.1)$$

We now formulate the Dirichlet problem for the Poisson equation in the region D with the boundary Γ:

$$\frac{\partial^2 u}{\partial x^2} + \frac{\partial^2 u}{\partial y^2} = -f(x,y), \quad 0 \leq x, y \leq 1 \qquad (4.2.2)$$

$$u(x, y) = \varphi(x, y), \quad (x, y) \in \Gamma \tag{4.2.3}$$

where $\varphi(x, y)$ is a given function. We introduce an uniform rectangular grid G_h in D with the aid of the formulas

$$x_j = jh_1, \ j = 0, 1, \ldots, M; \quad y_k = kh_2, \ k = 0, 1, \ldots, N$$

where $h_1 = 1/M$, $h_2 = 1/N$. The set of the internal points (x_j, y_k) of the mesh G_h, which lie inside the square D, will be denoted by G_h^0. The set of the G_h points lying on the boundary Γ will be denoted by Γ_h. Using the central differences of the form presented in Section 2.1 for the approximation of the second derivatives in Eq. (4.2.2) we can write the following difference scheme approximating Eq. (4.2.2) at points $(x_j, y_k) \in G_h^0$:

$$\Lambda_h u \equiv \frac{u_{j+1,k} - 2u_{j,k} + u_{j-1,k}}{h_1^2} + \frac{u_{j,k+1} - 2u_{j,k} + u_{j,k-1}}{h_2^2} =$$

$$-f(x_j, y_k), \quad j = 1, \ldots, M - 1; \quad k = 1, \ldots, N - 1 \tag{4.2.4}$$

Since the values of $u(x, y)$ are known on the boundary Γ in accordance with the boundary condition (4.2.3), we can write that

$$u_{j,k} = \varphi(x_j, y_k), \quad (x_j, y_k) \in \Gamma_h \tag{4.2.5}$$

Let us now consider the question of the approximation order of scheme (4.2.4). Assuming that the solution $u(x, y)$ of the problem (4.2.2), (4.2.3) has the bounded fourth derivatives we can establish the following equation with the aid of the Taylor formula:

$$\Lambda_h u = \frac{\partial^2 u}{\partial x^2} + \frac{\partial^2 u}{\partial y^2} + \frac{h^2}{24} \left[\frac{\partial^4 u(x + \theta_1 h_1, y)}{\partial x^4} + \frac{\partial^4 u(x - \theta_2 h_1, y)}{\partial x^4} \right.$$

$$\left. + \frac{\partial^4 u(x, y + \theta_3 h_2)}{\partial y^4} + \frac{\partial^4 u(x, y - \theta_4 h_2)}{\partial y^4} \right] \tag{4.2.6}$$

where $0 < \theta_\nu < 1$, $\nu = 1, 2, 3, 4$. It follows from Eq. (4.2.6) that the scheme (4.2.4) has the second order of approximation in space.

Let us now show that the five-point difference approximation of the Laplacian in the difference scheme (4.2.4) satisfies the *discrete maximum principle*, which can be formulated as[5,6]

Theorem 4.3. *If $\Lambda_h u \geq 0$ for $(x_j, y_k) \in G_h^0$, then the maximum value of u is attained on the boundary Γ_h. Similarly, if $\Lambda_h u \leq 0$ then the minimum value of u is attained on the boundary Γ_h.*

Proof. We prove the principle for the case when $h_1 = h_2$. Let us assume the contrary. Let us choose among the points of the mesh G_h

at which u achieves its maximum value some point (x_j, y_k) having the largest abscissa. By our assumption, (x_j, y_k) is an internal point, and $u_{j,k} > u_{j+1,k}$. Then we have at (x_j, y_k) point that

$$\Lambda_h u_{j,k} = [(u_{j+1,k} - u_{j,k}) + (u_{j,k+1} - u_{j,k}) + (u_{j-1,k} - u_{j,k})$$
$$+ (u_{j,k-1} - u_{j,k})]/(h_1^2) < 0$$

since the first expression in the brackets in the numerator is negative, and the remaining expressions are nonpositive. We arrive at a contradiction with the inequality $\Lambda_h u \geq 0$.

When $\Lambda_h u \leq 0$, by considering $\Lambda_h(-u) \geq 0$, this case reduces to the previous case.

It follows from the discrete maximum principle that the problem

$$\Lambda_h u_{j,k} = 0, \quad (x_j, y_k) \in G_h^0$$
$$u_{j,k} = 0, \quad (x_j, y_k) \in \Gamma_h$$

has only the zeroth solution $u = 0$, since the largest and the least values of this solution are attained on Γ_h, where $u_{j,k} = 0$. Consequently, the determinant of the system of linear equations (4.2.4) is different from zero and the difference boundary-value problem (4.2.4), (4.2.5) is uniquely solvable at an arbitrary right hand side $-f(x_j, y_k)$.

Despite the apparently simple structure of the difference equations (4.2.4) the solution of these equations, that is the determination of $(M - 1)(N - 1)$ values $u_{j,k}$, is not straightforward. The solution of the algebraic system (4.2.4) can be found iteratively, for example, with the aid of the pseudo-unsteady method, which we consider below in Section 4.3. Among other iterative methods for the solution of equations (4.2.4) we mention the Jacobi method,[5,7] the Gauss-Seidel method,[5,7] the successive over-relaxation method,[5,7] and the multigrid method.[7]

Exercise 4.1. Consider the difference equation

$$\Lambda_h u_{j,k} = 0, \quad j, k = 1, \ldots, M - 1, \ Mh = 1$$

Show that the solution u either takes equal values everywhere on G_h or the largest and the least values of the function u are attained at no internal point of the mesh G_h (the *enhanced maximum principle*[6]).

Exercise 4.2. Prove the discrete maximum principle on the unit square for the case with h_1 not equal to h_2.

4.3   PSEUDO-UNSTEADY METHODS

4.3.1   An Example of the Relation Between Stationary and Nonstationary Problems

For the computation of the solutions of many stationary problems one can consider the solution as a result of a temporal evolution of some transient process. The obtaining of the stationary solution in this way often proves to be simpler than the direct calculation of such a solution.

The time-dependent numerical methods, which are designed especially for the computation of the solutions to stationary problems, are often termed the *pseudo-unsteady* methods. The term "pseudo-unsteady" reflects the fact that the numerical solution obtained by the time-dependent method in the limit as the time $t \to \infty$ does indeed not depend on t.

Let us consider in a rectangular domain D the Dirichlet problem

$$\frac{\partial^2 u}{\partial x^2} + \frac{\partial^2 u}{\partial y^2} = 0 \tag{4.3.1}$$

$$u(x, y) = f(x, y), \quad (x, y) \in \Gamma \tag{4.3.2}$$

where Γ is the boundary of the domain D,

$$D = \{(x, y) \mid 0 \le x, y \le \pi\}$$

and $f(x, y)$ is a given function.

Let us consider along with the problem (4.3.1)-(4.3.2) the nonstationary problem

$$\frac{\partial U}{\partial t} = a^2 \left(\frac{\partial^2 U}{\partial x^2} + \frac{\partial^2 U}{\partial y^2} \right) \tag{4.3.3}$$

$$U(x, y, 0) = u_0(x, y) \tag{4.3.4}$$

$$U(x, y, t) = f(x, y), \ (x, y) \in \Gamma, \ t \ge 0 \tag{4.3.5}$$

where $a^2 = const > 0$, and the function $u_0(x, y)$ is arbitrary. The difference

$$v(x, y, t) = U(x, y, t) - u(x, y) \tag{4.3.6}$$

satisfies Eq. (4.3.3) with the initial conditions

$$v(x, y, 0) = v_0(x, y) = u_0(x, y) - U(x, y) \tag{4.3.7}$$

and the zero boundary conditions

$$v(x, y, t) = 0, \quad (x, y) \in \Gamma \tag{4.3.8}$$

The solution $v(x, y, t)$ may be presented in the form

$$v(x, y, t) = \sum_{k_1=1}^{\infty} \sum_{k_2=1}^{\infty} A_{k_1 k_2}(t) \sin k_1 x \sin k_2 y \tag{4.3.9}$$

where

$$A_{k_1 k_2}(t) = a_{k_1 k_2} e^{-a^2(k_1^2+k_2^2)t} \tag{4.3.10}$$

is the Fourier coefficient of the function $v(x, y, t)$; $a_{k_1 k_2}$ is the Fourier coefficient of the function $v_0(x, y)$ (see also Reference 8). The formulas (4.3.9) and (4.3.10) may be written in the operator form

$$v = S(t)v_0$$

The norm in the space $L_2(D)$ is defined as

$$\| v \| = \left(\int\!\!\int_D |v^2(x, y)| dx dy \right)^{1/2}$$

Note that

$$|A_{k_1 k_2}(t)| \le |a_{k_1 k_2}| e^{-2a^2 t} \quad \text{for } k_1 \ge 1, \ k_2 \ge 1$$

Therefore, the operator $S(t)$ has in the space $L_2(D)$ the norm

$$\| S(t) \| = e^{-2a^2 t} \tag{4.3.11}$$

From this it follows that

$$\| S(t) \| \to 0, \ t \to \infty \tag{4.3.12}$$

This means that

$$\| v(x, y, t) \| = \| U(x, y, t) - u(x, y) \| \to 0 \tag{4.3.13}$$

as $t \to \infty$, that is the solution of the nonstationary problem tends to the solution of the stationary problem under the same boundary conditions independently of the choice of the initial data.

4.3.2   General Construction of the Pseudo-Unsteady Method

Let us consider a PDE of the form

$$Au = f \qquad (4.3.14)$$

where the differential operator A involves the derivatives only with respect to the spatial variables. Assume that the solution of Eq. (4.3.14) is to be determined in a bounded spatial region D having the boundary Γ. Denote by Γ' the union of all those pieces of the Γ boundary, on which the boundary conditions

$$Bu|_{\Gamma'} = \varphi \qquad (4.3.15)$$

are specified, which ensure the existence and uniqueness of the solution of the boundary-value problem (4.3.14), (4.3.15). In Eq. (4.3.15), B is some given operator, which can involve, for example, the normal derivatives, and φ is a given function of the spatial coordinates. We now introduce the nonstationary equation

$$\sum_{i=1}^{m} C_i \frac{\partial^i v(x, y, t)}{\partial t^i} = Av(x, y, t) - f(x, y) \qquad (4.3.16)$$

where C_i are certain operators, which ensure the convergence of the pseudo-unsteady process:

$$\lim_{t \to \infty} v(x, y, t) = u(x, y)$$

Eq. (4.3.16) is solved at the initial conditions

$$\partial^k v / \partial t^k |_{t=0} = v_{0k}, \quad k = 0, 1, \ldots, m - 1$$

and at the *stationary* boundary conditions

$$Bv|_{\Gamma'} = \varphi$$

For example, the choice $C_1 = 1$, $c_2 = b^2$, $m = 2$ in Eq. (4.3.16), where $b^2 > 0$, ensures the decay of the difference $|v(x, y, t) - u(x, y)|$ as t increases[8] at any initial data $v(x, y, 0)$, $\partial v(x, y, 0)/\partial t$. Note that in a general case it is not necessary for the nonstationary equation (4.3.16) to describe some physical process. This equation can have only a formal meaning, it should only ensure the mathematical decay condition (4.3.12).

The acceleration of the convergence of the numerical (iterative) so-lution of Eq. (4.3.16) to the solution of the stationary problem (4.3.14), (4.3.15) can be achieved in a number of the ways:

(i) by the appropriate choice of the operators C_i in (4.3.16);

(ii) by the use of appropriate difference approximations of equation (4.3.16) (explicit, implicit schemes;[5-8] schemes of splitting,[6,8] etc.).

4.3.3 Application of the Pseudo-Unsteady Method to the Solution of Poisson Equation

Let us consider the boundary-value problem

$$\Delta u = -f(x, y) \tag{4.3.17}$$

$$u(x, y) = \varphi(x, y), \quad (x, y) \in \Gamma \tag{4.3.18}$$

on the unit square (4.2.1). Let us take the exact solution of this problem as in Reference 5 in the form

$$u(x, y) = \cos x \sin y \tag{4.3.19}$$

Substituting the function (4.3.17) into the left hand side of Eq. (4.3.17) we obtain the following expression for the function $f(x, y)$ on the right hand side of Eq. (4.3.17):

$$f(x, y) = 2 \cos x \sin y \tag{4.3.20}$$

Let us now approximate Eq. (4.3.17) on the set of mesh points G_h^0 constructed in Section 4.2 by the alternating direction implicit (ADI) scheme (see also Section 3.5)

$$\frac{\tilde{u}_{j,k} - u_{jk}^n}{\tau} = \frac{1}{2}[\Lambda_1 \tilde{u}_{j,k} + \Lambda_2 u_{j,k}^n + f(x_j, y_k)]$$

$$\frac{u_{j,k}^{n+1} - \tilde{u}_{jk}}{\tau} = \frac{1}{2}[\Lambda_1 \tilde{u}_{j,k} + \Lambda_2 u_{j,k}^{n+1} + f(x_j, y_k)]$$

$$j = 1, \ldots, M-1; \quad k = 1, \ldots, N-1 \tag{4.3.21}$$

$$u_{j,k}^{n+1}|_\Gamma = \tilde{u}_{j,k}|_\Gamma = \varphi(x_j, y_k), \quad (x_j, y_k) \in \Gamma_h \tag{4.3.22}$$

$$u_{j,k}^0| = \varphi_0(x_j, y_k), \quad (x_j, y_k) \in G_h \tag{4.3.23}$$

The difference operators Λ_1 and Λ_2 are defined as follows:

$$\Lambda_1 u_{j,k} = (u_{j+1,k} - 2u_{j,k} + u_{j-1,k})/(h_1^2)$$
$$\Lambda_2 u_{j,k} = (u_{j,k+1} - 2u_{j,k} + u_{j,k-1})/(h_2^2)$$

We will take the function $\varphi_0(x_j, y_k)$ in (4.3.23) in such a way that the equality

$$\varphi_0(x_j, y_k) = \varphi(x_j, y_k), \quad (x_j, y_k) \in \Gamma_h \tag{4.3.24}$$

is satisfied.

It was shown in Section 3.5 that the ADI scheme (4.3.21) is stable at any values of the time step τ. Therefore, one can try to find the optimal value of τ at which the speed of the convergence to the stationary solution will be maximum. The analytical solution of this optimization problem proves to be difficult for many difference schemes. As it was pointed out in Reference 9, the optimal value of τ depends on the mesh (i.e.; on M and N), on the geometric form of the spatial region D and on the type of the boundary conditions. In some cases, the analytical estimates for the optimal value of τ can be obtained from the Fourier stability analysis. Let us consider the difference

$$\varepsilon_{j,k}^n = u_{j,k}^n - u(x_j, y_k) \tag{4.3.25}$$

where $u(x_j, y_k)$ is the exact solution of the difference boundary-value problem

$$\Lambda_1 u_{j,k} + \Lambda_2 u_{j,k} = -f(x_j, y_k), \quad (x_j, y_k) \in G_h^0$$
$$u_{j,k} = \varphi(x_j, y_k), \quad (x_j, y_k) \in \Gamma_h \tag{4.3.26}$$

Then it is easy to show that the error ε^n satisfies the difference boundary-value problem

$$\frac{\tilde{\varepsilon}_{j,k} - \varepsilon_{j,k}^n}{\tau} = \frac{1}{2}[\Lambda_1 \tilde{\varepsilon}_{j,k} + \Lambda_2 \varepsilon_{j,k}^n]$$
$$\frac{\varepsilon_{j,k}^{n+1} - \tilde{\varepsilon}_{j,k}}{\tau} = \frac{1}{2}[\Lambda_1 \tilde{\varepsilon}_{j,k} + \Lambda_2 \varepsilon_{j,k}^{n+1}] \tag{4.3.27}$$
$$\tilde{\varepsilon}_{j,k} = \varepsilon_{j,k}^n = 0, \quad (x_j, y_k) \in \Gamma_h$$
$$\varepsilon_{j,k}^0 = \varphi_0(x_j, y_k) - u_{j,k}$$

Substituting the Fourier harmonic

$$\varepsilon_{j,k}^n = \lambda^n e^{i(jm_1 h_1 + km_2 h_2)}$$

into the difference scheme (4.3.27), we can obtain the representation for the solution $\varepsilon_{j,k}^n$ in the form

$$\varepsilon_{j,k}^n = \sum_{m_1, m_2} \lambda^n(m_1, m_2, \tau, h_1, h_2) e^{i(jm_1 h_1 + km_2 h_2)}$$

It can be seen from this formula that the optimum value of the step τ should be determined from the solution of the following optimisation problem:

$$\text{find } \min_{\tau}\left(\max_{m_1,m_2} |\lambda(m_1, m_2, \tau, h_1, h_2)| \right) \tag{4.3.28}$$

at fixed values of the steps h_1 and h_2. For the particular case of a square grid (i.e., $M = N$) on a unit square D the solution of the minimax problem (4.3.28) was found in the form[6]

$$\tau \approx \frac{1}{\sqrt{2}\pi M} \tag{4.3.29}$$

Since $h_1 = h_2 = h = 1/M$, the condition (4.3.29) can be rewritten as $\tau \approx h/(\sqrt{2}\pi)$. In this case

$$\max |\lambda| = 1 - \frac{\sqrt{2}\pi}{M} + O(\frac{1}{M^2}) \tag{4.3.30}$$

Let us assume that we want to reduce the initial value of the norm of the error $\| \varepsilon^0 \|$ by a factor of ν, $\nu > 1$. Then the needed number n of the time steps should be determined from the condition

$$\left(1 - \frac{\pi\sqrt{2}}{M} \right)^n \leq \frac{1}{\nu}$$

from where $n \approx M ln\nu/(\pi\sqrt{2}) = O(M)$. Each passage from u^n to u^{n+1} requires $O(M^2)$ arithmetic operations. Consequently the overall number of the arithmetic operations needed to reduce the error $\| \varepsilon^0 \|$ by a factor of ν will be of the order $O(M^3 ln\nu)$.

A more versatile technique of the convergence acceleration is the use of periodic sequences $\tau_1, \ldots, \tau_K$ ($K > 1$) instead of a fixed value of the time step τ. Such a sequence is repeated every K iterative cycles. Let us describe a method for the computation of the optimal sequence $\tau_1, \ldots, \tau_K$, which was proposed in Reference 10 for the case when $M = N$, that is the mesh G_h is a mesh of square cells. Then the amplification factor λ of scheme (4.3.27) for the errors ε^n_{jk} can be written in the form

$$\lambda(\tau) = \lambda_r(\tau)\lambda_s(\tau)$$

where

$$\lambda_\nu(\tau) = \frac{1 - 2\tau M^2 \sin^2 \frac{\pi\nu}{2M}}{1 + 2\tau M^2 \sin^2 \frac{\pi\nu}{2M}} \tag{4.3.31}$$

For the error ε^K_{jk} we obtain the expression[10]

$$\varepsilon^K_{jk} = \sum_{r,s=1}^{M-1} c^K_{rs} 2\sin\frac{r\pi j}{M} \sin\frac{s\pi k}{M} \tag{4.3.32}$$

where

$$c_{rs}^K = \left[\prod_{j=1}^K \lambda_r(\tau_j)\lambda_s(\tau_j)\right] c_{r,s}^0 \qquad (4.3.33)$$

It follows from (4.3.31) and (4.3.32) that the sequence $\tau_1, \ldots, \tau_K$ at which the quantity

$$\max_{r,s}\left|\prod_{j=1}^K \lambda_r(\tau_j)\lambda_s(\tau_j)\right|$$

takes the least value is optimal.

Let us show that by specifying an arbitrary positive q, $q < 1$, one can choose the iteration parameters $\tau_1, \tau_2, \ldots, \tau_K$ in such a way that the inequalities

$$|[\lambda_r(\tau_1)\lambda_s(\tau_1)][\lambda_r(\tau_2)\lambda_s(\tau_2)]\ldots[\lambda_r(\tau_K)\lambda_s(\tau_K)]| < q$$
$$r, s = 1, 2, \ldots, M-1 \qquad (4.3.34)$$

are satisfied. Then $\| \varepsilon^K \| \leq q \| \varepsilon^0 \|$. It is obvious that $|\lambda_i(\tau)| < 1$, $i = 1, \ldots, M-1$, $\tau > 0$. In order to satisfy the inequality (4.3.34) it is therefore sufficient that at least one of the K factors $\lambda_\nu(\tau_p)$, $p = 1, 2, \ldots, K$, satisfies the inequality (see also Eq. (4.3.32))

$$|\lambda_\nu(\tau_p)| = \left|\frac{1 - 2\tau_p M^2 \sin^2 \frac{\pi\nu}{2M}}{1 + 2\tau_p M^2 \sin^2 \frac{\pi\nu}{2M}}\right| \leq \sqrt{q} \qquad (4.3.35)$$

All the numbers $2M^2 \sin^2 \frac{\pi\nu}{2M}$, $\nu = 1, 2, \ldots, M-1$, belong to the interval

$$a = 0.5\pi^2 \leq \mu \leq 2M^2 = b \qquad (4.3.36)$$

Then it is sufficient for the satisfaction of inequality (4.3.34) that for each value of μ from the interval (4.3.36) the inequalities

$$-\sqrt{q} < \frac{1 - \tau_p \mu}{1 + \tau_p \mu} < \sqrt{q} \qquad (4.3.37)$$

are satisfied at least for one value of τ from the set $\{\tau_1, \tau_2, \ldots, \tau_K\}$. It is obvious that the inequalities

$$-\sqrt{q} \leq 1 - \tau_p \mu \leq \sqrt{q} \qquad (4.3.38)$$

imply the inequalities (4.3.37). We can rewrite the inequalities (4.3.38) as follows:

$$1 - \sqrt{q} \leq \tau_p \mu \leq 1 + \sqrt{q} \qquad (4.3.39)$$

Let us specify the μ_p and τ_p by the formulas

$$\mu_p = \Big(\frac{1+\sqrt{q}}{1-\sqrt{q}}\Big)^{p-1} a, \ \ p = 1, 2, \ldots, K$$

$$\tau_p = \frac{1-\sqrt{q}}{\mu_p}, \ \ \ p = 1, 2, \ldots, K \qquad (4.3.40)$$

Then the number $\tau_p \mu$ belongs to the interval (4.3.36) as μ increases from μ_p to μ_{p+1}. Let us choose the value of K from the condition $\mu_K \geq b$, that is

$$K \geq A ln\frac{b}{a} + 1 = A(2ln\ M + ln\frac{4}{\pi^2}) + 1, \ A = \Big[ln\frac{1+\sqrt{q}}{1-\sqrt{q}}\Big]^{-1}$$

Then we obtain the needed sequence $\tau_1, \tau_2, \ldots, \tau_K$ by formula (4.3.40).

Let us now briefly consider a question on the optimal choice of the cyclic sequences $\{\tau_1, \ldots, \tau_K\}$ in the case of using the *approximate factorization* schemes for the pseudo-unsteady numerical solution of two-dimensional elliptic problems. The optimal values of the time steps τ_p, $p = 1, \ldots, K$, are then computed by the formula

$$\tau = \tau_p = \alpha_h \Big(\frac{\alpha_l}{\alpha_h}\Big)^{(p-1)/(K-1)}, \ \ p = 1, 2, \ldots, K \qquad (4.3.41)$$

where $0 < \alpha < \alpha_h$, $K > 1$, and the constants α_l, α_h and K are specified by the user. The optimum value of K ranges between 6 and 20 depending on the grid used (i.e. depending on M and N)[11,12]; $2 \leq \alpha_h \leq 4$, $\alpha_l = ch$, $c > 0$.

4.4 THE FINITE ELEMENT METHOD

The finite element method (FEM) has gained widespread acceptance as an efficient method for the solution of the various problems of the mathematical physics and technology. Such a popularity of the method can be explained by the simplicity of its physical interpretation, of its mathematical form and versatile numerical algorithms facilitating the programming of complex problems.

The finite element method was considered at the initial stage of its development as a specialized engineering procedure for the computation of the stresses and displacements in the analysis of structures. First attempts to derive properties of finite elements of a continuum date back to 1941.[13,14] The traditional engineering interpretation of the finite element method is presented in.[15]

G.I. Marchuk[16] ranks the FEM in the class of the *variational difference methods*, to which the methods of Ritz and Galerkin also belong.

B.G. Galerkin had published his method in 1915.[17] The applications of the FEM to the fluid mechanics problems are considered in References 17 to 20. In the opinion of the authors of[21] the finite element method has the advantages over the finite difference methods when the tasks of the numerical integration of the elliptic PDEs are considered. The relations between the FEM and the theory of splines are discussed in Reference 22.

4.4.1   The Basic Notions

Let us illustrate the basic ideas of the finite element method at the example of the elliptic boundary-value problem

$$\Delta u = -f(x, y), \quad (x, y) \in D \qquad (4.4.1)$$

$$u|_\Gamma = 0 \qquad (4.4.2)$$

where

$$D = (a, b) \times (c, d), \quad -\infty < a < b < \infty, \quad -\infty < c < d < \infty \qquad (4.4.3)$$

Γ is the boundary of the spatial region D, and the Laplacian operator Δ is defined by the formula (4.1.2).

It is well known[23] that it is possible to consider instead of the problem (4.4.1), (4.4.2) an equivalent variational problem on the minimization of certain functional. Let us find the form of such a functional for the case of the problem (4.4.1), (4.4.2). Let us assume that the general form of this functional is as follows:

$$J(u) = \iint_D F(x, y, u, u_x, u_y) dx\, dy \qquad (4.4.4)$$

The necessary condition for the extremum of $J(u)$ is well known and consists of the fact that the function $u(x, y)$ should satisfy the Euler-Lagrange equation

$$\frac{\partial}{\partial x} F_{u_x} + \frac{\partial}{\partial y} F_{u_y} - F_u = 0 \qquad (4.4.5)$$

In accordance with Eq. (4.4.1) we now require that

$$\frac{\partial}{\partial x} F_{u_x} = \frac{\partial}{\partial x} u_x, \quad \frac{\partial}{\partial y} F_{u_y} = \frac{\partial}{\partial y} u_y, \quad F_u = -f(x, y)$$

We easily find from these conditions that the choice of the function F in (4.4.3) in the form $F = \frac{1}{2}(u_x^2 + u_y^2) - fu$ reverts Eq. (4.4.4) into the Poisson equation (4.4.1).

Let us introduce the Sobolev space $\overset{o}{W_2^1}(D)$ consisting of the functions satisfying the zero boundary conditions (4.4.2) and possessing the first derivatives whose squares are integrable over D. The norm in the space $\overset{o}{W_2^1}(D)$ is determined by the formula

$$\| u \|_{\overset{o}{W_2^1}(D)} = \left(\iint_D \left[u^2 + \left(\frac{\partial u}{\partial x}\right)^2 + \left(\frac{\partial u}{\partial y}\right)^2 \right] dx\, dy \right)^{1/2}$$

Then the boundary-value problem (4.4.1), (4.4.2) is equivalent to the following variational problem:

find the function $u \in \overset{o}{W_2^1}(D)$ such that

$$J(u) = \inf_{v \in \overset{o}{W_2^1}(D)} J(v) \tag{4.4.6}$$

where

$$J(v) = \iint_D \left[\left(\frac{\partial v}{\partial x}\right)^2 + \left(\frac{\partial v}{\partial y}\right)^2 - 2fv \right] dx\, dy \tag{4.4.7}$$

In order to construct an approximate solution of the minimization problem (4.4.6), (4.4.7) one must first of all perform the partitioning of the region D into a finite number of elementary cells in accordance with the general scheme of the finite element method. This can be done in different ways. Since in our case the region D given by the formula (4.4.3) is rectangular it can be partitioned into the rectangular and/or triangular elements.

The partitioning of a spatial region D into triangular cells has the advantage of a more accurate approximation of the curved boundaries in the case when the region D has geometrically complex boundary Γ.[14,20] In this connection we consider below the case when the rectangular region D is partitioned into a finite set of equal triangles. In order to obtain such an uniform triangular mesh let us at first introduce in $\bar{D} = D \bigcup \Gamma$ a rectangular uniform mesh by the formulas

$$\begin{aligned} x_i &= a + (i-1)h_1, \quad i = 1, \ldots, M_1 \\ y_j &= c + (j-1)h_2, \quad j = 1, \ldots, M_2 \end{aligned} \tag{4.4.8}$$

where

$$h_1 = (b-a)/(M_1 - 1), \quad h_2 = (d-c)/(M_2 - 1)$$

Now let us subdivide each elementary rectangle

$$d_{ij} = [x_i, x_{i+1}] \times [y_j, y_{j+1}], \quad i = 1, \ldots, M_1 - 1; \; j = 1, \ldots, M_2 - 1$$

into two triangles by the diagonal passing through the vertices (x_i, y_j) and (x_{i+1}, y_{j+1}), see Fig. 4.1.

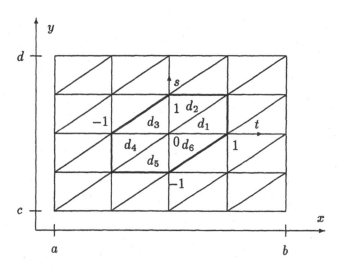

Figure 4.1: Triangular mesh in the rectangular region.

The approximate solution $u_n(x, y)$ of the boundary-value problem (4.4.1), (4.4.2) is determined in the finite element method in the form

$$u_n(x, y) = \sum_{i=2}^{M_1-1} \sum_{j=2}^{M_2-1} v_{ij} \psi_{ij}(x, y) \qquad (4.4.9)$$

where $n = (M_1 - 2)(M_2 - 2)$. The functions $\psi_{ij}(x, y)$ are called the *basis functions*. The coefficients v_{ij} represent the unknown values of the approximate solution $u_n(x, y)$ in the nodes of the mesh (4.4.8). The form of the basis functions $\psi_{ij}(x, y)$ in (4.4.9) is chosen from the following requirements:

(a) all the elements $\psi_{ij}(x, y)$ belong to the space $\overset{o}{W}_2^1 (D)$;

(b) at any $M_1 > 2$, $M_2 > 2$ the elements $\psi_{ij}(x, y)$ are linearly independent;

(c) the sequence $\{\psi_{ij}\}$ is complete in $\overset{o}{W}_2^1 (D)$, that is for any element $u \in \overset{o}{W}_2^1 (D)$ and $\varepsilon > 0$ there exists such a natural number N and the element (4.4.9) that

$$\| u - u_n \|_{\overset{o}{W}_2^1(D)} < \varepsilon$$

for all $n > N$.

There were proposed in the literature on the finite element method many specific forms of the basis functions. These choices affect the accuracy of the obtained FEM solutions and the form of an algebraic system for computing the coefficients v_{ij} of the linear combination (4.4.9). For example, in [22] the basis functions $\psi_{ij}(x, y)$ on an uniform rectangular mesh were taken as the products of the one-dimensional first-degree B-splines of the variables x and y. A detailed discussion of the various basis functions can be found in References 15,17,21.

In what follows we consider one of the simplest forms of the basis functions $\psi_{ij}(x, y)$, namely, the continuous piecewise linear functions of two variables, which are linear over any elementary triangle of the above introduced triangular mesh and which vanish on the boundary Γ of the region D. These basis functions have the form[22]

$$\psi_{ij}(x, y) = \psi\left(\frac{x - x_i}{h_1}, \frac{y - y_j}{h_2}\right), \quad i = 2, \ldots, M_1 - 1; \ j = 2, \ldots, M_2 - 1$$

$$(4.4.10)$$

where

$$\psi(t, s) = \begin{cases} 1 - t, & (t, s) \in d_1 \\ 1 - s, & (t, s) \in d_2 \\ t + 1 - s, & (t, s) \in d_3 \\ 1 + t, & (t, s) \in d_4 \\ 1 + s, & (t, s) \in d_5 \\ 1 - t + s, & (t, s) \in d_6 \\ 0, & (t, s) \notin d_j, \ j = 1, \ldots, 6 \end{cases} \quad (4.4.11)$$

The triangles d_j are shown in Fig. 4.1, and the form of the function $\psi(t, s)$ is shown in Fig. 4.2.

Let us write Eq. (4.4.1) in the operator form as

$$Au = f \qquad (4.4.12)$$

where $A = -\Delta$, and let us introduce the inner product

$$(u, v) = \iint_D uv \, dx \, dy$$

We can now observe that

$$\iint_D \left(-\frac{\partial^2 u}{\partial x^2}\right) u \, dx \, dy = -\oint_\Gamma u \frac{\partial u}{\partial x} dy + \iint_D \left(\frac{\partial u}{\partial x}\right)^2 dx \, dy = \iint_D \left(\frac{\partial u}{\partial x}\right)^2 dx \, dy$$

by virtue of the zero boundary condition (4.4.2). Therefore, we can write the functional (4.4.7) in the form

$$J(v) = (Av, v) - 2(f, v) \qquad (4.4.13)$$

Let us now substitute the expression (4.4.9) for $u_n(x, y)$ instead of v into the functional (4.4.13). Then we obtain a function of n dependent

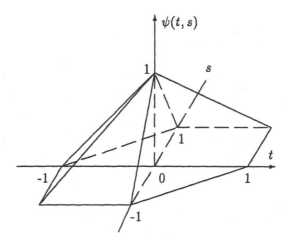

Figure 4.2: The piecewise linear function (4.4.11).

variables v_{ij}:

$$J(u_n) = \Big(\sum_{i=2}^{M_1-1} \sum_{j=2}^{M_2-1} v_{ij} A\psi_{ij}, \ \sum_{p=2}^{M_1-1} \sum_{q=2}^{M_2-1} v_{pq}\psi_{pq} \Big) -$$

$$2\Big(\sum_{p=2}^{M_1-1} \sum_{q=2}^{M_2-1} v_{pq}\psi_{pq}, \ f \Big) =$$

$$\sum_{i=2}^{M_1-1} \sum_{j=2}^{M_2-1} \Big[\sum_{p=2}^{M_1-1} \sum_{q=2}^{M_2-1} (A\psi_{ij}, \psi_{pq}) v_{pq} \Big] v_{ij} -$$

$$2 \sum_{p=2}^{M_1-1} \sum_{q=2}^{M_2-1} (\psi_{pq}, f) v_{pq} \qquad (4.4.14)$$

We now choose the coefficients v_{ij} in such a way that the function (4.4.14) takes the minimum value over all the v_{ij}, $i = 2, \ldots, M_1 - 1$, $j = 2, \ldots, M_2 - 1$. The necessary condition for the extremum of the function

$$J(u_n) = \tilde{J}(v_{22}, v_{23}, \ldots, v_{M_1-1,M_2-1})$$

is that all its partial derivatives with respect to these variables are equal to zero:

$$\frac{\partial J(u_n)}{\partial v_{ij}} = 0, \ i = 2, \ldots, M_1 - 1, \ j = 2, \ldots, M_2 - 1 \qquad (4.4.15)$$

It is easy to compute the partial derivatives in the left-hand side of Eq. (4.4.15) by using the expression (4.4.14). As a result we obtain from Eq. (4.4.15) the following system of the linear algebraic equations for determining v_{ij}:

$$\sum_{p=2}^{M_1-1} \sum_{q=2}^{M_2-1} (A\psi_{pq}, \psi_{ij})v_{pq} = (f, \psi_{ij})$$

$$i = 2, \ldots, M_1 - 1; \; j = 2, \ldots, M_2 - 1 \qquad (4.4.16)$$

In accordance with the definition (4.4.8) of the basis functions $\psi_{ij}(x, y)$ the entries $(A\psi_{pq}, \psi_{ij})$ of the matrix of the system (4.4.16) are equal to zero, if $|i-p| > 1$ and/or $|j-q| > 1$. Consequently, the matrix of system (4.4.16) will be a block tridiagonal matrix. Let us now compute directly the inner products in (4.4.16):

$$(A\psi_{pq}, \psi_{ij}) = \iint_D \left[\frac{\partial}{\partial x}\psi_{pq} \frac{\partial}{\partial x}\psi_{ij} + \frac{\partial}{\partial y}\psi_{pq} \frac{\partial}{\partial y}\psi_{ij} \right] dx\, dy \qquad (4.4.17)$$

As a result we arrive at the following five-point difference scheme:

$$(\delta_x^2 + \delta_y^2)v_{ij} = -g_{ij}, \; i = 2, \ldots, M_1 - 1, \; j = 2, \ldots, M_2 - 1 \quad (4.4.18)$$

$$v_{i1} = v_{i,M_2} = 0, \quad i = 1, \ldots, M_1$$
$$v_{1j} = v_{M_1,j} = 0, \quad j = 1, \ldots, M_2 \qquad (4.4.19)$$

where

$$\delta_x^2 v_{ij} = \frac{v_{i+1,j} - 2v_{ij} + v_{i-1,j}}{h_1^2}, \; \delta_y^2 v_{ij} = \frac{v_{i,j+1} - 2v_{ij} + v_{i,j-1}}{h_2^2}$$

$$g_{ij} = \iint_{d_{ij}} f(x, y)\psi_{ij}(x, y)dx\, dy \qquad (4.4.20)$$

and $d_{ij} = \bigcup_{k=1}^{6} d_k$, see Fig. 4.1. We can see that the obtained scheme (4.4.18) of the finite element method coincides with the accuracy up to the right hand sides with the conventional difference scheme (4.2.4). This enables one to use the efficient iterative schemes for the solution of the corresponding systems, for example, the ADI method (see Section 4.3.3), the splitting methods, the successive over-relaxation method, etc. A detailed discussion of the iterative methods may be found in Reference 18.

Let us rewrite the system (4.4.18) in the form

$$\frac{v_{i-1,j}}{h_1^2} - \left(\frac{2v_{ij}}{h_1^2} - \frac{v_{i,j-1} - 2v_{ij} + v_{i,j+1}}{h_2^2} \right) + \frac{v_{i+1,j}}{h_1^2} = -g_{ij}$$

$$i = 2, 3, \ldots, M_1 - 1; \; j = 2, 3, \ldots, M_2 - 1 \quad (4.4.21)$$

We now take into account the boundary conditions (4.4.19). Then we obtain the system of equations

$$\frac{v_{i-1,1}}{h_1^2} - \left(\frac{2v_{i,1}}{h_1^2} - \frac{-2v_{i,1} + v_{i,2}}{h_2^2}\right) + \frac{v_{i+1,1}}{h_1^2} = -g_{i1}$$

$$\frac{v_{i-1,j}}{h_1^2} - \left(\frac{2v_{ij}}{h_1^2} - \frac{v_{i,j-1} - 2v_{ij} + v_{i,j+1}}{h_2^2}\right) + \frac{v_{i+1,j}}{h_1^2} = -g_{ij}$$

$$j = 2, 3, \ldots, M_2 - 2$$

$$\frac{v_{i-1,M_2-1}}{h_1^2} - \left(\frac{2v_{i,M_2-1}}{h_1^2} - \frac{v_{i,M_2-2} - 2v_{i,M_2-1}}{h_2^2}\right) +$$

$$\frac{v_{i+1,M_2-1}}{h_1^2} = -g_{i,M_2-1}, \quad i = 2, 3, \ldots, M_1 - 1 \qquad (4.4.22)$$

Let us now introduce the vectors

$$\mathbf{V}_i = (v_{i1}, v_{i2}, \ldots, v_{i,M_2-1})^T, \quad \mathbf{F}_i = (g_{i1}, g_{i2}, \ldots, g_{i,M_2-1})^T$$

$$i = 2, 3, \ldots, M_1 - 1$$

and the tridiagonal $(M_2 - 2) \times (M_2 - 2)$ matrix

$$B = \frac{1}{h_2^2}\begin{pmatrix} -2 & 1 & 0 & 0 & \ldots & 0 \\ 1 & -2 & 1 & 0 & \ldots & 0 \\ 0 & 1 & -2 & 1 & \ldots & 0 \\ \ldots & \ldots & \ldots & \ldots & \ldots & 0 \\ \ldots & \ldots & \ldots & \ldots & 1 & -2 \end{pmatrix} \qquad (4.4.23)$$

Let I_{M_2-2} be the identity matrix of the dimension $(M_2 - 2) \times (M_2 - 2)$. Then the system of equations (4.4.21) can be written in the form

$$\frac{1}{h_1^2}I_{M_2-1}\mathbf{V}_{i-1} - \left(\frac{2}{h_1^2}I_{M_2-1} - B\right)\mathbf{V}_i + \frac{1}{h_1^2}I_{M_2-1}\mathbf{V}_{i+1} = -\mathbf{F}_i \quad (4.4.24)$$

$$i = 2, 3, \ldots, M_1 - 1$$

It can be seen from Eq. (4.4.23) that the matrix B possesses the property of the diagonal dominance. This implies that an iterative solution of Eq. (4.4.18) is possible. It can be constructed by first defining a *residual* R_h of the algebraic equation (4.4.18). At an arbitrary choice of the values v_{ml} in (4.4.18) R_h will not be zero. The solution of Eq. (4.4.18) will make $R_h = 0$. The residual R_h can be written as follows:

$$R_h = \frac{v_{i+1,j} - 2v_{ij} + v_{i-1,j}}{h_1^2} + \frac{v_{i,j+1} - 2v_{ij} + v_{i,j-1}}{h_2^2} + g_{ij} \quad (4.4.25)$$

Let us apply the method of *successive over-relaxation* (SOR) for the numerical solution of the equation $R_h = 0$. The SOR method is known to be one of the fastest iterative methods for the solution of block tridiagonal systems. In this method the iterative solution for the last level of iteration "$n + 1$" ($n = 0, 1, 2, \ldots$) is computed by the formula [18]

$$v_{ij}^{(n+1)} = \lambda v_{ij}^* + (1 - \lambda)v_{ij}^{(n)} \qquad (4.4.26)$$

where the solution v_{ij}^* is obtained with the aid of the *Gauss-Seidel method* [18]. In order to apply the Gauss-Seidel method to the solution of equation $R_h = 0$ let us at first solve this equation with respect to v_{ij}. From the formula (4.4.25) for R_h we easily obtain that

$$v_{ij} = \frac{1}{2(\frac{1}{h_1^2} + \frac{1}{h_2^2})} \left[\frac{v_{i-1,j} + v_{i+1,j}}{h_1^2} + \frac{v_{i,j-1} + v_{i,j+1}}{h_2^2} + g_{ij} \right]$$

The value v_{ij}^* is determined in accordance with the Gauss-Seidel scheme as

$$v_{ij}^* = \frac{1}{2(\frac{1}{h_1^2} + \frac{1}{h_2^2})} \left[\frac{v_{i-1,j}^{(n+1)} + v_{i+1,j}^{(n)}}{h_1^2} + \frac{v_{i,j-1}^{(n+1)} + v_{i,j+1}^{(n)}}{h_2^2} + g_{ij} \right] \qquad (4.4.27)$$

The value λ in Eq. (4.4.26) is the *relaxation parameter*, which is introduced to accelerate the convergence; $0 < \lambda < 2$. At $\lambda = 1$ the iterative scheme (4.4.26) reduces to the Gauss-Seidel method.

By using the boundary conditions $v_{i,1} = 0$ and $v_{t,j} = 0$ we can find from Eq. (4.4.27) the value $v_{2,2}^*$; after that we can compute at fixed $j = 2$ the value $v_{3,2}$, etc. When we have already computed the values $v_{i,1}^{n+1}$ for $i = 2, \ldots, M_1 - 1$, we can take the next value of j, $j = 3$, and find successively the values $v_{i,3}^{n+1}$ by using already found values $v_{i,2}^{n+1}$ and the boundary conditions (4.4.19), etc.

There exists for each specific problem an optimal value λ_{opt} of the relaxation parameter λ in Eq. (4.4.26), at which the convergence of the iterative process is achieved after the least number of iteration steps. This value depends, in particular, on the mesh refinement, that is on the values of M_1 and M_2.[18] The conventional criterion for the convergence of an iterative process like the process (4.4.26), (4.4.27) is as follows:[18] it is assumed that the convergence is achieved, if $\| R_h^{(n)} \|_{rms} < TOL$, where

$$\| R_h^{(n)} \|_{rms} = \left[\frac{1}{(M_1 - 2)(M_2 - 2)} \sum_{i=2}^{M_1-1} \sum_{j=2}^{M_2-1} (R_h)_{ij}^2 \right]^{\frac{1}{2}} \qquad (4.4.28)$$

The tolerance TOL is a user-specified small positive number, for example, $TOL = 10^{-5}$, $TOL = 10^{-6}$.

4.4.2   The Case of Inhomogeneous Dirichlet Boundary Condition

We have considered above the implementation of the finite element method in the case of the zero Dirichlet boundary condition (4.4.2). The zero boundary conditions are called the *homogeneous* boundary conditions, otherwise they are called *inhomogeneous* boundary conditions. In many applied problems one has to consider the inhomogeneous Dirichlet boundary conditions. Let us take the problem

$$\Delta u = -f(x, y), \quad (x, y) \in D \tag{4.4.29}$$

$$u|_\Gamma = \varphi(x, y) \tag{4.4.30}$$

where $\varphi(x, y)$ is a given function. In the case when $\varphi(x, y) \neq 0$ it is impossible to reduce the problem (4.4.29), (4.4.30) directly to the minimization of the functional (4.4.7). For this case there were developed in the FEM theory a number of the techniques, which enable one to apply the finite element method to the solution of inhomogeneous boundary-value problems.

One of these techniques consists of the reduction of the inhomogeneous boundary-value problem (4.4.29), (4.4.30) to a homogeneous boundary-value problem for the other function $v(x, y)$, which is related to the function $u(x, y)$. Let us briefly describe this technique following Reference 21.

Let us assume that the function $\varphi(x, y)$ in (4.4.29) admits an explicit analytic prolongation inside D, i.e. there exists such a smooth function $w(x, y)$ that

$$w|_\Gamma = \varphi(x, y) \tag{4.4.31}$$

and the function Δw is determined everywhere in D. In this case we can introduce the new dependent variable $v(x, y)$ by the formula

$$v(x, y) = u(x, y) - w(x, y), \quad (x, y) \in D \bigcup \Gamma$$

By virtue of Eqs. (4.4.30) and (4.4.31) we have that the function $v(x, y)$ satisfies the homogeneous Dirichlet boundary condition:

$$v(x, y)|_\Gamma = 0 \tag{4.4.32}$$

Let us now find the form of an elliptic PDE for the function $v(x, y)$. Since $u = v + w$, we have from Eq. (4.4.29) that

$$\Delta v = -f_1(x, y), \quad (x, y) \in D \tag{4.4.33}$$

where $f_1(x, y)$ is a given function,

$$f_1(x, y) = f(x, y) + \Delta w \qquad (4.4.34)$$

Example. Let

$$f(x, y) = 0, \quad \varphi(x, y) = -\frac{1}{2}(x^2 + y^2)$$

Let us take the function $w(x, y)$ in the form

$$w(x, y) = -\frac{1}{2}(x^2 + y^2), \quad -\infty < x, y < \infty$$

Then we arrive at the equation (4.4.33) with $f_1(x, y) = -2$.

In many practical problems the Dirichlet boundary conditions are insufficiently smooth to ensure the possibility of an explicit analytical prolongation. In such cases the inhomogeneous Dirichlet boundary conditions are added to the basic functional (4.4.7) in the form of the so-called *penalty functional*, and then instead of (4.4.7) the functional

$$J_\lambda(v) = J(v) + \lambda \int_\Gamma (v - \varphi)^2 d\Gamma$$

is minimized, where λ is a positive penalty coefficient[21]

Another approach to the treatment of the inhomogeneous boundary conditions is based on the use of the representation (4.4.9) for the approximate solution, where some of the coefficients v_{ij} are chosen in such a way that the approximate solution interpolates the boundary conditions. The remaining parameters v_{ij} corresponding to the internal nodes are computed with the aid of the Ritz method. The approximations of such type do not belong to the classical variational formulation of the finite element method. However, it was shown[21] that the accuracy of the approximate solution obtained with the aid of such a procedure is not worse than in the case of the classical FEM.

4.4.3 The Program fem1.ma

Let us consider the boundary-value problem (4.4.1), (4.4.2) with $f(x, y) = -2$. Let us take in (4.4.3)

$$a = -\frac{\pi}{2}, \quad b = \frac{\pi}{2}, \quad c = -\frac{\pi}{2}, \quad d = \frac{\pi}{2} \qquad (4.4.35)$$

Then the exact solution of the problem (4.4.1)-(4.4.3), (4.4.35) has the form[21]

$$u(x, y) = -\left(\frac{\pi}{2}\right)^2 + x^2 +$$

$$\frac{8}{\pi} \sum_{k=1}^{\infty} \frac{(-1)^{k+1}}{(2k-1)^3} \frac{\cosh((2k-1)y)}{\cosh((2k-1)\pi/2)} \cos((2k-1)x) \qquad (4.4.36)$$

We have implemented the FEM scheme (4.4.18)-(4.4.20) as applied to the numerical solution of the Poisson equation

$$u_{xx} + u_{yy} - 2 = 0 \qquad (4.4.37)$$

in the *Mathematica* program fem1.ma. The organization of the computation in this program is similar to that of the program mc2d.ma from Section 2.12. The complete listing of this program is presented in Appendix 2. The input parameters used in the program fem1.ma are described in Table 4.1.

Table 4.1. Parameters used in program fem1.ma

Parameter	Description
M1	the number M_1 of the grid nodes on the x-axis, $M_1 > 2$
M2	the number M_2 of the grid nodes on the y-axis, $M_2 > 2$
sor	the relaxation parameter λ entering the successive over-relaxation method, $0 < \lambda < 2$
TOL	the user-specified tolerance entering the criterion for the termination of the SOR iterations, $TOL > 0$
xview, yview, zview	the coordinates of the viewpoint for the surface $u = u(x, y)$. At least one of the coordinates xview, yview, zview should be greater than 1 in modulus

The main function of the file fem1.ma is

$$\text{fem1}\,[M_1, M_2, \lambda, xview, yview, zview].$$

The exact solution (4.4.36) was programmed in the form of the function

```
uex[x_, y_] :=
(* --- The function to compute the exact solution of the
       Poisson equation ------------------------------- *)
( z1 = zpi + x*x; z0 = -1.0; z2 = 0.0;
  Do[ z0 = -z0; k1 = 2*k - 1;
      zk = N[Cosh[k1*y]*Cos[k1*x]/Cosh[k1*Pi/2]];
  zk = z0*zk/(k1^3); z2 = z2 + zk, {k, nt}];
  z1 = z1 + zk1*z2; z1 )
```

The *Mathematica* function Cosh[x] computes the value of the hyperbolic cosine of x, that is

$$\cosh(x) = \frac{1}{2}(e^x + e^{-x})$$

The number of terms in the expansion (4.4.36) was determined from the requirement that the error produced by the truncation of the series (4.4.36) does not exceed 10^{-7}. This has led to the necessity of taking the first 108 terms of the series (the corresponding value nt = 108 is shown on the screen of the computer monitor as our program fem1.ma starts working).

The SOR method given by Eqs. (4.4.26), (4.4.27) was implemented in the lines

```
(* -------- Do-loop over the (i,j) nodes ---------- *)
 Do[
     Do[ vs = c1*(vp[[i-1, j]] + v[[i+1, j]]
 + cel2*(vp[[i, j-1]] + v[[i, j+1]]) - h12);
  vp[[i, j]] = sor*vs + (1-sor)*v[[i, j]],
   {i, 2, m1m}],
  {j, 2, m2m}];
(* ------------ Exchange of the lists ------------ *)
v = vp; Rhnorm = 0;
```

of the program fem1.ma. The relaxation parameter λ was taken to be equal to 1.50 for all computational examples presented below.

Since $f(x, y) = const$ in the case of Eq. (4.4.37), we have from Eq. (4.4.20) that

$$g_{ij} = -2 \iint_{d_{ij}} \psi_{ij}(x, y)dx\,dy \qquad (4.4.38)$$

We can calculate the integral in Eq. (4.4.38) with the aid of the *Mathematica* function

$$\text{Integrate}[f, \{x, xmin, xmax\}, \{y, ymin, ymax\}]$$

where $f(x, y)$ is the integrand. The y integral is done first. Its limits can depend on the value of x. For example, let us consider the item

$$s_6 = \iint_{d_6} \psi_{ij}(x, y)dx\,dy \qquad (4.4.39)$$

entering the integral in Eq. (4.4.38). With regard for Fig. 4.1 and formula (4.4.9) it is easy to determine the integration limits in the integral (4.4.39):

$$s_6 = \int_0^1 dt \int_{t-1}^0 (1 - t + s)ds \qquad (4.4.40)$$

Let us perform a small *Mathematica* session:

```
In[1]:= Integrate[1-t+s,{t,0,1}, {s,t-1,0}]
```

The result of the integration immediately appears on the monitor screen:

```
          1
Out[1]= ---
          6
```

In this way it easy to calculate that

$$\iint_{d_{ij}} \psi_{ij}(x, y)dx\, dy = 1$$

Therefore, we have simply that $g_{ij} = -2$ for all i, j. The use of the *Mathematica* function **Integrate** becomes especially efficient in cases when the integrals involved in the inner products (4.4.17) include the derivatives of more complex basis functions than the basis function of the form (4.4.9).

Table 4.2 The effect of the mesh refinement and of the value of TOL on the solution error δu

TOL	M_1	M_2	δu
10^{-3}	11	11	0.0076535
10^{-4}	11	11	0.0074898
10^{-5}	15	15	0.0037297
10^{-5}	17	17	0.0028364

In Table 4.2 we present some computational results obtained by the FEM scheme (4.4.18)-(4.4.20) at different values of M_1, M_2 and TOL. The value of the relaxation parameter λ in Eq. (4.4.26) was $\lambda = 1.50$ for all the runs. The mean-square solution error δu was computed by the formula

$$\delta u = \left(\frac{1}{(M_1 - 2)(M_2 - 2)} \sum_{i=2}^{M_1-1} \sum_{j=2}^{M_2-1} [v_{ij} - u_{ex}(x_i, y_j)]^2\right)^{0.5}$$

where $u_{ex}(x, y)$ is the exact solution (4.4.36). We can see from Table 4.2 that in the case of the 11×11 mesh the diminution of TOL from 10^{-3} to

10^{-4} reduces only slightly the solution error δu. This can be explained by the fact that for the given mesh steps h_1 and h_2 the approximation error of the FEM scheme used is $O(h_1^2) + O(h_2^2)$. Therefore, in order to improve further the solution accuracy one has to take smaller mesh steps h_1 and h_2 by increasing the values of M_1 and M_2. The needed computer time is proportional to the product $M_1 M_2$.

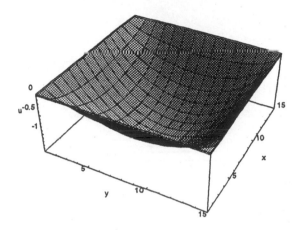

Figure 4.3: The numerical solution surface $v = v(x, y)$ obtained on the 15×15 mesh.

In Fig. 4.3 we present the numerical solution surface $v = v(x, y)$ obtained at $TOL = 10^{-5}$.

The *Mathematica* function `ListContourPlot[v]` makes a contour plot from an array of heights. In Fig. 4.4 we show the contour plot corresponding to the same numerical solution $v = v(x, y)$ as in Fig. 4.3.

We have also produced the color map of the obtained numerical solution (see Fig. 4.5). For this purpose we have augmented our program `fem1.m` by the color directives similarly to the program `mc2dc.ma` from Section 2.12. The corresponding file `fem2.ma` is available on the attached diskette.

Exercise 4.3. Let $\psi_{ij}(x, y) \in \overset{o}{W}{}_2^1 (D)$, $\psi_{pq}(x, y) \in \overset{o}{W}{}_2^1 (D)$, and $A = -(\partial^2/\partial x^2 + \partial^2/\partial y^2)$. Show that $(A\psi_{ij}, \psi_{pq}) = (A\psi_{pq}, \psi_{ij})$.

Exercise 4.4. Determine the optimal value λ_{opt} of the relaxation parameter λ in Eq. (4.4.22) at $M_1 = M_2 = 15$, $TOL = 10^{-5}$ with the aid of a number of numerical experiments.

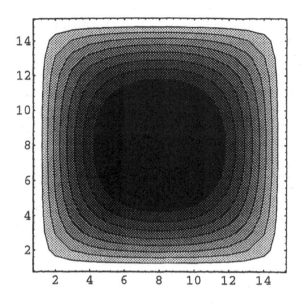

Figure 4.4: The contour plot of the numerical solution on the 15 × 15 mesh.

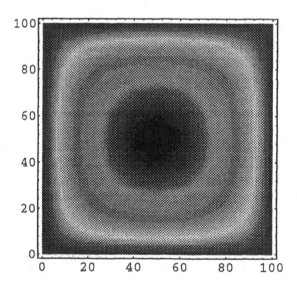

Figure 4.5: The color map of the numerical solution on the 15 × 15 mesh.

Hint. At first take λ in the interval $1.0 < \lambda < 1.5$, and after that try the values of λ from the interval $1.5 < \lambda < 2.0$. In this way one can organize a kind of a bisection process for determining λ_{opt}.

Exercise 4.5. Let us take an uniform rectangular mesh determined by Eqs. (4.4.8). Let us take the basis functions $\psi_{ij}(x, y)$ in the form

$$\psi_{ij}(x, y) = \varphi\left(\frac{x - x_i}{h_1}\right)\varphi\left(\frac{y - y_j}{h_2}\right)$$

where

$$\varphi(t) = \begin{cases} 1 + t, & t \in [-1, 0] \\ 1 - t, & t \in [0, 1] \\ 0, & t \notin [-1, 1] \end{cases}$$

that is $\varphi(t)$ is a piecewise linear function. Find the form of the algebraic system (4.4.16) for the determination of the values v_{pq} by using the *Mathematica* system for the computation of the integrals in Eq. (4.4.17).

4.5   NUMERICAL GRID GENERATION

We have presented earlier in Section 2.14 a simple algebraic method for the numerical grid generation. Now we want to present a grid generation technique, which is based on the numerical solution of certain elliptic PDEs. Let us again consider the problem of curvilinear grid generation inside a nozzle. In this case the spatial coordinates $x(\xi, \eta)$ and $y(\xi, \eta)$ are specified on all four segments of the rectangular region in the (ξ, η) plane, see Fig. 2.40.

The solution of such a boundary-value problem is a classical problem of partial differential equations, so that it is logical to take the coordinates $x(\xi, \eta)$, $y(\xi, \eta)$ to be solutions of a system of partial differential equations. If the coordinate points (and/or slopes) are specified on the entire closed boundary of the physical region, the equations must be elliptic. In cases when the specification is on only a portion of the boundary the equations are taken to be parabolic or hyperbolic.

We consider in this section the case of a completely specified boundary, which requires an elliptic partial differential system. The solutions of some elliptic systems satisfy the extremum principle, that is extremum of solutions cannot occur within the field. This property guarantees a one-to-one mapping between the physical and transformed regions. Since the variation of the curvilinear coordinate along a physical boundary segment must be monotonic, and is over the same range along facing boundary segments, it clearly follows that extrema of the curvilinear coordinates cannot be alllowed in the interior of the physical region, else overlapping of the coordinate system will occur. Therefore, it is clear that the governing PDEs for obtaining the curvilinear grid must be the equations containing the derivatives of ξ and η with respect to the Cartesian coordinates x and y.

Another important property of the elliptic grid generators is the inherent smoothness that prevails in the solution of elliptic systems. Furthermore, boundary slope discontinuities are not propagated into the field. Finally, the smoothing tendencies of elliptic operators, and the extremum principles, allow grids to be generated for any configurations without overlap of grid lines.

The simplest elliptic partial differential system, whose solutions satisfy the maximum principle, is the system of Laplace equations[2]

$$\frac{\partial^2 \xi}{\partial x^2} + \frac{\partial^2 \xi}{\partial y^2} = 0 \qquad (4.5.1)$$

$$\frac{\partial^2 \eta}{\partial x^2} + \frac{\partial^2 \eta}{\partial y^2} = 0 \qquad (4.5.2)$$

These equations can be obtained from the Euler-Lagrange equations for the minimization of the integrals

$$I_1 = \iint_D [(\partial \xi/\partial x)^2 + (\partial \xi/\partial y)^2] dx\, dy \qquad (4.5.3)$$

$$I_2 = \iint_D [(\partial \eta/\partial x)^2 + (\partial \eta/\partial y)^2] dx\, dy \qquad (4.5.4)$$

The grid in the rectqangular region in the (ξ, η) plane is covered by an uniform mesh with spacings $\Delta\xi$ and $\Delta\eta$. Since the coordinate lines are located at equal increments $\Delta\xi$ or $\Delta\eta$ of the curvilinear coordinate, the quantity

$$|\nabla\xi|^2 = (\partial\xi/\partial x)^2 + (\partial\xi/\partial y)^2$$

can be considered a measure of the grid point density along the coordinate line on which ξ varies, i.e., ξ must change rapidly in physical space where grid points are clustered. Minimization of the integral (4.5.3) thus leads to the smoothest coordinate line distribution over the field.

The curvilinear grid is generated in the domain D of the (x, y) plane with the aid of the numerical solution of equations (4.5.1), (4.5.2). Since the transformed region in the (ξ, η) plane has a simpler, rectangular, form, it is more convenient to do the computation in the rectangular transformed field, where the curvilinear coordinates, ξ and η, are the independent variables, with the Cartesian coordinates x and y as dependent variables. For this purpose let us perform the transformation of the equations (4.5.1) and (4.5.2). Let us at first express the derivatives

$$\partial\xi/\partial x, \ \partial\xi/\partial y, \ \partial\eta/\partial x, \ \partial\eta/\partial y$$

in terms of the derivatives

$$\partial x/\partial\xi, \ \partial x/\partial\eta, \ \partial y/\partial\xi, \ \partial y/\partial\eta$$

Consider the identity

$$x(\xi(x,y), \eta(x,y)) \equiv x \tag{4.5.5}$$

Differentiating the both sides of (4.5.5) with respect to x, we obtain the equation

$$\frac{\partial x}{\partial\xi}\frac{\partial\xi}{\partial x} + \frac{\partial x}{\partial\eta}\frac{\partial\eta}{\partial x} = 1 \tag{4.5.6}$$

The differentiation of the both sides of (4.5.5) with respect to y yields

$$\frac{\partial x}{\partial\xi}\frac{\partial\xi}{\partial y} + \frac{\partial x}{\partial\eta}\frac{\partial\eta}{\partial y} = 0 \tag{4.5.7}$$

Similarly, from the identity

$$y(\xi(x,y), \eta((x,y)) \equiv y$$

we find by differentiation the equations

$$\frac{\partial y}{\partial\xi}\frac{\partial\xi}{\partial x} + \frac{\partial y}{\partial\eta}\frac{\partial\eta}{\partial x} = 0 \tag{4.5.8}$$

$$\frac{\partial y}{\partial\xi}\frac{\partial\xi}{\partial y} + \frac{\partial y}{\partial\eta}\frac{\partial\eta}{\partial y} = 1 \tag{4.5.9}$$

Now take the system of the equations (4.5.6) and (4.5.8) for the derivatives $\partial\xi/\partial x$ and $\partial\eta/\partial x$. The determinant of this system is the Jacobian

$$J = \begin{vmatrix} \frac{\partial x}{\partial\xi} & \frac{\partial x}{\partial\eta} \\ \frac{\partial y}{\partial\xi} & \frac{\partial y}{\partial\eta} \end{vmatrix} = \frac{\partial x}{\partial\xi}\frac{\partial y}{\partial\eta} - \frac{\partial x}{\partial\eta}\frac{\partial y}{\partial\xi} \tag{4.5.10}$$

From the system (4.5.6), (4.5.8) we find the following expressions:

$$\frac{\partial\xi}{\partial x} = \frac{1}{J}\frac{\partial y}{\partial\eta}, \ \frac{\partial\eta}{\partial x} = -\frac{1}{J}\frac{\partial y}{\partial\xi} \tag{4.5.11}$$

Now take the equations (4.5.7) and (4.5.9). Solving this system with respect to $\partial\xi/\partial y$ and $\partial\eta/\partial y$, we obtain the expressions

$$\frac{\partial\xi}{\partial y} = -\frac{1}{J}\frac{\partial x}{\partial\eta}, \quad \frac{\partial\eta}{\partial y} = \frac{1}{J}\frac{\partial x}{\partial\xi} \tag{4.5.12}$$

Let us now find the expression for the Laplacian $(\partial^2/\partial x^2+\partial^2/\partial y^2)\xi(x,y)$ in terms of the derivatives

$$\partial x/\partial\xi,\ldots,\partial y/\partial\eta,\ \partial^2 x/\partial\xi^2,\ \partial^2 x/\partial\xi\partial\eta,\ \partial^2 x/\partial\eta^2,\ \partial^2 y/\partial\xi^2$$
$$\partial^2 y/\partial\xi\partial\eta,\ \partial^2 y/\partial\eta^2$$

For this purpose we differentiate the both sides of equation (4.5.6) with respect to x:

$$\frac{\partial x}{\partial\xi}\frac{\partial^2\xi}{\partial x^2} + \frac{\partial x}{\partial\eta}\frac{\partial^2\eta}{\partial x^2} = -\left[\frac{\partial^2 x}{\partial\xi^2}\left(\frac{\partial\xi}{\partial x}\right)^2 + 2\frac{\partial^2 x}{\partial\xi\partial\eta}\frac{\partial\xi}{\partial x}\frac{\partial\eta}{\partial x} + \frac{\partial^2 x}{\partial\eta^2}\left(\frac{\partial\eta}{\partial x}\right)^2\right] \tag{4.5.13}$$

Now differentiate the both sides of (4.5.8) with respect to x:

$$\frac{\partial y}{\partial\xi}\frac{\partial^2\xi}{\partial x^2} + \frac{\partial y}{\partial\eta}\frac{\partial^2\eta}{\partial x^2} = -\left[\frac{\partial^2 y}{\partial\xi^2}\left(\frac{\partial\xi}{\partial x}\right)^2 + 2\frac{\partial^2 y}{\partial\xi\partial\eta}\frac{\partial\xi}{\partial x}\frac{\partial\eta}{\partial x} + \frac{\partial^2 y}{\partial\eta^2}\left(\frac{\partial\eta}{\partial x}\right)^2\right] \tag{4.5.14}$$

Solving the equations (4.5.13) and (4.5.14) with respect to $\partial^2\xi/\partial x^2$ and $\partial^2\eta/\partial x^2$, we obtain the expressions

$$\frac{\partial^2\xi}{\partial x^2} = \frac{1}{J}\left[\frac{\partial x}{\partial\eta}\left(\frac{\partial^2 y}{\partial\xi^2}\left(\frac{\partial\xi}{\partial x}\right)^2 + \frac{\partial^2 y}{\partial\eta^2}\left(\frac{\partial\eta}{\partial x}\right)^2 + 2\frac{\partial^2 y}{\partial\xi\partial\eta}\frac{\partial\eta}{\partial x}\frac{\partial\xi}{\partial x}\right)\right.$$

$$\left. - \frac{\partial y}{\partial\eta}\cdot\left(\frac{\partial^2 x}{\partial\xi^2}\cdot\left(\frac{\partial\xi}{\partial x}\right)^2 + \frac{\partial^2 x}{\partial\eta^2}\cdot\left(\frac{\partial\eta}{\partial x}\right)^2 + 2\frac{\partial^2 x}{\partial\xi\partial\eta}\frac{\partial\eta}{\partial x}\frac{\partial\xi}{\partial x}\right)\right] \tag{4.5.15}$$

$$\frac{\partial^2\eta}{\partial x^2} = \frac{1}{J}\left[\frac{\partial y}{\partial\xi}\cdot\left(\frac{\partial^2 x}{\partial\xi^2}\left(\frac{\partial\xi}{\partial x}\right)^2 + \frac{\partial^2 x}{\partial\eta^2}\left(\frac{\partial\eta}{\partial x}\right)^2 + 2\frac{\partial^2 x}{\partial\xi\partial\eta}\frac{\partial\eta}{\partial x}\frac{\partial\xi}{\partial x}\right)\right.$$

$$\left. - \frac{\partial x}{\partial\xi}\cdot\left(\frac{\partial^2 y}{\partial\xi^2}\left(\frac{\partial\xi}{\partial x}\right)^2 + \frac{\partial^2 y}{\partial\eta^2}\left(\frac{\partial\eta}{\partial x}\right)^2 + 2\frac{\partial^2 y}{\partial\xi\partial\eta}\frac{\partial\eta}{\partial x}\frac{\partial\xi}{\partial x}\right)\right] \tag{4.5.16}$$

Let us now differentiate the both sides of the equation (4.5.7) with respect to y:

$$\frac{\partial x}{\partial\xi}\frac{\partial^2\xi}{\partial y^2} + \frac{\partial x}{\partial\eta}\frac{\partial^2\eta}{\partial y^2} = -\left[\frac{\partial^2 x}{\partial\xi^2}\left(\frac{\partial\xi}{\partial y}\right)^2 + 2\frac{\partial^2 x}{\partial\xi\partial\eta}\frac{\partial\xi}{\partial y}\frac{\partial\eta}{\partial y} + \frac{\partial^2 x}{\partial\eta^2}\left(\frac{\partial\eta}{\partial y}\right)^2\right] \tag{4.5.17}$$

Let us differentiate the both sides of the equation (4.5.9) with respect to y:

$$\frac{\partial y}{\partial \xi}\frac{\partial^2 \xi}{\partial y^2} + \frac{\partial y}{\partial \eta}\frac{\partial^2 \eta}{\partial y^2} = -\left[\frac{\partial^2 y}{\partial \xi^2}\left(\frac{\partial \xi}{\partial y}\right)^2 + 2\frac{\partial^2 y}{\partial \xi \partial \eta}\frac{\partial \xi}{\partial y}\frac{\partial \eta}{\partial y} + \frac{\partial^2 y}{\partial \eta^2}\left(\frac{\partial \eta}{\partial y}\right)^2\right] \quad (4.5.18)$$

Solving the equations (4.5.17) and (4.5.18) with respect to $\partial^2 \xi / \partial y^2$ and $\partial^2 \eta / \partial y^2$, we obtain the expressions

$$\frac{\partial^2 \xi}{\partial y^2} = \frac{1}{J}\left[\frac{\partial x}{\partial \eta}\left(\frac{\partial^2 y}{\partial \xi^2}\left(\frac{\partial \xi}{\partial y}\right)^2 + \frac{\partial^2 y}{\partial \eta^2}\left(\frac{\partial \eta}{\partial y}\right)^2 + 2\frac{\partial^2 y}{\partial \xi \partial \eta}\frac{\partial \eta}{\partial y}\frac{\partial \xi}{\partial y}\right)\right.$$

$$\left. - \frac{\partial y}{\partial \eta}\cdot\left(\frac{\partial^2 x}{\partial \xi^2}\cdot\left(\frac{\partial \xi}{\partial y}\right)^2 + \frac{\partial^2 x}{\partial \eta^2}\cdot\left(\frac{\partial \eta}{\partial y}\right)^2 + 2\frac{\partial^2 x}{\partial \xi \partial \eta}\frac{\partial \eta}{\partial y}\frac{\partial \xi}{\partial y}\right)\right] \quad (4.5.19)$$

$$\frac{\partial^2 \eta}{\partial y^2} = \frac{1}{J}\left[\frac{\partial y}{\partial \xi}\cdot\left(\frac{\partial^2 x}{\partial \xi^2}\left(\frac{\partial \xi}{\partial y}\right)^2 + \frac{\partial^2 x}{\partial \eta^2}\left(\frac{\partial \eta}{\partial y}\right)^2 + 2\frac{\partial^2 x}{\partial \xi \partial \eta}\frac{\partial \eta}{\partial y}\frac{\partial \xi}{\partial y}\right)\right.$$

$$\left. - \frac{\partial x}{\partial \xi}\cdot\left(\frac{\partial^2 y}{\partial \xi^2}\left(\frac{\partial \xi}{\partial y}\right)^2 + \frac{\partial^2 y}{\partial \eta^2}\left(\frac{\partial \eta}{\partial y}\right)^2 + 2\frac{\partial^2 y}{\partial \xi \partial \eta}\frac{\partial \eta}{\partial y}\frac{\partial \xi}{\partial y}\right)\right] \quad (4.5.20)$$

Using the formulas (4.5.15) and (4.5.20), we obtain that

$$\frac{\partial^2 \xi}{\partial x^2} + \frac{\partial^2 \xi}{\partial y^2} = \frac{1}{J}\left\{\frac{\partial x}{\partial \eta}\left[\frac{\partial^2 y}{\partial \xi^2}\left(\left(\frac{\partial \xi}{\partial x}\right)^2 + \left(\frac{\partial \xi}{\partial y}\right)^2\right) + \right.\right.$$

$$2\left(\frac{\partial \xi}{\partial x}\frac{\partial \eta}{\partial x} + \frac{\partial \eta}{\partial y}\frac{\partial \xi}{\partial y}\right)\cdot\frac{\partial^2 y}{\partial \xi \partial \eta} +$$

$$\left.\left(\left(\frac{\partial \eta}{\partial x}\right)^2 + \left(\frac{\partial \eta}{\partial y}\right)^2\right)\frac{\partial^2 y}{\partial \eta^2}\right] -$$

$$-\frac{\partial y}{\partial \eta}\cdot\left[\left(\left(\frac{\partial \xi}{\partial x}\right)^2 + \left(\frac{\partial \xi}{\partial y}\right)^2\right)\frac{\partial^2 x}{\partial \xi^2} + 2\frac{\partial^2 x}{\partial \xi \partial \eta}\left(\frac{\partial \eta}{\partial x}\frac{\partial \xi}{\partial x} + \frac{\partial \xi}{\partial y}\frac{\partial \eta}{\partial y}\right) + \right.$$

$$\left.\left.\left(\left(\frac{\partial \eta}{\partial x}\right)^2 + \left(\frac{\partial \eta}{\partial y}\right)^2\right)\frac{\partial^2 x}{\partial \eta^2}\right]\right\} \quad (4.5.21)$$

We can see that the expression (4.5.21) includes also the second derivatives of y with respect to ξ and η. It is possible to eliminate these terms. Let us indeed differentiate the both sides of equation (4.5.9) with respect to y:

$$\frac{\partial^2 y}{\partial \xi^2}\left(\frac{\partial \xi}{\partial y}\right)^2 + 2\frac{\partial^2 y}{\partial \xi \partial \eta}\frac{\partial \xi}{\partial y}\frac{\partial \eta}{\partial y} + \frac{\partial^2 y}{\partial \eta^2}\left(\frac{\partial \eta}{\partial y}\right)^2 = -\frac{\partial^2 \xi}{\partial y^2}\frac{\partial y}{\partial \xi} - \frac{\partial^2 \eta}{\partial y^2}\frac{\partial y}{\partial \eta} \quad (4.5.22)$$

Let us now differentiate the both sides of (4.5.8) with respect to x:

$$\frac{\partial^2 y}{\partial \xi^2}\left(\frac{\partial \xi}{\partial x}\right)^2 + 2\frac{\partial^2 y}{\partial \xi \partial \eta}\frac{\partial \xi}{\partial x}\frac{\partial \eta}{\partial x} + \frac{\partial^2 y}{\partial \eta^2}\left(\frac{\partial \eta}{\partial x}\right)^2 = -\frac{\partial^2 \xi}{\partial x^2}\frac{\partial y}{\partial \xi} - \frac{\partial^2 \eta}{\partial x^2}\frac{\partial y}{\partial \eta} \quad (4.5.23)$$

Adding the both sides of equations (4.5.22) and (4.5.23), we obtain the equation

$$\left[\left(\frac{\partial \xi}{\partial x}\right)^2 + \left(\frac{\partial \xi}{\partial y}\right)^2\right]\frac{\partial^2 y}{\partial \xi^2} + 2\left(\frac{\partial \xi}{\partial x}\frac{\partial \eta}{\partial x} + \frac{\partial \xi}{\partial y}\frac{\partial \eta}{\partial y}\right)\frac{\partial^2 y}{\partial \xi \partial \eta} +$$

$$\left[\left(\frac{\partial \eta}{\partial x}\right)^2 + \left(\frac{\partial \eta}{\partial y}\right)^2\right]\frac{\partial^2 y}{\partial \eta^2}$$

$$= -\frac{\partial y}{\partial \xi}\left(\frac{\partial^2 \xi}{\partial x^2} + \frac{\partial^2 \xi}{\partial y^2}\right) - \frac{\partial y}{\partial \eta}\left(\frac{\partial^2 \eta}{\partial x^2} + \frac{\partial^2 \eta}{\partial y^2}\right) = 0 \quad (4.5.24)$$

since the equations (4.5.1) and (4.5.2) hold. Thus we obtain from Eq. (4.5.21) the following nonlinear partial differential equation with regard to Eq. (4.5.24):

$$\left[\left(\frac{\partial \xi}{\partial x}\right)^2 + \left(\frac{\partial \xi}{\partial y}\right)^2\right]\frac{\partial^2 x}{\partial \xi^2} + 2\frac{\partial^2 x}{\partial \xi \partial \eta}\left(\frac{\partial \eta}{\partial x}\frac{\partial \xi}{\partial x} + \frac{\partial \xi}{\partial y}\frac{\partial \eta}{\partial y}\right) +$$

$$\left[\left(\frac{\partial \eta}{\partial x}\right)^2 + \left(\frac{\partial \eta}{\partial y}\right)^2\right]\frac{\partial^2 x}{\partial \eta^2} = 0 \quad (4.5.25)$$

Let us now substitute the expressions (4.5.11), (4.5.12) into (4.5.25):

$$(x_\eta^2 + y_\eta^2)x_{\xi\xi} - 2(x_\xi x_\eta + y_\xi y_\eta)x_{\xi\eta} + (x_\xi^2 + y_\xi^2)x_{\eta\eta} = 0 \quad (4.5.26)$$

where we have used the notations $x_\eta = \partial x/\partial \eta$, $x_{\xi\xi} = \partial^2 x/\partial \xi^2$, etc.

The equation (4.5.24) for $y(\xi, \eta)$ may be transformed in a similar way to yield the PDE for the determination of $y(\xi, \eta)$:

$$(x_\eta^2 + y_\eta^2)y_{\xi\xi} - 2(x_\xi x_\eta + y_\xi y_\eta)y_{\xi\eta} + (x_\xi^2 + y_\xi^2)y_{\eta\eta} = 0 \quad (4.5.27)$$

Now consider the question of the numerical solution of the equations (4.5.26) and (4.5.27). Let us introduce an uniform rectangular mesh in the (ξ, η) plane (see Fig. 2.40,b). Let $\Delta\xi$ and $\Delta\eta$ be the steps of this mesh along the ξ-axis and η-axis, respectively. Let the grid nodes ξ_i and η_j be as follows:

$$\xi_i = (i-1)\Delta\xi, \ i = 1, \ldots, I; \quad \eta_j = (j-1)\Delta\eta, \ j = 1, \ldots, J \quad (4.5.28)$$

Then the inlet section $x = 0$ (Fig. 2.40,a) is mapped onto the line $\xi = 0$. The nozzle wall "23" of Fig. 2.40,a is mapped onto the line $\eta = \eta_J =$

$(J-1)\Delta\eta$. The exit section "34" of Fig. 2.40,a is mapped onto the line $\xi = \xi_I = (I-1)\Delta\xi$. And finally the nozzle axis $y = 0$ is mapped onto the line $\eta = \eta_1 = 0$. This means that the values of the functions $x(\xi, \eta)$ and $y(\xi, \eta)$ are given on the boundary of the computational domain in the (ξ, η) plane:

$$x_{1,j} \equiv x(\xi_1, \eta_j) = 0; \quad x_{I,J} = x_{ex.}, \quad j = 1, \ldots, J$$

$$x_{i,1} = X_{i,1}, \quad x_{i,J} = X_{i,J}, \quad i = 1, \ldots, I \qquad (4.5.29)$$

$$y_{1,j} \equiv y(\xi_1, \eta_j) = Y_{1,j}; \quad y_{I,j} = Y_{I,j}, \quad j = 1, \ldots, J$$

$$y_{i,1} = 0, \quad y_{i,J} = f(X_{i,J}), \quad i = 1, \ldots, I \qquad (4.5.30)$$

Equations (4.5.29) represent the difference discretization of the Dirichlet boundary conditions for the function $x(\xi, \eta)$. The values $X_{i,1}$ and $X_{i,J}$ should be specified by the user in such a way that the numbers $X_{1,1}, X_{2,1}, \ldots, X_{I,1}$ and $X_{1,J}, X_{2,J}, \ldots, X_{I,J}$ form the monotone sequences. The abscissas $X_{i,J}$ can be specified in such a way that the grid in the (x, y) plane becomes more condensed in the neighborhood of the nozzle throat to increase the numerical solution accuracy in this subregion. For this purpose the points $(X_{i,J}, f(X_{i,J}))$ can be placed more densely in the neighborhood of the nozzle throat. Here $y = f(x)$ is the equation of the nozzle wall, and $f(x)$ is a given function.

The ordinates $Y_{1,j}$ and $Y_{I,j}$ in (4.5.30) are also user-specified, so that

$$0 = Y_{1,1} < Y_{1,2} < \ldots Y_{1,J} = f(0)$$

$$0 = Y_{I,1} < Y_{I,2} < \ldots < Y_{I,J} = f(x_{ex.})$$

where $x_{ex.}$ is the abscissa of the exit section of the nozzle (section "34" in Fig. 2.40,a).

By virtue of (4.5.29) and (4.5.30) we need the values x_{ij} and y_{ij} only for $2 \le i \le I - 1$ and $2 \le j \le J - 1$. Let $f(\xi, \eta)$ be any of the functions $x(\xi, \eta)$ or $y(\xi, \eta)$. Since we can use the arbitrary stretching of ξ- and η-coordinates, we will assume that $\Delta\xi = \Delta\eta = 1$. Then we can use the following second-order approximations for the partial derivatives of the function $f(\xi, \eta)$:

$$(f_\xi)_{ij} = \frac{1}{2}(f_{i+1,j} - f_{i-1,j}), \quad (f_\eta)_{ij} = \frac{1}{2}(f_{i,j+1} - f_{i,j-1})$$

$$(f_{\xi\xi})_{ij} = f_{i+1,j} - 2f_{ij} + f_{i-1,j}, (f_{\eta\eta})_{ij} = f_{i,j+1} - 2f_{ij} + f_{i,j-1} \quad (4.5.31)$$

$$(f_{\xi\eta})_{ij} = \frac{1}{4}(f_{i+1,j+1} - f_{i+1,j-1} - f_{i-1,j+1} + f_{i-1,j-1})$$

If we replace all derivatives in equations (4.5.26) and (4.5.27) by central difference expressions (4.5.31), we obtain two systems of nonlinear algebraic equations for determining x_{ij} and y_{ij}, $2 \le i \le I - 1$, $2 \le j \le J - 1$. A number of different algorithms can be used for the solution of these equations, including point and line SOR (Successive Over-Relaxation),[24] ADI (Alternating Direction Implicit), and multigrid iteration.[24] The ADI method is known to possess a faster convergence than the SOR method when applied to the numerical solution of the Thompson-Thames-Mastin grid equations (4.5.26), (4.5.27).[25]

We now describe the implementation of an ADI difference scheme for the numerical solution of the PDEs (4.5.26) and (4.5.27). Let us introduce the following difference operators:

$$\Lambda_\xi^n f_{ij} = \left[(x_\eta^2)_{ij}^n + (y_\eta^2)_{ij}^n\right](f_{i+1,j} - 2f_{ij} + f_{i-1,j})$$

$$\Lambda_\eta^n f_{ij} = \left[(x_\xi^2)_{ij}^n + (y_\xi^2)_{ij}^n\right](f_{i,j+1} - 2f_{ij} + f_{i,j-1})$$

$$\Lambda_{\xi\eta}^n f_{ij} = -\frac{1}{2}\left[(x_\xi x_\eta)_{ij}^n + (y_\xi y_\eta)_{ij}^n\right](f_{i+1,j+1} - f_{i+1,j-1} - f_{i-1,j+1} + f_{i-1,j-1})$$

Here the superscript n denotes the number of iteration. Then the ADI difference scheme for the iterative solution of equation (4.5.26) may be written as follows:

$$\frac{\bar{x}_{ij} - x_{ij}^n}{0.5\tau} = \Lambda_\xi^n \bar{x}_{ij} + \Lambda_{\xi\eta}^n x_{ij}^n + \Lambda_\eta^n x_{ij}^n \quad (4.5.32)$$

$$\frac{x_{ij}^{n+1} - \bar{x}_{ij}}{0.5\tau} = \Lambda_\xi^n \bar{x}_{ij} + \Lambda_{\xi\eta}^n \bar{x}_{ij} + \Lambda_\eta^n x_{ij}^{n+1} \quad (4.5.33)$$

The ADI scheme for equation (4.5.27) may be written down in a similar way:

$$\frac{\bar{y}_{ij} - y_{ij}^n}{0.5\tau} = \Lambda_\xi^n \bar{y}_{ij} + \Lambda_{\xi\eta}^n y_{ij}^n + \Lambda_\eta^n y_{ij}^n$$

$$\frac{y_{ij}^{n+1} - \bar{y}_{ij}}{0.5\tau} = \Lambda_{\xi}^n \bar{y}_{ij} + \Lambda_{\xi\eta}^n \bar{y}_{ij} + \Lambda_{\eta}^n y_{ij}^{n+1} \qquad (4.5.34)$$

The parameter τ in (4.5.32)-(4.5.34) may be regarded as a pseudo-time step. That is the schemes (4.5.32)-(4.5.33) and (4.5.34) may be considered as the difference schemes approximating the parabolic equations of the form

$$\partial f / \partial t = Lf$$

where L is the differential operator of the left-hand sides of equations (4.5.20) and (4.5.27). We have discussed the pseudo-unsteady method above in Section 4.3.

The algebraic system (4.5.32) is solved along each line $\eta = \eta_j$, $2 \leq j \leq J$. It is easy to see that it has the form

$$A\bar{x}_{i-1,j} - B\bar{x}_{ij} + C\bar{x}_{i+1,j} = D_{ij}, \ i = 2,\dots,I-1 \qquad (4.5.35)$$

where $A, B, C,$ and D are known coefficients. It can easily be shown that these coefficients satisfy the condition of the diagonal dominance. Therefore, an efficient Gaussian elimination type method presented above in Section 2.9 can be used for obtaining the solution of the system (4.5.35) under the boundary conditions (4.5.29).

The algebraic system (4.5.33) is solved along each line $\xi = \xi_i$, $2 \leq i \leq I-1$. It is easy to see that it has the form

$$A_1 x_{i,j-1}^{n+1} - B_1 x_{ij}^{n+1} + C_1 x_{i,j+1}^{n+1} = D_1 \qquad (4.5.36)$$

and $|B_1| > |A_1| + |C_1|$. Therefore, the Gaussian elimination method of Section 2.9 can be used also for the solution of the system (4.5.36).

The difference schemes (4.5.32)-(4.5.33) and (4.5.34) need the start values x_{ij}^0 and y_{ij}^0 for $2 \leq i \leq I-1$ and $2 \leq j \leq J-1$. These values can be computed with the aid of the linear unidirectional interpolation between the facing sides of a rectangle in the (ξ, η) plane by using the boundary values (4.5.29) and (4.5.30). For example, the values y_{ij}^0 may be specified with the aid of the formula

$$y_{ij}^0 = f(X_{i,J}) \cdot (j-1)/(J-1), \quad i = 1,\dots,I, \ j = 1,\dots,J$$

Since the systems (4.5.32)-(4.5.34) are nonlinear, the convergence (as $n \to \infty$) depends on the initial guess x_{ij}^0 and y_{ij}^0 in iterative solutions.

We have implemented the above elliptic grid generator in the *Mathematica* program gridttm.ma. Difference equations (4.5.32)-(4.5.34) were solved numerically under the Dirichlet boundary conditions (4.5.29), (4.5.30) by an efficient variant of the Gauss elimination method described in Section 2.9. The input parameters used in the program gridttm.ma are described in Table 4.3.

Table 4.3. Parameters used in program `gridttm.ma`

Parameter	Description
hc	the nozzle wall height h_c in the critical section of the nozzle, $h_c > 0$
cy0	the ordinate of the nozzle wall in the inlet section $x = 0$ related to h_c
cR0	the radius of the curvature of the nozzle wall in its subsonic part related to h_c
cR	the nozzle wall radius of the curvature in the critical section related to h_c
cx0	the abscissa of the critical nozzle section related to h_c
cx4	the nozzle length related to h_c, cx4 > cx0
thet1	the inclination angle ϑ_1 of the nozzle wall in the subsonic part of the nozzle; the angle is specified in radians
thet2	the inclination angle ϑ_2 of the nozzle wall in the supersonic part of the nozzle; the angle is specified in radians
ix	the number i_x of grid points on the ξ-axis, $i_x > 1$
jy	the number j_y of grid points on the η-axis, $j_y > 1$
Saf	the safety factor θ in the formula (4.5.37) for the computation of the pseudo time step τ; $\theta > 0$
eps	a small positive number TOL entering the criterion (4.5.38) for the termination of the iterations by the ADI method

The main function of the file `gridttm.ma` is

$$\texttt{grid2}[hc, cy0, cR0, cR, cx0, cx4, \vartheta_1, \vartheta_2, i_x, j_y, \theta, TOL].$$

The time step τ was specified by the formula

$$\tau = \theta x_{ex}/(I - 1) \tag{4.5.37}$$

where θ is a positive safety factor chosen empirically from the requirement of the fastest convergence of the iterative solution by the pseudo unsteady difference equations (4.5.32)-(4.5.34). The initial guess (x_{ij}^0, y_{ij}^0), $1 \le i \le I$, $1 \le j \le J$, was specified with the aid of simple algebraic formulas (2.14.3), (2.14.4) as in the program `grid1.ma` from Section 2.14. Since the coefficients of the difference equations (4.5.32), (4.5.33) and (4.5.34) for determining the x- and y- coordinates of the grid nodes coincide, it was convenient to write a function implementing the computation by these ADI schemes. This function has the name `ADI[nxy]` in our program `gridttm.ma` and is as follows:

```
(* ------ The ADI solver in the form of a function ----- *)
ADI[nxy_] :=
( xg = x; yg = x; xn1 = xx; xn2 = xn1;
(* ----- The first half-step
        (passage from t = n*dt to t = (n+0.5)*dt) ----- *)
xg[[1]] = 0.0;
   Do[ yg[[1]] = 0.0;
       If[nxy != 1, yg[[1]] = yw[[1]]*(j-1)/(jy-1)];
       xn1[[i, j]] = yg[[1]]; xn2[[1, j]] = yg[[1]];
Print["nxy = ", nxy,
      ", yy[[",ifix,", ", j, "]] = ", yy[[ifix,j]] ];
(* --- The scalar sweeps along the lines eta = const --- *)
       Do[ xet = (xx[[i, j+1]] - xx[[i, j-1]])/2;
           yet = (yy[[i, j+1]] - yy[[i, j-1]])/2;
   xks = (xx[[i+1,j]] - xx[[i-1,j]])/2;
   yks = (yy[[i+1,j]] - yy[[i-1,j]])/2;
If[nxy == 1,
fkset = xx[[i+1, j+1]] - xx[[i+1, j-1]] -
xx[[i-1, j+1]] + xx[[i-1, j-1]],
           fkset = yy[[i+1, j+1]] - yy[[i+1, j-1]] -
           yy[[i-1, j+1]] + yy[[i-1, j-1]] ];
a1 = xet*xet + yet*yet;  b1 = xks*xks + yks*yks;
If[nxy == 1,
 fet = xx[[i, j+1]] - 2*xx[[i,j]] + xx[[i, j-1]],
        fet = yy[[i, j+1]] - 2*yy[[i,j]] + yy[[i, j-1]] ];
g1 = b1*fet - (1/2)*(xks*xet + yks*yet)*fkset;
a2 = dt2*a1;   b2 = -1 - dt*a1;
   If[nxy == 1, fcij = xx[[i, j]], fcij = yy[[i, j]] ];
f2 = -fcij - dt2*g1;   den = b2 + a2*xg[[i - 1]];
xg[[i]] = -a2/den;  yg[[i]] = (f2 - a2*yg[[i - 1]])/den,
     {i, 2, ix1}];
If[nxy == 1, xn1[[ix, j]] = x4,
   xn1[[ix, j]] = yw[[ix]]*(j - 1)/jy1];
   xn2[[ix,j]] = xn1[[ix, j]];
  Do[ i1 = ix - i + 1;
     xn1[[i1, j]] = xg[[i1]]*xn1[[i1+1, j]] + yg[[i1]],
 {i, 2, ix1}],
                       {j, 2, jy1}];
(* ----- The second half-step
(passage from t = (n + 0.5)*dt to t = (n + 1)*dt) ----- *)
If[nxy == 1, xn2[[1, jy]] = 0; xn2[[ix, jy]] = x4;
           xn2[[ix, 1]] = x4,
```

```
       xn2[[1, jy]] = yw[[1]]; xn2[[ix, 1]] = 0;
       xn2[[ix, jy]] = yw[[ix]] ];

Do[ xg[[1]] = 0;
    If[nxy == 1, yg[[1]] = x[[i]], yg[[1]] = 0];
       xn2[[i,1]] = yg[[1]];
(* --- The scalar sweeps along the lines xi = const --- *)
         Do[ xet = (xx[[i, j+1]] - xx[[i, j-1]])/2;
             yet = (yy[[i, j+1]] - yy[[i, j-1]])/2;
  xks = (xx[[i+1,j]] - xx[[i-1,j]])/2;
  yks = (yy[[i+1,j]] - yy[[i-1,j]])/2;
fkset = xn1[[i+1, j+1]] - xn1[[i+1, j-1]] -
xn1[[i-1, j+1]] + xn1[[i-1, j-1]];
a1 = xet*xet + yet*yet;  b1 = xks*xks + yks*yks;
g1 = a1*(xn1[[i+1, j]] - 2*xn1[[i,j]] + xn1[[i-1,j]]);
g1 = g1 - (1/2)*(xks*xet + yks*yet)*fkset;
a2 = dt2*b1;   b2 = -1 - dt*b1;
  f2 = -xn1[[i, j]] -dt2*g1;  den = b2 + a2*xg[[j-1]];
 xg[[j]] = -a2/den;  yg[[j]] = (f2 - a2*yg[[j-1]])/den,
 {j, 2, jy1}];
xn2[[i, jy]] = x[[i]];
If[nxy != 1, xn2[[i, jy]] = yw[[i]] ];
  Do[ j1 = jy - j + 1;
  xn2[[i, j1]] = xg[[j1]]*xn2[[i, j1+1]] + yg[[j1]],
     {j, 2, jy1}],
{i, 2, ix1}];
  Return[xn2] );
```

At **nxy** = 1 the above function computes the abscissas x_{ij}^{n+1} of the grid nodes, and at **nxy** = 2 the ordinates y_{ij}^{n+1}. The convergence criterion was as follows:

$$\max_{(i,j) \in G_h} |x_{ij}^{n+1} - x_{ij}^n| < TOL, \quad \max_{(i,j) \in G_h} |y_{ij}^{n+1} - y_{ij}^n| < TOL \quad (4.5.38)$$

where

$$G_h = \{(i,j)| \ (2 \le i \le I-1) \times (2 \le j \le J-1)\},$$

and TOL is a user-specified tolerance.

In Fig. 4.6-b we show the curvilinear grid obtained with the aid of the above ADI solver at $\theta = 40.0$ in Eq. (4.5.37). It can be seen that the lines $\xi = const$ do not change in the result of the iterations and remain the straight line segments parallel with the y-axis. This means that only vertical positions of the grid nodes have changed as

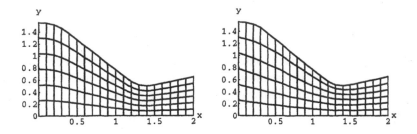

Figure 4.6: The curvilinear grid in the nozzle. (a) the initial grid; (b) the grid obtained after 67 ADI iterations.

a result of the solution of the elliptic grid equations, so that a one-dimensional grid construction problem was indeed solved along each grid line $\xi = \xi_i$, $i = 2, 3, \ldots, I - 1$. Similar examples of curvilinear grid construction in nonconvex regions can be found in the Appendix C of the book.[24]

The tolerance $TOL = 0.0003$ for the example of Fig. 4.6. At this value of TOL, 67 iterations were needed to achieve the satisfaction of the convergence criterion (4.5.38).

It can be seen from Fig. 4.6-b that the curvilinear grid obtained is non-orthogonal near the nozzle wall. We will not discuss here the questions of the grid control near the boundary, we refer the reader to the Reference 24.

Exercise 4.6. Show with the aid of the Euler-Lagrange equation from the variational calculus that the equation (4.5.1) can be obtained from the functional (4.5.3).

Exercise 4.7. Write down the equation (4.5.32) and find the expressions for the coefficients A, B, C and D in (4.5.35). Show that $|B| > |A| + |C|$.

Exercise 4.8. Find the condition for the orthogonality of the curvilinear grid lines at any point of intersection of these lines.
Hint. At first find the tangent vectors to each of the two grid lines intersecting at a grid node, and then equal the inner product of these vectors to zero.

Exercise 4.9. Consider the particular coordinate transformation (2.14.1). Find the condition for the function $f_w(x)$, at which the curvilinear grid will be orthogonal.

Exercise 4.10. Find the form of the functions $x(\xi, \eta)$, $y(\xi, \eta)$ entering the transformation (2.14.2), which maps the spatial region determined by the inequality

$$x^2 + y^2 \leq R^2 \tag{4.5.39}$$

where $R > 0$ and R is given, onto a rectangular region in the (ξ, η) plane. Determine whether the families of grid lines $\xi = const$ and $\eta = const$ form an orthogonal curvilinear grid in the (x, y) plane.

Exercise 4.11. Let a closed planar curve be given by the equation

$$x^2 + 2y^2 - 2x + 12y + 15 = 0 \tag{4.5.40}$$

Find the form of a one-to-one transformation (2.14.2), which enables one to map the interior of the curve (4.5.40) onto a rectangular region in the (ξ, η) plane. Determine whether the families of grid lines $\xi = const$ and $\eta = const$ form an orthogonal curvilinear grid in the (x, y) plane. Make a *Mathematica* program, which determines the form of the functions $x(\xi, \eta)$ and $y(\xi, \eta)$ in (2.14.2) at least for some types of planar closed curves.

Exercise 4.12. Modify the program `gridttm.ma` by considering the cyclic sequences of the time steps in accordance with Eq. (4.3.41) instead of a constant time step. Determine the constants α_l, α_h and K in Eq. (4.3.41) with the aid of numerical experiments from the requirement of the fastest convergence of the iterative solution. Compare the obtained number of the iterations needed to satisfy the convergence criterion (4.5.38) at $TOL = 0.0003$ with the 67 iterations mentioned earlier.

4.6 LOCAL APPROXIMATION STUDY OF FINITE VOLUME OPERATORS ON ARBITRARY GRIDS

We have already presented above in Sections 2.14, 3.4 and 4.5 a number of difference approximations to various partial derivative operators on curvilinear grids of quadrilateral cells and on triangular grids. These grids can generally have irregular shapes to be adapted to geometrically complex boundaries. It is nevertheless necessary to be sure of the fact that the approximation of the original differential operators takes place also in these cases. The work[10] appears to represent the first attempt at the automation of the local approximation study of the spatial differencing operators on an arbitrary curvilinear nonorthogonal grid of quadrilateral cells. The symbolic computations were performed in Reference 26 with the aid of the Soviet language REFAL. In a more recent work[27] the procedure for determining the local approximation order of difference operators on two-dimensional curvilinear grids was implemented with the aid of the computer algebra system REDUCE.[28] The method of the work[27] was extended for the case of any number of the spatial dimensions in Reference 29. According Reference 29, this method can be applied also for the computer-aided investigation of the local approximation of difference operators on a mesh of Dirichlet cells.

The method of[27] enables one to solve the following problem: let a difference operator $P_h u$ be given, which approximates certain operator Pu on a spatial grid; for example, $Pu = grad(u)$. Find the value k in the formula

$$P_h u = Pu + O(h^k) + o(h^k)$$

where $O(h^k)$ is the leading term of the approximation error, and h is a quantity, which characterizes the local dimensions of the grid cells in the neighborhood of the spatial point under consideration. Thus the authors of References 26 and 27 have not considered the problem on the computation of an explicit form of the leading term of approximation error. However, it is well known that information of such kind is important for understanding the structure of the approximation "viscosity" of the numerical method[30,31] (see also Section 2.3), since this structure affects the computational stability, the presence of oscillations, dispersion.

We describe below a symbolic algorithm, which enables us to establish the local approximation order of a finite-volume spatial differencing operator on an arbitrary spatial grid and to obtain an explicit form of the leading term of the approximation error. The basic steps of this algorithm are the same as in References 27 and 29.

4.6.1 Finite Volume Discretization on Quadrilateral Cells

Let us at first describe the symbolic algorithm for the local approximation study of the finite volume discretization of the partial derivatives $\partial F(u)/\partial x$, $\partial G(u)/\partial y$ on a curvilinear non-orthogonal grid of quadrilateral cells. It follows from Eq. (3.4.11) that the finite volume approximation of the partial derivatives $\partial F(u)/\partial x$ and $\partial G(u)/\partial y$ has the form

$$[\partial F(u)/\partial x]_{i,j} = [(F_{i,j-1} + F_{i,j})\Delta y_{AB} +$$
$$(F_{i,j} + F_{i+1,j})\Delta y_{BC} + (F_{i,j} + F_{i,j+1})\Delta y_{CD} +$$
$$(F_{i,j} + F_{i-1,j})\Delta y_{DA}]/(2A_{ij}) \qquad (4.6.1)$$

$$[\partial G(u)/\partial y]_{i,j} = -[(G_{i,j-1} + G_{i,j})\Delta x_{AB} +$$
$$(G_{i,j} + G_{i+1,j})\Delta x_{BC} + (G_{i,j} + G_{i,j+1})\Delta x_{CD} +$$
$$(G_{i,j} + G_{i-1,j})\Delta x_{DA}]/(2A_{ij}) \qquad (4.6.2)$$

The area of a quadrilateral cell can be computed with the aid of the formula (cf. (3.4.20))

$$A_{ij} = \oint_{ABCDA} = \frac{1}{2}(x_A + x_B)\Delta y_{AB} + \frac{1}{2}(x_B + x_C)\Delta y_{BC} +$$

$$\frac{1}{2}(x_C + x_D)\Delta y_{CD} + \frac{1}{2}(x_D + x_A)\Delta y_{DA}] \quad (4.6.3)$$

Let us now present the steps of the symbolic algorithm for the local approximation study of the differencing operators (4.6.1), (4.6.2), which we have implemented with the aid of the *Mathematica* system.

Step 1. The input data are input into the *Mathematica* program. These are the following data:

a) the difference operator under consideration;

b) the order kt> 1 of the Taylor expansions;

Step 2. All the grid functions $F_{m,k} = F(u_{m,k})$ in the difference operator (4.6.1) are expanded into truncated Taylor series up to the above specified order kt in accordance with the formula

$$F(x, y) = F(x_c, y_c) +$$

$$\sum_{v=1}^{kt} \sum_{i_1+i_2=v} \frac{1}{i_1! i_2!} \frac{\partial^{i_1+i_2} F}{\partial x^{i_1} \partial y^{i_2}} (x - x_c)^{i_1}(y - y_c)^{i_2} + R \quad (4.6.4)$$

where (x_c, y_c) are the coordinates of the center of the Taylor expansions, and R is the remainder term of the Taylor formula.

Step 3. The specification of the center (xc, yc) of the Taylor expansion. In the case of Eqs. (4.6.1)-(4.6.2) it is natural to take

$$x_c = x_{i,j}, \quad y_c = y_{i,j}$$

see Fig. 3.10-a.

Step 4. The formula (4.6.1) involves the spatial coordinates

$$(x_A, y_A), \ (x_B, y_B), \ (x_C, y_C), \ (x_D, y_D), \ (x_{i,j-1}, y_{i,j-1}),$$

$$(x_{i,j+1}, y_{i,j+1}), \ (x_{i-1,j}, y_{i-1,j}), \ (x_{i+1,j}, y_{i+1,j}) \quad (4.6.5)$$

The coordinates (x_A, y_A), (x_B, y_B), (x_C, y_C), (x_D, y_D) are now represented as the deviations from the center (x_c, y_c) of the Taylor expansion:

$$\begin{aligned}
x_A &= x_c + C_A^x \cdot h, & y_A &= y_c + C_A^y \cdot h \\
x_B &= x_c + C_B^x \cdot h, & y_B &= y_c + C_B^y \cdot h \\
x_C &= x_c + C_C^x \cdot h, & y_C &= y_c + C_C^y \cdot h \\
x_D &= x_c + C_D^x \cdot h, & y_D &= y_c + C_D^y \cdot h
\end{aligned} \quad (4.6.6)$$

Here h is a quantitative measure, which locally characterizes the density of a grid of quadrilateral cells in the neighborhood of the (x_c, y_c) point.

Following[4] let us denote by $S(i,j)$ the stencil of the node (i,j), that is all those points which are involved in the computation by formula (4.6.1). In accordance with (4.6.5) we can write that

$$S(i,j) = \{A, B, C, D, (i,j), (i,j-1), (i,j+1), (i-1,j), (i+1,j)\} (4.6.7)$$

The stencil $S(i,j)$ thus involves 9 points. Then h may be computed by the formula

$$h = \frac{1}{8}\{ \sum_{(k,m)\in S(i,j)} [(x_{k,m} - x_c)^2 + (y_{k,m} - y_\bullet)^2]\}^{0.5} \qquad (4\ 6\ 8)$$

The constants $C_A^y, C_B^y, \ldots, C_D^y$ in (4.6.6) do not depend on h. Their specific values are generally not known, which enables us to consider arbitrary distortions of quadrilateral cells. Fig. 3.10-a gives some idea on how the quadrilateral cells can be distorted to take into account the geometrically complex boundaries of the spatial computational domain.

The formula (4.6.3) for the area A_{ij} involves only the coordinates (4.6.6). Since the quantity A_{ij} enters the denominator in Eqs. (4.6.1) and (4.6.2), it is desirable to express the coordinates

$$(x_{i,j-1}, y_{i,j-1}), \quad (x_{i,j+1}, y_{i,j+1}), \quad (x_{i-1,j}, y_{i-1,j}), \quad (x_{i+1,j}, y_{i+1,j}) \ (4.6.9)$$

in (4.6.5) in terms of the coordinates (4.6.6) to ensure the proper cancellation of similar terms in the numerator of the expanded expression (4.6.1), so that the coefficient of $\partial F/\partial x$ becomes equal to 1.

Let us at first consider the point $(x_{i-1,j}, y_{i-1,j})$. We now assume that

(i) the $(i-1,j)$ point deviates little from the line passing through the points C' and D', see Fig. 3.10-a;

(ii) the doubled length of the straight line segment connecting the points $(i-1,j)$ and D' is approximately equal to the length of the line segment $C'D'$.

Then it is easy to find that

$$x_2 = x_{D'} - \frac{1}{2}(x_{C'} - x_{D'}), \quad y_2 = y_{D'} - \frac{1}{2}(y_{C'} - y_{D'}).$$

The coordinates $x_{D'}$ and $y_{D'}$ are easily computed as

$$x_{D'} = \frac{1}{2}(x_A + x_D), \quad y_{D'} = \frac{1}{2}(y_A + y_D).$$

The coordinates $(x_{C'}, y_{C'})$ are computed in a similar way as the arithmetic means of (x_B, y_B) and (x_C, y_C). The remaining points in the set (4.6.7) are considered in a similar way. As a result we obtain the following formulas:

$$
\begin{aligned}
x_{i-1,j} &= (3x_A + 3x_D - x_B - x_C)/4 \\
y_{i-1,j} &= (3y_A + 3y_D - y_B - y_C)/4 \\
x_{i,j-1} &= (3x_A + 3x_B - x_C - x_D)/4 \\
y_{i,j-1} &= (3y_A + 3y_B - y_C - y_D)/4 \\
x_{i+1,j} &= (3x_B + 3x_C - x_A - x_D)/4 \\
y_{i+1,j} &= (3y_B + 3y_C - y_A - y_D)/4 \\
x_{i,j+1} &= (3x_C + 3x_D - x_A - x_B)/4 \\
y_{i,j+1} &= (3y_C + 3y_D - y_A - y_B)/4
\end{aligned}
\tag{4.6.10}
$$

We can see that all the coordinates in (4.6.5) are now expressed with the aid of formulas (4.6.10) in terms of the coordinates (4.6.6) only.

In the particular case of an uniform rectangular grid it is possible to specify the constants $C_A^x, \ldots, C_D^y$ from the geometric considerations (see Fig. 3.10-a):

$$
\begin{aligned}
C_A^x &= -\tfrac{1}{2}; & C_A^y &= -\tfrac{1}{2}; & C_B^x &= \tfrac{1}{2}; & C_B^y &= -\tfrac{1}{2} \\
C_C^x &= \tfrac{1}{2}; & C_C^y &= \tfrac{1}{2}; & C_D^x &= -\tfrac{1}{2}; & C_D^y &= \tfrac{1}{2}
\end{aligned}
\tag{4.6.11}
$$

The expressions (4.6.6) are substituted for the spatial coordinates in Eq. (4.6.1). The substitution of these expressions into Eqs. (4.6.10) yields the formulas like

$$
x_{i-1,j} = x_c + C_2^x \cdot h
$$

where

$$
C_2^x = (3C_A^x + 3C_D^x - C_B^x - C_C^x)/4
$$

etc. (see also the listing of the program `appr2.ma` below).

Step 5. Let us denote the operator appearing on the right hand side of Eq. (4.6.1) as a result of the foregoing steps as follows:

$$
\left[\frac{\partial F(u)}{\partial x}\right]_{i,j} = H\left(F, \frac{\partial F}{\partial x}, \frac{\partial F}{\partial y}, \frac{\partial^2 F}{\partial x^2}, \ldots, \frac{\partial^{kt} F}{\partial y^{kt}}, h\right)
\tag{4.6.12}
$$

Now let us group the terms in the expression $H(\cdot)$ in accordance with their powers of h. Let

$$
H(\cdot) = \sum_{\nu=0}^{kt} r_\nu h^\nu
\tag{4.6.13}
$$

If $r_0 = \partial F/\partial x$, then the approximation of the differential operator $\partial F/\partial x$ by the difference formula (4.6.1) takes place.

We have implemented the above symbolic algorithm in the *Mathematica* program `appr2.ma` whose listing we present as follows.

———————————— File appr2.m ————————————

```
kt = 3;
(* kt = order of expansion into the Taylor series, kt >1 *)

(* --- The expansion of the function F(x,y) into the
        Taylor series with respect to (xc, yc) point ---- *)

    F[x_, y_]:=
(sum = Fc + Fx*(x - xc) + Fy*(y - yc) +
1/2*Fxx*(x - xc)^2 + Fxy*(x - xc)*(y - yc) +
1/2*Fyy*(y - yc)^2;
If[kt > 2, sum = sum +
1/6*Fxxx*(x - xc)^3 + 1/2*Fxxy*(x - xc)^2*(y - yc) +
1/2*Fxyy*(x - xc)*(y - yc)^2 +1/6*Fyyy*(y - yc)^3]; sum )

(* --- The difference operator under consideration --- *)
         dyAB = yB - yA;    dyBC = yC - yB;
         dyCD = yD - yC;    dyDA = yA - yD;
(* ----------- The numbering of points:
  1 = (i, j);   2 = (i - 1, j);   3 = (i,j - 1);
  4 = (i + 1,j);   5 = (i,j + 1) -------------------- *)
(* -- Substitution of the Taylor series expansions --- *)
sc1 = (F[x3, y3] + F[x1, y1])*dyAB +
                        (F[x1, y1] + F[x4, y4])*dyBC +
       (F[x1, y1] + F[x5, y5])*dyCD +
(F[x1, y1] + F[x2, y2])*dyDA;
(* --------- Aij is the area of the cell ABCD -------- *)
Aij = ((xA + xB)*dyAB + (xB + xC)*dyBC + (xC + xD)*dyCD +
       (xD + xA)*dyDA)/2;
  Aij = Simplify[Aij];
  OutputForm[sc1] >> ap2.m;
                        OutputForm[Aij] >>> ap2.m;

                   sc1 = sc1;   sc1 = Expand[sc1];
  sc2 = sc1/(2*Aij);
Print["The difference operator expanded into Taylor",
" series has the form"];
 Print[sc2];

(* -- Representation of the spatial coordinates of points
      A, B, C, D, 1, 2, 3, 4, 5 as the deviations from the
      center (xc, yc) of the Taylor expansions -------- *)
      xA = xc + cax*h;    yA = yc + cay*h;
```

```
     xB = xc + cbx*h;    yB = yc + cby*h;
     xC = xc + ccx*h;    yC = yc + ccy*h;
     xD = xc + cdx*h;    yD = yc + cdy*h;
               x1 = xc;               y1 = yc;
               x2 = xc + c2x*h;       y2 = yc + c2y*h;
               x3 = xc + c3x*h;       y3 = yc + c3y*h;
               x4 = xc + c4x*h;       y4 = yc + c4y*h;
               x5 = xc + c5x*h;       y5 = yc + c5y*h;
  sc1 = sc1;    sc1 = Expand[sc1];
  OutputForm[sc1] >>> ap2.m;
  Aij = Aij;    Aij = Expand[Aij];  Aij = Simplify[Aij];
  OutputForm[Aij] >>> ap2.m;
  a2 = Coefficient[Aij, h, 2];
OutputForm[a2] >>> ap2.m;
  r0 = Coefficient[sc1, h, 2];
                              OutputForm[r0] >>> ap2.m;
  r1 = Coefficient[sc1, h, 3];
                              OutputForm[r1] >>> ap2.m;
  r2 = Coefficient[sc1, h, 4];
                              OutputForm[r2] >>> ap2.m;


(* The expressions for (x2,y2), (x3,y3), (x4,y4),(x5,y5)
in terms of (xA,yA), (xB,yB), (xC,yC), (xD,yD)    *)
    c2x = (3*cax + 3*cdx - cbx - ccx)/4;
    c2y = (3*cay + 3*cdy - cby - ccy)/4;
        c3x = (3*cax + 3*cbx - ccx - cdx)/4;
        c3y = (3*cay + 3*cby - ccy - cdy)/4;
            c4x = (3*cbx + 3*ccx - cax - cdx)/4;
            c4y = (3*cby + 3*ccy - cay - cdy)/4;
                c5x = (3*ccx + 3*cdx - cax - cbx)/4;
                c5y = (3*ccy + 3*cdy - cay - cby)/4;
  r0 = r0;  r0 = Simplify[r0];
Print["The coefficient of h^0 in the operator sc2 is"];
      r0 = r0/(2*a2);
Print[r0];
  OutputForm[r0] >>> ap2.m;
  r1 = r1;  r1 = Simplify[r1];
Print["The coefficient of h^1 in the operator sc2 is"];
      r1 = r1/(2*a2);
Print[r1];
  OutputForm[r1] >>> ap2.m;
  r2 = r2;  r2 = Simplify[r2];
Print["The coefficient of h^1 in the operator sc2 is"];
      r2 = r2/(2*a2);
```

```
Print[r2];
  OutputForm[r2] >>> ap2.m;

(* ----------- The particular case of a square
              grid in the (x,y) plane ------------- *)
   cax = -1/2;   cay = -1/2;   cbx = 1/2;   cby = -1/2;
   ccx = 1/2;    ccy = 1/2;    cdx = -1/2;  cdy = 1/2;
 r1 = r1; r1 = Simplify[r1];
Print["The coefficient of h in the operator sc2"];
Print["in the case of a square grid is"];
Print[r1];
          OutputForm[r1] >>> ap2.m;
 r2 = r2; r2 = Simplify[r2];
Print["The coefficient of h^2 in the operator sc2"];
Print["in the case of a square grid is"];
Print[r2];
          OutputForm[r2] >>> ap2.m;
```

When writing the program appr2.ma we have used some elements of the program appr1.ma from Section 2.5. In particular, we have at first written the function F[x_,y_] similarly to the function u[k_,m_] as follows:

```
F[x_, y_]:= Block[
(* ---- xs = xc, ys = yc ---- *)
sf = f[xs, ys];
Do[
   Do[i2 = ktj - i1;
     dr1 = If[i1 == 0, 1, (x - xc)^i1];
     dr2 = If[i2 == 0, 1, (y - yc)^i2];
   sf = sf + Derivative[i1, i2][f][xs, ys]*dr,
{i1, 0, ktj}],
     {ktj, kt}];
 sf ]
```

But it turned out that the expressions for the leading terms of the truncation error in case of a curvilinear grid are much longer than in the case of an uniform rectangular grid. The internal representation of the partial derivative $\partial^{i_1+i_2} F/\partial x^{i_1} \partial y^{i_2}$ is relatively long, see the body of the above function F[x_,y_]. We have denoted by F[x_,y_] the function $\tilde{F}(x,y) = F(u(x,y))$. To reduce the length of the expressions for the Taylor expansions of the difference operators we have written the function F[x_,y_] as follows:

```
    F[x_, y_]:=
(sf = Fc + Fx*(x - xc) + Fy*(y - yc) +
1/2*Fxx*(x - xc)^2 + Fxy*(x - xc)*(y - yc) +
1/2*Fyy*(y - yc)^2;
If[kt > 2, sf = sf +
1/6*Fxxx*(x - xc)^3 + 1/2*Fxxy*(x - xc)^2*(y - yc) +
1/2*Fxyy*(x - xc)*(y - yc)^2 +1/6*Fyyy*(y - yc)^3]; sf )
```

We have introduced the following notations here: $Fc= \tilde{F}(x_c, y_c)$, $Fx= \partial\tilde{F}/\partial x$, $Fxyy= \partial^3\tilde{F}/\partial x\partial y^2$, etc. This has enabled us to reduce substantially the length of the symbolic expressions produced by the program appr2.ma.

The results of symbolic computations are stored in the file ap2.m. These are:

(1) the expression sc1 for the numerator of the rational expression (4.6.1) obtained as a result of the execution of Step 2;

(2) the expression Aij for the area A_{ij} of the cell $ABCD$ obtained as a result of the execution of Step 2;

(3) the expression sc1 for the numerator of the rational expression (4.6.1) obtained as a result of the execution of Step 4;

(4) the expression Aij for the area A_{ij} of the cell $ABCD$ obtained as a result of the execution of Step 4;

(5) the expressions r0, r1, r2 for the r_j, $j = 0, 1, 2$, in the representation (4.6.13);

(6) the expressions r1, r2 for the particular case of an uniform rectangular grid in the (x, y) plane.

In the process of the work of the program appr2.ma the following message appears on the screen of the computer monitor:

```
The coefficient of h^0 in the operator sc2 is
Fx
```

This means that the approximation (4.6.1) of the partial derivative $\partial F/\partial x$ by the finite volume method indeed takes place. Let us now look at the leading term r_1 of the truncation error as obtained by the program appr2.ma:

$r1 = (-3\ cax\ cay\ cbx\ Fxx - cay\ cbx^2\ Fxx + cax^2\ cby\ Fxx +$

$3\ cax\ cbx\ cby\ Fxx + cay\ cbx\ ccx\ Fxx - 3\ cbx\ cby\ ccx\ Fxx -$

$cby\ ccx^2\ Fxx - cax\ cbx\ ccy\ Fxx + cbx^2\ ccy\ Fxx +$

$3\ cbx\ ccx\ ccy\ Fxx + 3\ cax\ cay\ cdx\ Fxx$

$-\ cax\ cby\ cdx\ Fxx - cay\ ccx\ cdx\ Fxx + cby\ ccx\ cdx\ Fxx$

$+\ cax\ ccy\ cdx\ Fxx - 3\ ccx\ ccy\ cdx\ Fxx + cay\ cdx^2\ Fxx -$

$ccy\ cdx^2\ Fxx - cax^2\ cdy\ Fxx + cax\ cbx\ cdy\ Fxx -$

$cbx\ ccx\ cdy\ Fxx + ccx^2\ cdy\ Fxx - 3\ cax\ cdx\ cdy\ Fxx +$

$3\ ccx\ cdx\ cdy\ Fxx - 3\ cay^2\ cbx\ Fxy - cax\ cay\ cby\ Fxy$

$+\ cay\ cbx\ cby\ Fxy + 3\ cax\ cby^2\ Fxy + cay\ cby\ ccx\ Fxy$

$-\ 3\ cby^2\ ccx\ Fxy - cax\ cby\ ccy\ Fxy - cbx\ cby\ ccy\ Fxy +$

$cby\ ccx\ ccy\ Fxy + 3\ cbx\ ccy^2\ Fxy +$

$3\ cay^2\ cdx\ Fxy - cay\ cby\ cdx\ Fxy + cby\ ccy\ cdx\ Fxy -$

$3\ ccy^2\ cdx\ Fxy + cax\ cay\ cdy\ Fxy + cay\ cbx\ cdy\ Fxy -$

$cay\ ccx\ cdy\ Fxy + cax\ ccy\ cdy\ Fxy - cbx\ ccy\ cdy\ Fxy -$

$ccx\ ccy\ cdy\ Fxy - cay\ cdx\ cdy\ Fxy + ccy\ cdx\ cdy\ Fxy -$

$3\ cax\ cdy^2\ Fxy + 3\ ccx\ cdy^2\ Fxy - 2\ cay^2\ cby\ Fyy +$

$2\ cay\ cby^2\ Fyy - 2\ cby^2\ ccy\ Fyy + 2\ cby\ ccy^2\ Fyy +$

$2\ cay^2\ cdy\ Fyy - 2\ ccy^2\ cdy\ Fyy - 2\ cay\ cdy^2\ Fyy +$

$2\ ccy\ cdy^2\ Fyy)\ /\ (4\ (-(cay\ cbx) + cax\ cby - cby\ ccx +$

$cbx\ ccy + cay\ cdx - ccy\ cdx - cax\ cdy + ccx\ cdy))$

It follows from this formula that the approximation of the operator $\partial F / \partial x$ by the finite volume method has only the first order of approximation in the case of a general curvilinear grid.

The expression for the next term, r_2, of the truncation error was obtained in the form

r2= (ccy (cax + cbx - 3 ccx - 3 cdx)^3 Fxxx +

cby (3 cax + 3 cbx - ccx - cdx)^3 Fxxx +

ccy (-cax + 3 cbx + 3 ccx - cdx)^3 Fxxx +

cby (cax - 3 cbx - 3 ccx + cdx)^3 Fxxx +

cay (-3 cax - 3 cbx + ccx + cdx)^3 Fxxx +

cay (3 cax - cbx - ccx + 3 cdx)^3 Fxxx +

(-3 cax + cbx + ccx - 3 cdx)^3 cdy Fxxx +

(-cax - cbx + 3 ccx + 3 cdx)^3 cdy Fxxx)/384

+ ccy (-cax - cbx + 3 ccx + 3 cdx)^2 (cay + cby -

3 ccy - 3 cdy) Fxxy/128 +

cby (-3 cax - 3 cbx + ccx + cdx)^2 (3 cay + 3 cby

- ccy - cdy) Fxxy/128 +

ccy (cax - 3 cbx - 3 ccx + cdx)^2 (-cay + 3 cby +

3 ccy - cdy) Fxxy/128 +

(3 cax - cbx - ccx + 3 cdx)^2 (-3 cay + cby + ccy -

3 cdy) cdy Fxxy/128 +

cby (cax - 3 cbx - 3 ccx + cdx)^2 (cay - 3 cby -

3 ccy + cdy) Fxxy/128 +

cay (-3 cax - 3 cbx + ccx + cdx)^2 (-3 cay - 3 cby

+ ccy + cdy) Fxxy/128 +

cay (3 cax − cbx − ccx + 3 cdx)^2 (3 cay − cby −

ccy + 3 cdy) Fxxy/128 +

(−cax − cbx + 3 ccx + 3 cdx)^2 cdy (−cay − cby +

3 ccy + 3 cdy) Fxxy/128 +

(−3 cax + cbx + ccx − 3 cdx) (−3 cay + cby + ccy

 3 cdy)^2 cdy Fxyy/128 +

ccy (−cax + 3 cbx + 3 ccx − cdx) (cay − 3 cby −

3 ccy + cdy)^2 Fxyy/128 +

cby (cax − 3 cbx − 3 ccx + cdx) (cay − 3 cby −

3 ccy + cdy)^2 Fxyy/128 +

cby (3 cax + 3 cbx − ccx − cdx) (−3 cay − 3 cby +

ccy + cdy)^2 Fxyy/128 +

cay (−3 cax − 3 cbx + ccx + cdx) (−3 cay − 3 cby +

ccy + cdy)^2 Fxyy/128 +

cay (3 cax − cbx − ccx + 3 cdx) (3 cay − cby − ccy

+ 3 cdy)^2 Fxyy/128 +

ccy (cax + cbx − 3 ccx − 3 cdx) (−cay − cby +

3 ccy + 3 cdy)^2 Fxyy/128 +

(−cax − cbx + 3 ccx + 3 cdx) cdy (−cay − cby +

3 ccy + 3 cdy)^2 Fxyy/128 +

(ccy (cay + cby − 3 ccy − 3 cdy)^3 Fyyy +

cby (3 cay + 3 cby − ccy − cdy)^3 Fyyy +

```
ccy (-cay + 3 cby + 3 ccy - cdy)^3 Fyyy +

(-3 cay + cby + ccy - 3 cdy)^3  cdy Fyyy +

cby (cay - 3 cby - 3 ccy + cdy)^3  Fyyy +

cay (-3 cay - 3 cby + ccy + cdy)^3  Fyyy +

cay (3 cay - cby - ccy + 3 cdy)^3  Fyyy +

cdy (-cay - cby + 3 ccy + 3 cdy)^3  Fyyy)/384)/

(-(cay cbx) + cax cby - cby ccx + cbx ccy + cay cdx -

ccy cdx - cax cdy + ccx cdy)
```

In the particular case of an uniform rectangular spatial grid, when the equalities (4.6.11) are satisfied, the above expressions for **r1** and **r2** revert into the following very short expressions:

```
0
Fxxx
----
 6
```

It follows from here that in this particular case the finite volume differencing formula (4.6.1) has the second order of approximation in space. The obtained leading term of the truncation error $r_2 h^2 = F_{xxx}/6$ coincides with that obtained in Section 2.1 in the case of an uniform grid.

4.6.2   Finite Volume Discretization on Triangular Grids

Let us consider the nodal scheme of Fig. 3.10-b. It follows from Section 3.4 that the finite volume approximation of the partial derivative $\partial F(u)/\partial x$ on the mesh of triangular cells has in the case of the nodal scheme the form

$$\left[\frac{\partial F(u)}{\partial x}\right]_i = \frac{1}{A_i}\left[\frac{1}{2}\sum_{k=1}^{6}(F_k + F_{k+1})\Delta y_{k,k+1}\right] \qquad (4.6.14)$$

where

$$\Delta y_{k,k+1} = y_{k+1} - y_k, \quad k = 1,\ldots,6; \quad y_7 = y_1$$

$$A_i = \frac{1}{2}\sum_{k=1}^{6}(x_k + x_{k+1})\Delta y_{k,k+1}$$

We now take the coordinates of the center of Taylor expansions as $x_c = x_i$, $y_c = y_i$. Similarly to Eqs. (4.6.6) we present the coordinates of points (x_j, y_j), $j = 1, \ldots, 6$, as the deviations from the (x_c, y_c) point:

$$x_j = x_c + C_j^x \cdot h, \; y_j = x_c + C_j^y \cdot h, \; j = 1, \ldots, 6 \qquad (4.6.15)$$

We have implemented the steps 1–5 of the above algorithm as applied to the operator (4.6.14) in the *Mathematica* program appr3.ma whose listing we present as follows.

```
────────────── File appr3.ma ──────────────

kt = 2;
(* kt = order of expansion into the Taylor series, kt>1  *)


(* --- The expansion of the function F(x,y) into the
       Taylor series with respect to (xc, yc) point ---- *)

   F[x_, y_]:=
(sum = Fc + Fx*(x - xc) + Fy*(y - yc) +
1/2*Fxx*(x - xc)^2 + Fxy*(x - xc)*(y - yc) +
1/2*Fyy*(y - yc)^2;
If[kt > 2, sum = sum +
1/6*Fxxx*(x - xc)^3 + 1/2*Fxxy*(x - xc)^2*(y - yc) +
1/2*Fxyy*(x - xc)*(y - yc)^2 +1/6*Fyyy*(y - yc)^3]; sum )

(* --- The difference operator under consideration --- *)
          dy12 = y2 - y1;    dy23 = y3 - y2;
          dy34 = y4 - y3;    dy45 = y5 - y4;
          dy56 = y6 - y5;    dy67 = y1 - y6;
(* -- Substitution of the Taylor series expansions --- *)
Print["Substitution of the Taylor series expansions"];
sc1 = (F[x1, y1] + F[x2, y2])*dy12 +
                         (F[x2, y2] + F[x3, y3])*dy23 +
       (F[x3, y3] + F[x4, y4])*dy34 +
(F[x4, y4] + F[x5, y5])*dy45 +
       (F[x5, y5] + F[x6, y6])*dy56 +
(F[x6, y6] + F[x1, y1])*dy67;
(* --------- Ai is the area of the hexagon -------- *)
Ai = ((x1 + x2)*dy12 + (x2 + x3)*dy23 + (x3 + x4)*dy34 +
       (x4 + x5)*dy45 + (x5 + x6)*dy56 + (x6 + x1)*dy67)/2;
  Ai = Simplify[Ai];
  OutputForm[sc1] >> ap3.m;
                         OutputForm[Ai] >>> ap3.m;
```

```
                         sc1 = sc1;   sc1 = Expand[sc1];
     sc2 = sc1/(2*Ai);
Print["The difference operator expanded into Taylor",
" series has the form"];
 Print[sc2];


(* -- Representation of the spatial coordinates of points
      1, 2, 3, 4, 5, 6 as the deviations from the
      center (xc, yc) of the Taylor expansions -------- *)
                    x1 = xc + c1x*h;   y1 = yc + c1y*h;
                    x2 = xc + c2x*h;   y2 = yc + c2y*h;
                    x3 = xc + c3x*h;   y3 = yc + c3y*h;
                    x4 = xc + c4x*h;   y4 = yc + c4y*h;
                    x5 = xc + c5x*h;   y5 = yc + c5y*h;
     x6 = xc + c6x*h;   y6 = yc + c6y*h;
Print["Substitution of the coordinates of points 1,...,6"];
Print["into the expanded difference operator as the"];
Print["deviations from the (xc, yc) point."];
    sc1 = sc1;   sc1 = Expand[sc1];
    OutputForm[sc1] >>> ap3.m;
    Ai = Ai;   Ai = Expand[Ai];  Ai = Simplify[Ai];
    OutputForm[Ai] >>> ap3.m;
    a2 = Coefficient[Ai, h, 2];
 OutputForm[a2] >>> ap3.m;
Print["Determination of the coefficient of h^2 in the"];
Print["numerator of the expanded difference operator."];
    r0 = Coefficient[sc1, h, 2];   r0 = Simplify[r0];
                              OutputForm[r0] >>> ap3.m;


Print["Determination of the coefficient of h^3 in the"];
Print["numerator of the expanded difference operator."];
    r1 = Coefficient[sc1, h, 3];   r1 = Simplify[r1];
                              OutputForm[r1] >>> ap3.m;


Print["Determination of the coefficient of h^4 in the"];
Print["numerator of the expanded difference operator."];
    r2 = Coefficient[sc1, h, 4];   r2 = Simplify[r2];
                              OutputForm[r2] >>> ap3.m;


Print["The coefficient of h^0 in the operator sc2 is"];
        r0 = r0/(2*a2);
Print[r0];
  OutputForm[r0] >>> ap3.m;
```

```
Print["The coefficient of h^1 in the operator sc2 is"];
      r1 = r1/(2*a2);
Print[r1];
  OutputForm[r1] >>> ap3.m;
Print["The coefficient of h^1 in the operator sc2 is"];
      r2 = r2/(2*a2);
Print[r2];
  OutputForm[r2] >>> ap3.m;


(* ----------- The particular case of equilateral
              triangular cells ---------------- *)
Print["The particular case of equilateral triangles"];
 cc = Table[N[Cos[(4 + j)*Pi/3], 20], {j, 6}];
 ss = Table[N[Sin[(4 + j)*Pi/3], 20], {j, 6}];
         Print[cc];          Print[ss];
c1x = cc[[1]]; c1y = ss[[1]]; c2x = cc[[2]]; c2y = ss[[2]];
c3x = cc[[3]]; c3y = ss[[3]]; c4x = cc[[4]]; c4y = ss[[4]];
c5x = cc[[5]]; c5y = ss[[5]]; c6x = cc[[6]]; c6y = ss[[6]];
 r1 = r1; r1 = Simplify[r1];
Print["The coefficient of h in the operator sc2"];
Print["in the case of equilateral triangular cells is"];
Print[r1];
          OutputForm[r1] >>> ap3.m;
 r2 = r2; r2 = Simplify[r2];
Print["The coefficient of h^2 in the operator sc2"];
Print["in the case of equilateral triangular cells is"];
Print[r2];
          OutputForm[r2] >>> ap3.m;
```

The structure of this program is similar to that of the program appr2.ma. To perform the computation by this program one must at first specify the integer value kt. Then one must input the difference operator. Then one can perform a computation by the program appr2.ma. The results of symbolic computations are stored in the file ap3.m, their list is the same as in the case of the program appr2.ma.

In the process of the work of the program appr3.ma the following message appears on the screen of the computer monitor:

```
The coefficient of h^0 in the operator sc2 is
Fx
```

This means that the approximation (4.6.14) of the partial derivative $\partial F/\partial x$ on the grid of arbitrary triangular cells indeed takes place. Let us now consider the leading term r_1 of the truncation error as obtained by the program appr3.ma:

```
r1=(-(c1y c2x^2 Fxx)+c1x^2 c2y Fxx-c2y c3x^2 Fxx +

c2x^2 c3y Fxx - c3y c4x^2 Fxx+ c3x^2 c4y Fxx -

c4y c5x^2 Fxx + c4x^2 c5y Fxx + c1y c6x^2  Fxx -

c5y c6x^2 Fxx- c1x^2 c6y Fxx + c5x^2 c6y Fxx +

2 c1x c1y c2y Fxy- 2 c1y c2x c2y Fxy+ 2 c2x c2y c3y Fxy -

2 c2y c3x c3y Fxy+ 2 c3x c3y c4y Fxy- 2 c3y c4x c4y Fxy +

2 c4x c4y c5y Fxy- 2 c4y c5x c5y Fxy- 2 c1x c1y c6y Fxy +

2 c5x c5y c6y Fxy+ 2 c1y c6x c6y Fxy- 2 c5y c6x c6y Fxy +

c1y^2 c2y Fyy- c1y c2y^2 Fyy+ c2y^2 c3y Fyy -

c2y c3y^2 Fyy + c3y^2 c4y Fyy- c3y c4y^2 Fyy+

c4y^2 c5y Fyy - c4y c5y^2 Fyy - c1y^2 c6y Fyy+

c5y^2 c6y Fyy+ c1y c6y^2 Fyy- c5y c6y^2 Fyy) /

(2 (-(c1y c2x) + c1x c2y - c2y c3x + c2x c3y - c3y c4x +

c3x c4y - c4y c5x + c4x c5y + c1y c6x - c5y c6x -

c1x c6y + c5x c6y))
```

It follows from this formula that the approximation of the operator $\partial F/\partial x$ by the finite volume method on a triangular grid has only the first order of approximation. Interestingly, the coordinates (x_c, y_c) of the center of Taylor expansions do not enter the formula for r_1.

We will not present here the expression for the next term, r_2, of the truncation error because of its bulky form.

Let us consider the particular case of a grid of equilateral triangles. Let us take as h the length of a side of an individual cell. Then we have that

$$x_j = x_i + h \cos \varphi_j, \quad y_j = y_i + h \sin \varphi_j$$
$$\varphi_j = \left(\frac{3\pi}{2} + \frac{\pi}{6}\right) + (j-1)\frac{\pi}{3}, \quad j = 1, 2, \ldots, 6$$

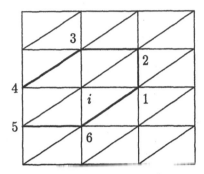

Figure 4.7: Uniform triangular grid.

Therefore, we can assume that

$$C_j^x = \cos\varphi_j, \quad C_j^y = \sin\varphi_j, \quad j = 1,\dots,6 \qquad (4.6.16)$$

in Eqs. (4.6.15). The substitution of the numerical values (4.6.16) into the above presented expression for r_1 reverts this expression into zero: $r_1 = 0$ (see also the output file ap3.m). The obtained expression for r_2 has the form

$$r_2 = 0.125h^2(F_{xxx} + F_{xyy})$$

This means that in the particular case of the mesh of equilateral triangles the finite volume approximation (4.6.14) of the spatial derivative $\partial F/\partial x$ has the second order of approximation.

Exercise 4.13. Modify the program appr2.ma to find the expressions r0, r1, r2 corresponding to the difference operator (4.6.2).

Exercise 4.14. Assume that a rectangular region in the (x, y) plane is discretized by an uniform rectangular grid with the steps h_1 along the x-axis and h_2 along the y-axis. Let us now draw one diagonal in each cell at the angle $\alpha = \arctan(h_2/h_1)$ to the positive direction of the x-axis. Then any inner node i of the obtained triangular mesh is a common vertex of the six triangular elements (see Fig. 4.7). Determine the order of the local approximation of the operator $\partial F(x, y)/\partial x$ in the node i.

Hint. Replace the lines of the program appr3.ma

```
c1x = cc[[1]]; c1y = ss[[1]]; c2x = cc[[2]]; c2y = ss[[2]];
c3x = cc[[3]]; c3y = ss[[3]]; c4x = cc[[4]]; c4y = ss[[4]];
c5x = cc[[5]]; c5y = ss[[5]]; c6x = cc[[6]]; c6y = ss[[6]];
```

by the lines

```
c1x = 1;      c1y = 0;      c2x = 1;      c2y = c;
c3x = 0;      c3y = c;      c4x = -1;     c4y = 0;
c5x = -1;     c5y = -c;     c6x = 0;      c6y = -c;
```

Here c is the cell aspect ratio, $c = h_2/h_1$.

Exercise 4.15. Modify the program appr3.ma to find the expressions r0, r1, r2 corresponding to the difference approximation of the spatial derivative $\partial G(u)/\partial y$ on a mesh of triangular cells of Fig. 4.7.

4.7 LOCAL APPROXIMATION STUDY OF DIFFERENCE SCHEMES ON LOGICALLY RECTANGULAR GRIDS

In this section we want to describe an algorithm for the local approximation study of difference operators and difference schemes on logically rectangular grids. These are the curvilinear grids, which with the aid of a one-to-one mapping

$$x = x(\xi, \eta), \quad y = y(\xi, \eta)$$

can be transformed to an uniform rectangular grid in the (ξ, η) plane. Such grids now have a widespread acceptance.[2,24]

4.7.1 Scalar Differencing Operator

Let us take some differentiable function $f(x, y)$. Let

$$\tilde{f}(\xi, \eta) = f(x(\xi, \eta), y(\xi, \eta)) \qquad (4.7.1)$$

Then we can write that

$$\frac{\partial \tilde{f}}{\partial \xi} = \frac{\partial f}{\partial x}\frac{\partial x}{\partial \xi} + \frac{\partial f}{\partial y}\frac{\partial y}{\partial \xi}, \quad \frac{\partial \tilde{f}}{\partial \eta} = \frac{\partial f}{\partial x}\frac{\partial x}{\partial \eta} + \frac{\partial f}{\partial y}\frac{\partial y}{\partial \eta} \qquad (4.7.2)$$

Solving this system with respect to $\partial f/\partial x$ and $\partial f/\partial y$, we can find the following expressions:

$$\frac{\partial f}{\partial x} = \frac{1}{J}\left(\frac{\partial \tilde{f}}{\partial \xi}\frac{\partial y}{\partial \eta} - \frac{\partial \tilde{f}}{\partial \eta}\frac{\partial y}{\partial \xi}\right), \quad \frac{\partial f}{\partial y} = \frac{1}{J}\left(\frac{\partial \tilde{f}}{\partial \eta}\frac{\partial x}{\partial \xi} - \frac{\partial \tilde{f}}{\partial \xi}\frac{\partial x}{\partial \eta}\right) \quad (4.7.3)$$

where J is the Jacobian (4.5.10). Let us now construct a finite difference approximation of the partial derivative $\partial f/\partial x$ basing on its expression (4.7.3). For this purpose we approximate the partial derivatives $\partial \tilde{f}/\partial \xi$, $\partial \tilde{f}/\partial \eta$, $\partial x/\partial \xi$, $\partial x/\partial \eta$, $\partial y/\partial \xi$, $\partial y/\partial \eta$ in (4.7.3) by the central differences as in (4.5.31):

$$\begin{aligned}
(x_\xi)_{jk} &= (x_{j+1,k} - x_{j-1,k})/2, \quad &(x_\eta)_{jk} &= (x_{j,k+1} - x_{j,k-1})/2 \\
(y_\xi)_{jk} &= (y_{j+1,k} - y_{j-1,k})/2, \quad &(y_\eta)_{jk} &= (y_{j,k+1} - y_{j,k-1})/2 \quad (4.7.4) \\
(\tilde{f}_\xi)_{jk} &= (\tilde{f}_{j+1,k} - \tilde{f}_{j-1,k})/2, \quad &(\tilde{f}_\eta)_{jk} &= (\tilde{f}_{j,k+1} - \tilde{f}_{j,k-1})/2
\end{aligned}$$

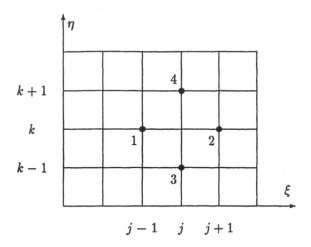

Figure 4.8: Numbering of grid nodes in the vicinity of the (j, k) node.

Replacing the derivatives $\partial \tilde{f}/\partial \xi$, $\partial \tilde{f}/\partial \eta$, $\partial x/\partial \xi$, $\partial x/\partial \eta, \partial y/\partial \xi$, $\partial y/\partial \eta$ entering the differential operator $\partial f/\partial \xi$ by the corresponding central differences (4.7.4) we obtain the difference operator

$$(D_{xh}f)_{jk} = (1/J_{jk})[(\tilde{f}_\xi)_{jk}(y_\eta)_{jk} - (\tilde{f}_\eta)_{jk}(y_\xi)_{jk}] \qquad (4.7.5)$$

where

$$J_{jk} = (x_\xi)_{jk}(y_\eta)_{jk} - (x_\eta)_{jk}(y_\xi)_{jk} \qquad (4.7.6)$$

Let us consider in more detail the central difference for $(\tilde{f}_\xi)_{jk}$ and $(\tilde{f}_\eta)_{jk}$. Since Eq. (4.7.1) holds, we can write that

$$\begin{aligned}
(\tilde{f}_\xi)_{jk} &= (f(x_{j+1,k}, y_{j+1,k}) - f(x_{j-1,k}, y_{j-1,k}))/2 \\
(\tilde{f}_\eta)_{jk} &= (f(x_{j,k+1}, y_{j,k+1}) - f(x_{j,k-1}, y_{j,k-1}))/2
\end{aligned} \qquad (4.7.7)$$

We now apply the Steps 1-5 of the algorithm presented in Section 4.6 for the local approximation study of the difference operator (4.7.5). The stencil St of the difference operator (4.7.5) includes four points:

$$St = \{(j-1, k), (j+1, k), (j, k-1), (j, k+1)\}$$

For convenience of further notation, these points are numbered by 1,2,3,4 as shown in Fig. 4.8. We now represent the coordinates (x_m, y_m), $m = 1, \ldots, 4$, as the deviations from the center (x_c, y_c) of the Taylor expansion similarly to Eqs. (4.6.6):

$$x_m = x_c + C_m^x \cdot h, \quad y_m = x_c + C_m^y \cdot h, \ m = 1, \ldots, 4 \qquad (4.7.8)$$

In accordance with Eq. (4.7.5) we take $x_c = x_{j,k}$, $y_c = y_{j,k}$.

The expansions of the grid values $f(x_{m,l}, y_{m,l})$, $(m, l) \in St$ into the Taylor series are performed similarly to formula (4.6.4).

We can present the obtained Taylor expansion for the difference operator $D_{xh}f$ similarly to Eq. (4.6.13) in the form

$$D_{xh}f = \sum_{\nu=0}^{kt} r_\nu h^\nu \qquad (4.7.9)$$

where the coefficients r_ν, $\nu = 0, 1, \ldots, kt$, do not depend on h. If $r_0 = \partial f / \partial x$, then the approximation of the derivative $\partial f / \partial x$ by the difference formula (4.7.5) takes place.

We have implemented the above symbolic algorithm in the *Mathematica* program appr4.ma whose listing we present below.

```
──────────────────  File appr4.ma  ──────────────────


kt = 3;
(* kt = order of expansion into the Taylor series, kt >1 *)
(* - The program for the local approximation study of
      difference operators on logically rectangular grids *)

(* --- The expansion of the function F(x,y) into the
        Taylor series with respect to (xc, yc) point ---- *)

    f[x_, y_]:=
(sum = fc + fx*(x - xc) + fy*(y - yc) +
1/2*fxx*(x - xc)^2 + fxy*(x - xc)*(y - yc) +
1/2*fyy*(y - yc)^2;
If[kt > 2, sum = sum +
1/6*fxxx*(x - xc)^3 + 1/2*fxxy*(x - xc)^2*(y - yc) +
1/2*fxyy*(x - xc)*(y - yc)^2 +1/6*fyyy*(y - yc)^3]; sum )

(* --- The difference operator under consideration --- *)
       xks = (x2 - x1)/2;      yks = (y2 - y1)/2;
       xet = (x4 - x3)/2;      yet = (y4 - y3)/2;
    (* -- Jac is the jacobian of the transformation -- *)
            Jac = xks*yet - xet*yks;
(* -- Substitution of the Taylor series expansions --- *)
Print["Substitution of the Taylor series expansions"];
Dxhfnum = (1/2)*(f[x2, y2] - f[x1, y1])*yet -
                (1/2)*(f[x4, y4] - f[x3, y3])*yks;

    Jac = Simplify[Jac];
```

```
    OutputForm[Dxhfnum] >> ap4.m;
                                OutputForm[Jac] >>> ap4.m;


                    Dxhfnum = Expand[Dxhfnum];
       Dxhf = Dxhfnum/Jac;
Print["The difference operator expanded into Taylor",
" series has the form"];
 Print[Dxhf];


(* -- Representation of the spatial coordinates of points
      1, 2, 3, 4 as the deviations from the
      center (xc, yc) of the Taylor expansions -------- *)
                    x1 = xc + c1x*h;    y1 = yc + c1y*h;
                    x2 = xc + c2x*h;    y2 = yc + c2y*h;
                    x3 = xc + c3x*h;    y3 = yc + c3y*h;
                    x4 = xc + c4x*h;    y4 = yc + c4y*h;
Print["Substitution of the coordinates of points 1,...,4"];
Print["into the expanded difference operator as the"];
Print["deviations from the (xc, yc) point."];
   Dxhfnum = Dxhfnum;   Dxhfnum = Expand[Dxhfnum];
   OutputForm[Dxhfnum] >>> ap4.m;
   Jac = Jac;    Jac = Expand[Jac];  Jac = Simplify[Jac];
   OutputForm[Jac] >>> ap4.m;
   a2 = Coefficient[Jac, h, 2];
 OutputForm[a2] >>> ap4.m;
Print["Determination of the coefficient of h^2 in the"];
Print["numerator of the expanded difference operator."];
   r0 = Coefficient[Dxhfnum, h, 2];  r0 = Simplify[r0];
                                OutputForm[r0] >>> ap4.m;


Print["Determination of the coefficient of h^3 in the"];
Print["numerator of the expanded difference operator."];
   r1 = Coefficient[Dxhfnum, h, 3];  r1 = Simplify[r1];
                                OutputForm[r1] >>> ap4.m;


Print["Determination of the coefficient of h^4 in the"];
Print["numerator of the expanded difference operator."];
   r2 = Coefficient[Dxhfnum, h, 4];  r2 = Simplify[r2];
                                OutputForm[r2] >>> ap4.m;


Print["The coefficient of h^0 in the operator Dxhf is"];
      r0 = r0/a2;
Print[r0];
   OutputForm[r0] >>> ap4.m;
```

```
Print["The coefficient of h^1 in the operator Dxhf is"];
        r1 = r1/a2;
Print[r1];
   OutputForm[r1] >>> ap4.m;
Print["The coefficient of h^1 in the operator Dxhf is"];
        r2 = r2/a2;
Print[r2];
   OutputForm[r2] >>> ap4.m;


(* ------- The particular case of square cells ------- *)
Print["The particular case of square cells"];
        c1x = -1; c1y = 0; c2x = 1; c2y = 0;
        c3x = 0; c3y = -1; c4x = 0; c4y = 1;
  r1 = r1; r1 = Simplify[r1];
Print["The coefficient of h in the operator Dxhf"];
Print["in the case of square cells is"];
Print[r1];
            OutputForm[r1] >>> ap4.m;
  r2 = r2; r2 = Simplify[r2];
Print["The coefficient of h^2 in the operator Dxhf"];
Print["in the case of square cells is"];
Print[r2];
            OutputForm[r2] >>> ap4.m;
```

When writing the program appr4.ma we have used some elements of the program appr3.ma from Section 4.6. The results of symbolic computations are stored in the file ap4.m. Their list is the same as in the case of the program appr3.ma.

In the process of the work of the program appr4.ma the following message appears on the screen of the computer monitor:

```
The coefficient of h^0 in the operator Dxhf is
fx
```

This means that the difference operator (4.7.5) indeed ensures the approximation of the partial derivative $\partial f/\partial x$ at the (j, k) point.

The leading term r_1 of the truncation error was obtained by the program appr4.ma in the form

```
r1=(-(c1y c3x^2 fxx) +c2y c3x^2 fxx +c1x^2 c3y fxx -

c2x^2 c3y fxx + c1y c4x^2 fxx - c2y c4x^2 fxx -
```

```
c1x^2 c4y fxx +c2x^2 c4y fxx +

2 c1x c1y c3y fxy -2 c2x c2y c3y fxy -2 c1y c3x c3y fxy +

2 c2y c3x c3y fxy -2 c1x c1y c4y fxy +2 c2x c2y c4y fxy +

2 c1y c4x c4y fxy - 2 c2y c4x c4y fxy + c1y^2 c3y fyy -

c2y^2 c3y fyy -2 c1y c3y^2 fyy + c2y c3y^2 fyy -

c1y^2 c4y fyy + c2y^2 c4y fyy + 2 c1y c4y^2 fyy -

c2y c4y^2 fyy) /(2 (-(c1y c3x) + c2y c3x + c1x c3y -

c2x c3y + c1y c4x - c2y c4x - c1x c4y + c2x c4y))
```

This means that the approximation of the operator $\partial f/\partial x$ by the central differences on a logically rectangular grid has only the first order of approximation in space in the case of a general curvilinear grid.

We will not present here the expression for the next term, r_2, of the truncation error because of its bulky form. The interested reader can find it in the output file ap4.m.

Let us consider the particular case of an uniform rectangular grid in the (x, y) plane. In this case we should take in (4.7.8) the following values of the constants C_m^x and C_m^y (see Fig. 4.8):

$$C_1^x = -1, \quad C_1^y = 0, \quad C_2^x = 1, \quad C_2^y = 0$$
$$C_3^x = 0, \quad C_3^y = -1, \quad C_4^x = 0, \quad C_4^y = 1$$

At these values, the program appr4.ma printed the following messages on the screen of the computer monitor:

```
The particular case of square cells
The coefficient of h in the operator Dxhf
in the case of square cells is
0
The coefficient of h^2 in the operator Dxhf
in the case of square cells is
fxxx
----
6
```

It follows from here that in this particular case the difference operator (4.7.5) has the second order of approximation in space. The obtained leading term of the truncation error $r_2 h^2 = h^2 f_{xxx}/6$ coincides with that obtained in Section 2.1 in the case of an uniform grid.

4.7.2   Local Approximation Study of a Difference Scheme for the Euler Equations

We consider the system of Euler equations governing the two-dimensional nonstationary flow of an inviscid compressible non-heat-conducting fluid

$$\frac{\partial \mathbf{u}}{\partial t} + \frac{\partial \mathbf{F(u)}}{\partial x} + \frac{\partial \mathbf{G(u)}}{\partial y} = 0 \qquad (4.7.10)$$

where

$$\mathbf{u} = \begin{pmatrix} \rho \\ \rho u \\ \rho v \\ \rho E \end{pmatrix}, \; \mathbf{F(u)} = \begin{pmatrix} \rho u \\ p + \rho u^2 \\ \rho u v \\ pu + \rho u E \end{pmatrix}, \; \mathbf{G(u)} = \begin{pmatrix} \rho v \\ \rho u v \\ p + \rho v^2 \\ pv + \rho v E \end{pmatrix} \quad (4.7.11)$$

Here ρ is the gas density, p is the pressure, u and v are the components of the gas velocity vector along the x- and y- axes, respectively. $E = \varepsilon + (u^2 + v^2)/2$, ε is the specific internal energy. We assume in what follows that the system (4.7.10), (4.7.11) is completed by the ideal gas equation of state $p = (\gamma - 1)\rho\varepsilon$.

In the case when a stationary gas dynamic problem is considered the solution $\mathbf{u}$ does not depend on the time t: $\mathbf{u} = \mathbf{u}(x, y)$. Then $\partial \mathbf{u}/\partial t = 0$, and the system (4.7.10) reverts to the form

$$\frac{\partial \mathbf{F(u)}}{\partial x} + \frac{\partial \mathbf{G(u)}}{\partial y} = 0 \qquad (4.7.12)$$

where the vector columns $\mathbf{u}$, $\mathbf{F(u)}$ and $\mathbf{G(u)}$ are given by formulas (4.7.11). In the case of an irrotational isentropic gas flow it is possible to introduce the velocity potential $\varphi(x, y)$ by the formulas

$$u = \partial\varphi/\partial x, \quad v = \partial\varphi/\partial y$$

Then the following equation can be obtained from the Euler equations (4.7.12) for the velocity potential[7,32]:

$$(u^2 - c^2)\varphi_{xx} + (v^2 - c^2)\varphi_{yy} + 2uv\varphi_{xy} = 0 \qquad (4.7.13)$$

where c is the sound speed. In the regions of subsonic flow velocities Eq. (4.7.13) is of *elliptic* type, and in the regions of supersonic flow velocities it is of *hyperbolic* type.[32]

Let us now transform the nonstationary system (4.7.10), so that the transformed PDE system ivolves the $\xi-$ and $\eta-$derivatives instead of the $x-$ and $y-$derivatives of the original system. In the case of a differentiable function $f(x, y)$ the equations (4.7.3) hold. Let us now take another function $g(x, y)$ and let $\tilde{g}(\xi, \eta) = g(x(\xi, \eta), y(\xi, \eta))$. Taking into account the formulas (4.7.3) we can write that

$$\frac{\partial f}{\partial x} + \frac{\partial g}{\partial y} = \frac{1}{J}\left[\left(\frac{\partial \tilde{f}}{\partial \xi}\frac{\partial y}{\partial \eta} - \frac{\partial \tilde{f}}{\partial \eta}\frac{\partial y}{\partial \xi}\right) + \left(\frac{\partial \tilde{g}}{\partial \eta}\frac{\partial x}{\partial \xi} - \frac{\partial \tilde{g}}{\partial \xi}\frac{\partial x}{\partial \eta}\right)\right] \qquad (4.7.14)$$

The formula (4.7.14) has the shortcoming that it is in a nondivergence form, that is it does not have the form

$$\partial f/\partial x + \partial g/\partial y = (1/J)(\partial A/\partial \xi + \partial B/\partial \eta) \qquad (4.7.15)$$

where A and B are some functions of ξ and η. Let us show the validity of the following formula:

$$\frac{\partial f}{\partial x} + \frac{\partial g}{\partial y} = \frac{1}{J}\left[\frac{\partial}{\partial \xi}\left(\tilde{f}\frac{\partial y}{\partial \eta} - \tilde{g}\frac{\partial x}{\partial \eta}\right) + \frac{\partial}{\partial \eta}\left(\tilde{g}\frac{\partial x}{\partial \xi} - \tilde{f}\frac{\partial y}{\partial \xi}\right)\right] \qquad (4.7.16)$$

We can obviously write that

$$\frac{\partial}{\partial \xi}\left(\tilde{f}\frac{\partial y}{\partial \eta} - \tilde{g}\frac{\partial x}{\partial \eta}\right) + \frac{\partial}{\partial \eta}\left(\tilde{g}\frac{\partial x}{\partial \xi} - \tilde{f}\frac{\partial y}{\partial \xi}\right) =$$

$$\frac{\partial \tilde{f}}{\partial \xi}\frac{\partial y}{\partial \eta} + \tilde{f}\frac{\partial^2 y}{\partial \eta \partial \xi} - \frac{\partial \tilde{f}}{\partial \eta}\frac{\partial y}{\partial \xi} - \tilde{f}\frac{\partial^2 y}{\partial \xi \partial \eta} + \frac{\partial \tilde{g}}{\partial \eta}\frac{\partial x}{\partial \xi} + \tilde{g}\frac{\partial^2 x}{\partial \eta \partial \xi} - \frac{\partial \tilde{g}}{\partial \xi}\frac{\partial x}{\partial \eta} -$$

$$\tilde{g}\frac{\partial^2 x}{\partial \xi \partial \eta} = \left(\frac{\partial \tilde{f}}{\partial \xi}\frac{\partial y}{\partial \eta} - \frac{\partial \tilde{f}}{\partial \eta}\frac{\partial y}{\partial \xi}\right) + \left(\frac{\partial \tilde{g}}{\partial \eta}\frac{\partial x}{\partial \xi} - \frac{\partial \tilde{g}}{\partial \xi}\frac{\partial x}{\partial \eta}\right) +$$

$$\tilde{f}\left(\frac{\partial^2 y}{\partial \eta \partial \xi} - \frac{\partial^2 y}{\partial \xi \partial \eta}\right) + \tilde{g}\left(\frac{\partial^2 x}{\partial \xi \partial \eta} - \frac{\partial^2 x}{\partial \eta \partial \xi}\right)$$

The functions $y(\xi, \eta)$ and $x(\xi, \eta)$ are assumed to be sufficiently smooth, therefore, $y_{\xi\eta} = y_{\eta\xi}$, $x_{\xi\eta} = x_{\eta\xi}$, and we easily see that the right hand side of formula (4.7.16) coincides with the right hand side of formula (4.7.14). Let us now return to the Euler equation system (4.7.10). Let us set in (4.7.16) $f(x, y) = \mathbf{F(u)}$, $g(x, y) = \mathbf{G(u)}$. We also assume in the following that the curvilinear grid in the (x, y) plane is fixed, that is $x = x(\xi, \eta)$, $y = y(\xi, \eta)$. Then we may rewrite the Euler equation system in the divergence form as follows:

$$\frac{\partial \mathbf{u}J}{\partial t} + \frac{\partial}{\partial \xi}\left(\mathbf{F}\frac{\partial y}{\partial \eta} - \mathbf{G}\frac{\partial x}{\partial \eta}\right) + \frac{\partial}{\partial \eta}\left(\mathbf{G}\frac{\partial x}{\partial \xi} - \mathbf{F}\frac{\partial y}{\partial \xi}\right) = 0 \qquad (4.7.17)$$

Using the formulas (4.5.11) and (4.5.12) we can rewrite equation (4.7.17) also in the form

$$\frac{\partial \mathbf{u}J}{\partial t} + \frac{\partial}{\partial \xi}\left(J\xi_x\mathbf{F} + J\xi_y\mathbf{G}\right) + \frac{\partial}{\partial \eta}\left(J\eta_x\mathbf{F} + J\eta_y\mathbf{G}\right) = 0 \qquad (4.7.18)$$

Let us introduce for brevity of notation the following vector columns:

$$\tilde{\mathbf{F}} = J\xi_x\mathbf{F} + J\xi_y\mathbf{G}, \quad \tilde{\mathbf{G}} = J\eta_x\mathbf{F} + J\eta_y\mathbf{G} \qquad (4.7.19)$$

Then we can rewrite the system (4.7.18) in the form

$$\frac{\partial \mathbf{u}J}{\partial t} + \frac{\partial \tilde{\mathbf{F}}}{\partial \xi} + \frac{\partial \tilde{\mathbf{G}}}{\partial \eta} = 0 \qquad (4.7.20)$$

We will illustrate the algorithm for the local approximation study at the example of the MacCormack scheme[33] applied to the system (4.7.20). We can write this scheme similarly to (2.12.3):

$$(\mathbf{u}J)_{j,k}^{(1)} = (\mathbf{u}J)_{j,k}^n - \frac{\tau}{\Delta\xi}(\tilde{\mathbf{F}}_{j+1,k}^n - \tilde{\mathbf{F}}_{j,k}^n) - \frac{\tau}{\Delta\eta}(\tilde{\mathbf{G}}_{j,k+1}^n - \tilde{\mathbf{G}}_{j,k}^n)$$

$$(\mathbf{u}J)_{j,k}^{n+1} = \frac{1}{2}\{(\mathbf{u}J)_{j,k}^{(1)} + (\mathbf{u}J)_{j,k}^n - \frac{\tau}{\Delta\xi}[\tilde{\mathbf{F}}_{j,k}^{(1)} - \tilde{\mathbf{F}}_{j-1,k}^{(1)}] - \frac{\tau}{\Delta\eta}[\tilde{\mathbf{G}}_{j,k}^{(1)} - \tilde{\mathbf{G}}_{j,k-1}^{(1)}]\}$$

$$(4.7.21)$$

where we have used the notations

$$\tilde{\mathbf{F}}_{j,k}^{(1)} = \tilde{\mathbf{F}}(\mathbf{u}_{j,k}^{(1)}), \quad \tilde{\mathbf{G}}_{j,k}^{(1)} = \tilde{\mathbf{G}}(\mathbf{u}_{j,k}^{(1)})$$

etc. Note that the stencil of the MacCormack scheme (4.7.21) is non-symmetric with respect to the (j,k) point (see Fig. 4.9) and includes thirteen points. Let $\Omega_{j,k}$ be the stencil of scheme (4.7.21). In the case under consideration it looks as

$$\Omega_{j,k} = \{(j-2,k+1),(j-1,k+1),(j,k+1),(j+1,k+1),$$
$$(j-2,k),(j-1,k),(j,k),(j+1,k),(j-1,k-1),(j,k-1),$$
$$(j+1,k-1),(j,k-2),(j+1,k-2)\} (4.7.22)$$

Note that in the case of an uniform rectangular grid in the (x,y) plane the stencil of scheme (4.7.21) would include only seven points:

$$\Omega_{j,k}^{unif.} = \{(j,k),(j+1,k),(j-1,k),(j,k+1),$$
$$(j,k-1),(j-1,k+1),(j+1,k-1)\}$$

The additional points in the stencil in the case of a general curvilinear grid are due to the fact that we have to approximate the derivatives y_η, x_η, x_ξ and y_ξ appearing in (4.7.17). The additional stencil points, which have appeared in (4.7.22) because of the central difference approximations of the ξ-derivatives of x and y in (4.7.17), are shown in Fig. 4.9 by $\otimes$. The additional stencil points due to the central difference approximation of the η-derivatives of x and y in (4.7.17) are shown in Fig. 4.9 by $\oplus$.

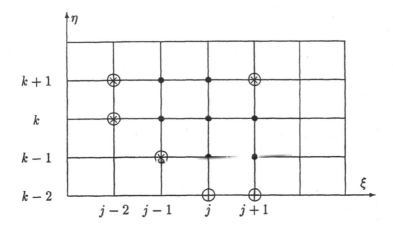

Figure 4.9: The points of the stencil of scheme (4.7.21) at $t = t_n$

Since the algebraic computations needed for the approximation study of scheme (4.7.21) are rather laborious (although the underlying idea of these computations is very simple), it is reasonable to use the symbolic manipulations on a computer (the computer algebra) for the implementation of this study. We mention here the computer algebra systems REDUCE, Mathematica and MAPLE among the most popular systems. Now we describe the steps of the algorithm for the local approximation study of difference schemes approximating the Euler equation system (4.7.20).

Step 1. If the difference scheme under consideration involves the intermediate steps (as the MacCormack scheme (4.7.21), we at first eliminate the intermediate grid quantities to obtain a difference scheme involving only the grid values at the time levels t_n and t_{n+1}. In the case of scheme (4.7.21) we substitute the expressions for $u_{j,k}^{(1)}$ from the first equation of scheme (4.7.21) into the second equation of the scheme. As a result we obtain the difference equation of the form

$$(uJ)_{j,k}^{n+1} = \sum_{m,l \in \Omega_{j,k}} \mathbf{D}_{j,k}^{m,l}(u_{m,l}^n) \qquad (4.7.23)$$

Here the functions $\mathbf{D}_{jk}^{ml}$ involve also the coordinates of nodes comprised in the stencil.

Step 2. All the functions $u_{m,l}^n$ in the right hand side of (4.7.23) as

well as the function $\mathbf{u}_{j,k}^{n+1}$ in the left hand side of (4.7.23) are expanded
into truncated Taylor series at some point (x^*, y^*, t^*), which will be
called the center of the Taylor expansion.

$$\mathbf{u}_{kl}^{\nu} = \mathbf{u}(x^*, y^*, t^*)+$$

$$\sum_{i_1+i_2+i_3=1}^{\mathbf{kt}} \frac{1}{i_1! i_2! i_3!} \frac{\partial^{i_1+i_2+i_3} \mathbf{u}}{\partial x^{i_1} \partial y^{i_2} \partial t^{i_3}} \cdot (x_{kl} - x^*)^{i_1} (y_{kl} - y^*)^{i_2} (t_{\nu} - t^*)^{i_3} + R$$

$$(4.7.24)$$

where $\mathbf{kt}$ is the user-specified order of Taylor expansion (for the Mac-
Cormack scheme we take $\mathbf{kt} \geq 3$), and R is the remainder term of the
Taylor formula. The expressions of the form (4.7.24) are substituted
into the difference scheme (4.7.23). Let us denote the result of these
substitutions by $\mathbf{Q}_1 = 0$. The expression $\mathbf{Q}_1$ is a ratio of two poly-
nomials depending on the derivatives $\partial^{i_1+i_2+i_3} \mathbf{u}/\partial x^{i_1} \partial y^{i_2} \partial t^{i_3}$ and the
coordinates $x^*, y^*, t^*, x_{kl}, y_{kl}, t_n, t_{n+1} = t_n + \tau$, where the derivatives
enter only the numerator.

Step 3. The specification of the point (x^*, y^*, t^*). In the case of the
MacCormack scheme (4.7.21) it is natural to take

$$x^* = x_j, \quad y^* = y_k, \quad t^* = t_n$$

In addition, we make in $\mathbf{Q}_1$ the substitution $t_{n+1} = t_n + \tau$.

Step 4. Representation of the 26 nodes coordinates (x_{ml}, y_{ml}),
$m, l \in \Omega_{j,k}$, in the form of the deviations from the center (x^*, y^*):

$$x_{ml} = x^* + C_{ml}^x h, \quad y_{ml} = y^* + C_{ml}^y h, \quad (m, l) \in \Omega_{j,k} \quad (4.7.25)$$

Here C_{ml}^x, C_{ml}^y are constants, which do not depend on h and τ. The
quantity h characterizes as in Section 4.6 the local density of the curvilinear mesh in the neighborhood of the (j, k) node; it can be given by
the formula

$$h = (1/(N_s - 1)) \left\{ \sum_{(m,l) \in \Omega_{j,k}} [(x_{ml} - x^*)^2 + (y_{ml} - y^*)^2] \right\}^{0.5}$$

in the case when $x^* = x_j, y^* = y_k$. Here N_s is the number of points in
the stencil of the scheme. For example, $N_s = 13$ in the case of the Mac-
Cormack scheme (4.7.21). Let us denote the result of the substitutions
(4.7.25) into the expression $\mathbf{Q}_1 = 0$ by $\mathbf{Q}_2 = 0$.

Step 5. The original equation system (4.7.10) is resolved with respect to the time derivative, and for this equation system a table is calculated which contains the differential consequencies of equation (4.7.10)
up to the order $\mathbf{kt}$ (see (4.7.24)). By the differential consequence of
equation (4.7.10) is meant a differential equation obtained as a result of

one or multiple differentiation of the original equation with respect to independent variables. For example, let us find one of the second-order differential consequences of equation (4.7.10):

$$\frac{\partial}{\partial t}\left(\frac{\partial \mathbf{u}}{\partial t}\right) = -\frac{\partial}{\partial t}\left[\frac{\partial \mathbf{F}(\mathbf{u})}{\partial x} + \frac{\partial \mathbf{G}(\mathbf{u})}{\partial y}\right] =$$

$$-\frac{\partial}{\partial x}\frac{\partial \mathbf{F}(\mathbf{u})}{\partial t} - \frac{\partial}{\partial y}\frac{\partial \mathbf{G}(\mathbf{u})}{\partial t} = -\frac{\partial}{\partial x}\left(A(\mathbf{u})\frac{\partial \mathbf{u}}{\partial t}\right) - \frac{\partial}{\partial y}\left(B(\mathbf{u})\frac{\partial \mathbf{u}}{\partial t}\right) =$$

$$\frac{\partial}{\partial x}\left[A(\mathbf{u})\left(\frac{\partial \mathbf{F}(\mathbf{u})}{\partial x} + \frac{\partial \mathbf{G}(\mathbf{u})}{\partial y}\right)\right] + \frac{\partial}{\partial y}\left[B(\mathbf{u})\left(\frac{\partial \mathbf{F}(\mathbf{u})}{\partial x} + \frac{\partial \mathbf{G}(\mathbf{u})}{\partial y}\right)\right] \quad (4.7.26)$$

where $A(\mathbf{u})$ and $B(\mathbf{u})$ are the Jacobi matrices,

$$A(\mathbf{u}) = \partial \mathbf{F}/\partial \mathbf{u}, \ \ B(\mathbf{u}) = \partial \mathbf{G}/\partial \mathbf{u}$$

The second-order and the higher-order derivatives of $\mathbf{u}$, which contain the differentiation with respect to t, are expressed with the aid of the differential consequences like (4.7.26) in terms of the spatial derivatives of $\mathbf{u}$. Let us denote the result of the corresponding substitutions by $\mathbf{Q}_3 = 0$.

Step 6. The original differential equation (4.7.10) is extracted in $\mathbf{Q}_3$, so that $\mathbf{Q}_3$ takes the form

$$\mathbf{Q}_3 = \frac{\partial \mathbf{u}}{\partial t} + \frac{\partial \mathbf{F}(\mathbf{u})}{\partial x} + \frac{\partial \mathbf{G}(\mathbf{u})}{\partial y} + R_3$$

where R_3 is the remainder term. The right hand side of equation

$$L\mathbf{u} = \frac{\partial \mathbf{u}}{\partial t} + \frac{\partial \mathbf{F}(\mathbf{u})}{\partial x} + \frac{\partial \mathbf{G}(\mathbf{u})}{\partial y} = -R_3$$

is written down in the form of a polynomial in τ and h:

$$L\mathbf{u} = -R_3 = \sum_{k=k_0}^{\mathbf{k_t}-1} \sum_{\alpha_0 + \alpha_1 = k} c_{\alpha_0 \alpha_1} \tau^{\alpha_0} h^{\alpha_1} \quad (4.7.27)$$

where $c_{\alpha_0 \alpha_1}$ are certain 4×4 matrices, whose entries do not depend on h and τ; $k_0 \geq 0$. The summation in k in (4.7.27) is performed up to the value $k = M - 1$. The reason for this is that the derivatives with respect to independent variables are usually approximated by the divided differences.

If it turns out for a specific difference scheme that $c_{0,0} = 0$ in (4.7.27), then this means that the difference scheme approximates the original equation system (4.7.10). The MacCormack scheme (4.7.21) is known to have the second order of approximation both in space and time in the particular case of an uniform rectangular grid in the (x, y) plane.[33,31] It follows from the consideration of the local approximation of the operator (4.7.5) that the approximation order of scheme (4.7.21) in space reduces by one in the case of a general curvilinear grid, so that the order of approximation in space is $O(h)$.

The above algorithm can be used also for the local approximation study of the difference schemes approximating the stationary Euler equation system (4.7.12), if the latter is solved numerically by the pseudo-unsteady method. In this case the time derivatives in the obtained expression (4.7.27) should be equalled to zero, because the converged numerical stationary solution does not depend on t.

The Euler equations are nonlinear; they involve, for example, the product $\rho v u^2$. The substitution of the Taylor series expansions for ρ, u, v in such a product leads to tremendous intermediate expressions. This can become the reason for the memory exhaustion in the process of symbolic computations. In this connection we enumerate in what follows a number of the techniques for the solution of the problem of the memory exhaustion.

1°. *Separate consideration of approximations of the four Euler equations* (4.7.10), (4.7.11). In this case, for the first run on a computer, the momentum and energy equations are eliminated from the further consideration, and only the local approximation of the first equation of system (4.7.10)-(4.7.11), the continuity equation, is studied. The needed amount of computer memory thus reduces by a factor of about four. For the second run, only the second equation of system (4.7.10), (4.7.11) is taken, etc. This can lead to the needed complete result after the four runs on a computer.

2°. *The introduction of intermediate notations.* In the interactive process of a symbolic computation, one can find the expressions for the coefficients of various powers of h and immediately make their denotations. For example, a large expression for the coefficient r2 in the product r2 h^2 can be replaced by a temporary name r20. The substitution r20 = r2 can be made after obtaining the final expression of the form (4.7.27).

Exercise 4.16. Let the walls of a plane channel in the (x, y) plane be given by the equations $y = f_1(x)$ (the lower wall) and $y = f_2(x)$ (the upper wall), where $a \le x \le b$, a and b are the given constants.

Find the form of the transformation (2.14.2) at which the flow region inside the channel is mapped onto a rectangle in the (ξ, η) plane. Derive the transformed Euler equations in the variables ξ, η, t.

ANSWERS TO THE EXERCISES

4.5.

$$\left(\delta_x^2 + \delta_y^2 + \frac{h_1^2 + h_2^2}{6}\delta_x^2\delta_y^2\right)v_{ij} = -g_{ij}$$

$$i = 2, \ldots, M_1 - 1; \quad j = 2, \ldots, M_2 - 1$$

$$\delta_x^2 = \frac{v_{i+1,j} - 2v_{ij} + v_{i-1,j}}{h_1^2}, \quad \delta_y^2 = \frac{v_{i,j+1} - 2v_{ij} + v_{i,j-1}}{h_2^2}$$

$$g_{ij} = \iint_{d_{ij}} f(x,y)\psi_{ij}(x,y)dx\,dy$$

$$d_{ij} = [x_{i-1}, x_{i+1}] \times [y_{j-1}, y_{j+1}]$$

See also[22].

4.8.

$$x_\xi x_\eta + y_\xi y_\eta = 0 \qquad (4.7.28)$$

4.9. Let us at first solve the system (2.14.1) with respect to x and y:

$$x = \frac{\xi \cdot x_{nz}}{i_x - 1}, \quad y = \frac{\eta \cdot f_w(x)}{j_y - 1} \qquad (4.7.29)$$

Then we calculate the partial derivatives

$$\partial x(\xi, \eta)/\partial\xi, \ \partial x(\xi, \eta)/\partial\eta, \ \partial y(\xi, \eta)/\partial\xi, \ \partial y(\xi, \eta)/\partial\eta$$

as this is required by the orthogonality condition (4.7.28):

$$x_\xi = x_{nz}/(i_x - 1), \quad x_\eta = 0$$

$$y_\xi = \frac{\eta}{j_y - 1}\frac{df_w(x)}{dx}\cdot\frac{x_{nz}}{i_x - 1}; \ y_\eta = \frac{f_w(x)}{j_y - 1}$$

Therefore,

$$x_\xi x_\eta + y_\xi y_\eta = \frac{\eta x_{nz}f_w'(x)f_w(x)}{(j_y - 1)^2(i_x - 1)} = 0 \qquad (4.7.30)$$

Since it is assumed that $f_w(x) > 0$, we obtain from (4.7.30) that

$$df_w(x)/dx = 0$$

or $f_w(x) = const$. This means that the channel wall should be parallel to the x-axis to ensure that the curvilinear grid produced with the aid of the transformation (2.14.1) is orthogonal.

4.10. Let us introduce the polar coordinates r, φ by the formulas

$$x = r\cos\varphi, \quad y = r\sin\varphi \qquad (4.7.31)$$

where

$$0 \le r \le R, \quad 0 \le \varphi \le 2\pi \qquad (4.7.32)$$

It follows from the inequalities (4.7.32) that the transformation (4.7.31) maps the interior of the circle (4.5.37) onto a rectangular region in the (φ, r) plane.

In order to check whether the orthogonality condition (4.7.28) is satisfied let us set $\xi \equiv \varphi$, $\eta \equiv r$. Then we have that

$$x_\xi = -r\sin\varphi, \ x_\eta = \cos\varphi, \ y_\xi = r\cos\varphi, \ y_\eta = \sin\varphi$$

therefore, the condition (4.7.28) is satisfied.

4.11. Let us at first find the following equivalent form for the equation (4.5.40):

$$\frac{(x-1)^2}{4} + \frac{(y+3)^2}{2} = 1$$

This means that the (x, y) points satisfying the inequality

$$\frac{(x-1)^2}{4} + \frac{(y+3)^2}{2} \le 1 \qquad (4.7.33)$$

belong to the interior of the region bounded by the ellipse (4.7.33). We now introduce the polar coordinates r, φ as follows:

$$(x-1)/2 = r\cos\varphi, \ (y+3)/\sqrt{2} = r\sin\varphi$$

where in accordance with (4.7.33)

$$0 \le r \le 1, \ 0 \le \varphi \le 2\pi$$

From the orthogonality condition (4.7.28) we easily obtain the equation

$$\sin 2\varphi = 0$$

It follows from here that the curvilinear grid lines in the (x, y) plane will be orthogonal only at $\varphi = 0$, $\varphi = \frac{\pi}{2}$, $\varphi = \pi$, $\varphi = \frac{3\pi}{2}$.

4.14.

$$(F_x)_h = F_x + \frac{h^2}{6}(F_{xxx} + cF_{xxy} + c^2 F_{xyy})$$

that is the finite volume approximation of the operator F_x is second-order accurate in this particular case. The *Mathematica* program appr31.ma, which solves this task, is available on the attached diskette.

4.16. Similarly to formulas (2.14.1) let us specify the curvilinear coordinates ξ and η by the formulas

$$\xi = (i_x - 1)\frac{x - a}{b - a}, \quad 0 \le \xi \le i_x - 1 \qquad (4.7.34)$$

$$\eta = (j_y - 1)\frac{y - f_1(x)}{f_2(x) - f_1(x)}, \quad 0 \le \eta \le j_y - 1 \qquad (4.7.35)$$

It can be seen that the region inside the channel is mapped onto the rectangular region

$$0 \le \xi \le i_x - 1, \quad 0 \le \eta \le j_y - 1$$

in the (ξ, η) plane. Let us now express x and y in terms of ξ and η by using Eqs. (4.7.34) and (4.7.35):

$$\begin{cases} x = a + c\xi \\ y = f_1(a + c\xi) + \frac{\eta}{j_y - 1}[f_2(a + c\xi) - f_1(a + c\xi)] \end{cases} \qquad (4.7.36)$$

where $c = (b - a)/(i_x - 1)$. We obtain the following expression for the Jacobian of the transformation (4.7.36):

$$J = x_\xi y_\eta - x_\eta y_\xi = \frac{c}{j_y - 1}(f_2 - f_1)$$

Then we have from Eqs. (4.7.34) and (4.7.35) that

$$\xi_x = 1/c, \quad \xi_y = 0$$

$$\eta_x = -(j_y - 1)\frac{f_1'(f_2 - f_1) + (y - f_1)(f_2' - f_1')}{(f_2 - f_1)^2}, \eta_y = \frac{j_y - 1}{f_2 - f_1}$$

where $f_j' = df_j(x)/dx$, $j = 1, 2$. Therefore, the transformed Euler equations in the variables (ξ, η, t) have the form (4.7.20), where

$$\tilde{F} = \frac{f_2 - f_1}{j_y - 1}\mathbf{F}$$

$$\tilde{G} = -c\frac{f_1'(f_2 - f_1) + (y - f_1)(f_2' - f_1')}{f_2 - f_1}\mathbf{F} + c\mathbf{G}$$

REFERENCES

1. **Tikhonov, A.N. and Samarskii, A.A.,** *Equations of the Mathematical Physics,* Fifth Edition (in Russian), Nauka, Moscow, 1977.
2. **Thompson, J.F., Warsi, Z.U.A. and Mastin, C.W.,** *Numerical Grid Generation-Foundations and Applications,* Elsevier Science Publishing Company, New York, 1985.
3. **Samarskii, A.A. and Andreev, V.B.,** *Difference Methods for Elliptic Equations* (in Russian), Nauka, Moscow, 1976.
4. **Konovalov, A.N.,** *Problems of Multiphase Fluid Filtration,* World Scientific, Singapore, 1994.
5. **Strikwerda, J.C.,** *Finite Difference Schemes and Partial Differential Equations,* Wadsworth & Brooks/Cole Advanced Books & Software, Pacific Grove, California, 1989.
6. **Godunov, S.K. and Ryabenkii, V.S.,** *Difference Schemes.. An Introduction to the Theory* (in Russian). Nauka, Moscow, 1977. [English transl.: **Godunov, S.K. and Ryabenkii, V.S.,** *Difference Schemes. An Introduction to the Underlying Theory.* (Studies in Mathematics and its Applications, 19). Elsevier Science Publishing Co., Inc., New York, 1987]
7. **Fletcher, C.A.J.,** *Computational Techniques for Fluid Dynamics 1. Fundamental and General Techniques,* Springer-Verlag, Berlin, Heidelberg, New York, 1988.
8. **Yanenko, N.N.,** *The Method of Fractional Steps.* English transl. edited by M. Holt. Springer-Verlag, New York, 1971.
9. **Roache, P.J.,** *Computational Fluid Dynamics,* Hermosa, Albuquerque, New Mexico, 1976.
10. **Douglas, J. and Rachford, H.H.,** On the numerical solution of heat conduction problems in two and three space variables, *Transactions of the American Mathematical Society,* 82, 421, 1956.
11. **Catherall, D.,** Optimum approximate-factorization schemes for two-dimensional steady potential flows, *AIAA Journal,* 20, 1057, 1982.
12. **Catherall, D.,** The optimisation of approximate-factorisation schemes for solving elliptic partial differential equations, *Journal of Computational Physics,* 78, 138, 1988.
13. **Hrenikoff, A.,** Solutions of problems in elasticity by the framework method, *Journal of Applied Mechanics,* 8, 169, 1941.

14. **Zienkiewicz, O.C.**, Finite elements-the background story. *The Mathematics of Finite Elements and Applications. Proc. Brunel University Conference of the Institute of Mathematics and Its Applications Held in April 1972* / Ed. J.R. Whiteman, Academic Press, London, 1973, p. 1-35.
15. **Zienkiewicz, O.C.**, *The Finite Element Method*, Third Edition, McGraw-Hill, London, 1977.
16. **Marchuk, G.I.**, *Methods of Numerical Mathematics* (in Russian), Nauka, Moscow, 1977.
17. **Fletcher, C.A.J.**, *Computational Galerkin Methods*, Springer-Verlag, Berlin, 1984.
18. **Fletcher, C.A.J.**, *Computational Techniques for Fluid Dynamics*, Springer-Verlag, Berlin, 1988.
19. **Baker, A.J.**, *Finite Element Computational Fluid Mechanics*, McGraw-Hill, New York, 1983.
20. **Oran, E.S. and Boris, J.P.**, *Numerical Simulation of Reactive Flow*, Elsevier Science Publishing Co., New York, 1987.
21. **Mitchell, A.R. and Wait, R.**, *The Finite Element Method in Partial Differential Equations*, Wiley-Interscience, New York, 1977.
22. **Zavyalov, Yu.S., Kvasov, B.I. and Miroshnichenko, V.L.**, *The Methods of Spline Functions* (in Russian), Nauka, Moscow, 1980.
23. **Mikhlin, S.G.**, *Variational Methods in the Mathematical Physics* (in Russian), Nauka, Moscow, 1970.
24. **Knupp, P. and Steinberg, S.**, *Fundamentals of Grid Generation*, CRC Press, Boca Raton, 1994.
25. **Steinberg, S.**, personal communication, 1995.
26. **Ganzha, V.G., Shashkov, M.Yu and Shapeev, V.P.**, *Local Approximation study of difference operators in symbolic form.* Preprint of the Keldysh Institute of Applied Mathematics of the USSR Academy of Sciences (in Russian), No. 79, Moscow, 1987, p. 1-12.
27. **Ganzha, V.G. and Shashkov, M.Yu.**, Local approximation study of difference operators by means of REDUCE system. In: *ISSAC'90 Proceedings* /Eds. S. Watanabe and M. Nagata. Addison-Wesley Publishing Company, Reading, 1990, p. 185.
28. **Hearn, A.C.**, *REDUCE 3.5, User's Manual.* Rand Corporation, Santa Monica, Cal., 1993.
29. **Ganzha, V.G., Solovyov, A.V. and Shashkov, M.Yu.**, A new algorithm for the symbolic computation of the analyses of local approximation of difference operators. In: *Advances in Computer Methods for Partial Differential Equations – VII. Proceedings of the Seventh IMACS Internat. Conf. on Comp-*

uter Methods for Partial Differential Equations, New Brunswick, New Jersey, USA, June 22-24, 1992. / Eds. R. Vichnevetsky, D. Knight and G.Richter. IMACS Publ., New Brunswick, NJ 08903, 1992, p. 280-286.

30. **Shokin, Yu.I.,** *The Method of Differential Approximation.* (Translated from the Russian by K.G. Roesner). Springer-Verlag, Berlin, 1983.

31. **Shokin, Yu.I. and Yanenko, N.N.,** *The Method of Differential Approximation. Application to Gas Dynamics.* (in Russian). Nauka, Siberian Division, Novosibirsk, 1985.

32. **Ovsiannikov, L.V.,** Lectures on the Fundamentals of Gas Dynamics (in Russian). Nauka, Moscow, 1981.

33. **MacCormack, R.W.,** *The effect of Viscosity in Hypervelocity Impact Cratering.* AIAA Paper 69-354, 1969.

APPENDIX GLOSSARY OF PROGRAMS

This book is supplied with a disk containing all the programs which were described in the book. These programs were adjusted for the work with *Mathematica* in the Microsoft Windows environment. The manual

T.Williams and J. Walsh. User's Guide for Microsoft Windows. *Mathematica*. A System for Doing Mathematics by Computer. Version 2.2. Second edition. Wolfram Research, 1993

discusses in detail the features of *Mathematica* that are specific to the Windows version.

This version of *Mathematica* as well as the later Version 2.2.1 Enhanced Version for Microsoft Windows require that the computer has

- A 386 - or 386SX-class computer or higher

- A numeric coprocessor for the Enhanced version of *Mathematica*

- 4 megabytes RAM (6 megabytes RAM is recommended)

- Windows Version 3.0 or higher in enhanced mode (Windows 3.1 is strongly recommended)

- Approximately 14 megabytes of free storage for a full installation

We have made and tested our programs on a personal computer working under the monitoring system DOS. In accordance with the rules of DOS the length of the name of each *Mathematica* program available on the disk does not exceed 8 symbols. In order to use any specific program you can simply copy the needed files on your computer.

The *Mathematica* kernel uses the PostScript language to represent all the graphics it produces. The *Mathematica* for Windows front end includes a PostScript interpreter to translate PostScript code from the kernel into screen images.

Mathematica 2.2 automatically generates the auxiliary files *.mb ("binary parts") which are needed to store the information about the graphics objects. In this connection we have recorded on the disk also the auxiliary files *.mb for all 25 *Mathematica* programs available on the disk. The size of the *.mb files, in bytes, can be quite large in cases when many graphics objects are to be stored. In this connection we have stored in our files on the disk only a few graphics examples to ensure that all our *.ma and *.mb files are placed on a single 1.44 MB disk.

One of the features of *Mathematica* 2.2 or 2.2.1 is that the *Mathematica* documents are called Notebooks. Notebooks can contain ordinary text and graphics as well as *Mathematica* input and output. The disk that comes with this book contains 25 *Mathematica* notebooks.

The information in a Notebook is stored in cells, whose characteristics vary with the cell's function and the kind of information it holds. The brackets on the right side of the Notebook window indicate the extent of each cell.

The typical structure of the Notebooks for this book is as follows. Each Notebook begins with a title, for example, **Fourier Method**. The contents of a Notebook are subdivided into following Sections:

Impressum. This Section contains the text:
This Mathematica-Notebook is part of the book entitled

V.G. Ganzha and E.V. Vorozhtsov. Numerical Solutions For Partial Differential Equations: Problem Solving Using Mathematica. ©CRC Press

Initialisation. This Section indicates the *Mathematica* packages like `Algebra'ReIm'`, `Algebra'Trigonometry'` which are needed for the work of a specific *Mathematica* program contained in the Notebook.

General Description. This Section explains the purpose of the Notebook, the method implemented in a specific program, the limitations of the program, and refers the reader to a specific Section of the book, where the implemented method is described in detail.

User's Guide. This Section represents an internal instruction of the Notebook for the practical use of a *Mathematica* program contained in the Notebook. This instruction can include several steps.

Program Listing. This Section contains the program listing and usually includes the major part of the *Mathematica* program.

Parameters Used in Program *.ma. This Section explains in detail the physical or mathematical meaning of each input parameter, gives the limitations on the numerical values of parameters if necessary.

Examples of the Input Data. This Section gives one or several examples of the specification of the input parameters. These examples enable the user to reproduce all the numerical, analytical or graphics examples that are presented in our book. At the end of this Appendix we present Tables 1-3, which help the interested reader to easily reproduce any of the Figures whose numbers are indicated in the Tables.

If you want to introduce some change into the input data, please always save the original version of the file. This will help you to find an error in your version of input data if our program fails at your data.

One of the possible errors in the process of the preparation of the input data is that one of the entries (divided by commas from each other) is missing, so that the total number of the entries is less than the number needed by our program. Then the *Mathematica* system shows your file of input data on the screen of the computer monitor, for example, as follows:

```
advbc[1., 0., 2., 4, 1., 0.2, 80, 1.08]
```

and thus does not proceed to the computation. In this case we recommend you to look more attentively at your prepared variant of input data.

The Structure of the Output. This Section explains the meaning of each of the data that are printed out in the process of the work of the program.

The names of our *Mathematica* notebooks are given below in the alphabetic order. After the name of each notebook (given by bold face letters) we reproduce the meaning of each entry of the input data by using the corresponding cell of the Notebook. If the given program uses some functions, we also describe briefly the purpose of these functions (the names of the functions are printed in Italics).

We will present below the listings of a number of *.ma programs, which implement different symbolic or numerical algorithms.

advbac.ma The main program of this file,

```
advbc[a_,a1_,b1_,jf0_,ul_,ur_,M_, cap_, Npic_]
```

solves numerically the advection equation (2.1.1) under the initial condition (2.2.6) of a step form (the square wave test) by the explicit one-sided difference scheme (2.1.9). The meaning of the input data is explained in the following cell of the notebook `advbac.ma`:

Parameter	Description
a	the advection speed in eqn (2.1.1), a > 0;
a1	the abscissa of the left end of the interval on the x-axis;
b1	the abscissa of the right end of the above interval, a1 < b1;
jf0	the number of the grid node, which coincides with the discontinuity position at t = 0; it should satisfy the inequalities 3 < jf0 < M;

ul,ur   the constant solution values at t = 0 to the left and
        to the right of the discontinuity;
M   the number of the grid node coinciding with the right
        boundary x = b1 of the spatial integration interval
        (see also (2.1.5));
cap   the Courant number, cap = a*dt/h, where dt is the
        time step of the difference scheme, and h is the step
        of the uniform grid in the interval a1 <=x <= b1;
Npic   the number of the pictures of the difference solution
        graphs for Npic moments of time, Npic > 1.

Below we present the listing of the *Mathematica* program advbac.ma.

```
ClearAll[advbc];
advbc[a_,a1_,b1_,jf0_,ul_,ur_,M_, cap_, Npic_]:=
Module[
(* --- Definition of the local variables --- *)
{uj,up,ug,ut,tnt,y,j,k,n,M1,h,t,dt,xf0,Tmax,N},
(* - uj = the difference solution at time level n,
up = the difference solution at time level n+1 - *)
      h = (b1 - a1)/M;   M1 = M + 1;
        dt = cap*h/a;    xf0 = a1 + jf0*h;
  Tmax = (b1 - 5.0*h - xf0)/a;
uj = Table[ul, {j,0,M}];   up = uj;
               ug = Table[ur,{j, 1, M1}];
          tnt = Table[Tmax, {j, Npic}];
          mt = Table[j, {j, Npic}];
(* - Tmax is the right end of the time interval - *)
          N = Floor[Tmax/dt];
          N = If[N < 1, 1, N];
   nplot = Floor[N/Npic]; nplot = Max[1, nplot];
(* -- The solution graphs u(x, tk) will be given at
     tk = (k-1)*nplot*dt, k = 1,...,Npic ------ *)
       um = Max[Abs[ul], Abs[ur]];
       Print["um = ",um];
     t= 0.0;  n2 = nplot;  nt = 1;  tnt[[1]] = 0.0;
(* -- Computation of the initial values u(xj,0)
                                at grid points -- *)
            Do[ uj[[j]] = ul, {j, 0, jf0}];
            Do[ uj[[j]] = ur, {j, jf0+1, M, 1}];
  (* - Generation of the two-dimensional table ut - *)
       ut = Table[ur, {k, Npic}, {j, M1}];
(* Storing the initial profile u(x,0) in ut[[j, 1]] *)
```

```
         Do[ ut[[1,j + 1]] = uj[[j]], {j, 0, M}];
(* ---- Begin of the Do-loop for time steps ----- *)
         Do[
             t = t + dt;
             Print["n = ", n, ", t = ", t];
(* --- Do-loop over grid nodes on the x-axis ---- *)
             Do[
(* -- The first-order scheme (2.1.9)
                           for equation (2.1.1) -- *)
             up[[j]] = uj[[j]] - cap*(uj[[j]] - uj[[j-1]]),
               {j, 1, M }];
(* -- The boundary condition on the left
                           boundary x = a1 -- *)
                 up[[0]] = ul;
(* ---------- Exchange of the lists ------------- *)
                 uj = up;
(* ------- Find Max[Abs[u(j,n+1) - um]----------- *)
             umax = 0.0;
       Do[ y = Abs[uj[[j]] -um]; umax = Max[y, umax],
           {j, 0, M, 1}];
             If[n == n2, Goto[store]];
         If[umax > 2.*um, Goto [endsteps], Goto [endn]];
Label[endsteps]; Tmax = t;
Label[store]; n2 = n2 + nplot;
       nt = nt + 1;
                     If[nt <= Npic,
         tnt[[nt]] = t;
(* -- Storing the solution u(x,t) in ut[[nt,j]] - *)
         Do[ ut[[nt, j + 1]] = uj[[j]], {j, 0, M}]];
         If[umax > 2.0*um||nt == Npic, Break[]];
 Label[endn],
(* ---------- End of the Do-loop in n ----------- *)
             {n, 0, N, 1}];
(* Plotting the solution graphs for several
                                  moments of time *)
(* ---- Determining min(ut) and max(ut) --------- *)
             umin = 10^6;  umax = -umin;
       nt = If[nt > Npic, Npic, nt];
       Do[
           y = ut[[k, j]];
             umin = If[y < umin, y, umin];
               umax = If[y > umax, y, umax],
           {k, nt}, {j, M1}];
         Print["umin = ", umin, ", umax = ", umax];
```

```
(* -- Storing the pictures in the elements mt[[j]],
      j = 1,...,Npic -- *)
    umin = If[umin > 0, 0.0, umin];
    umax = If[umax < 0, 0.0, umax];
    Do[
        ug = ut[[j]];
        t  = tnt[[j]];
      Print["The solution graph at t = ", t];
  mt[[j]] = ListPlot[ug,
                AxesLabel -> {"x", "u"},
                PlotJoined -> True,
                PlotRange -> {{0, M1}, {umin, umax}}],
          {j, nt}];
If[nt == 2,
gr = Show[GraphicsArray[{mt[[1]], mt[[2]]}]]  ];
If[nt == 3,
gr = Show[GraphicsArray[{{mt[[1]], mt[[2]]},
                        {mt[[3]]}}]]];
If[nt == 4,
gr = Show[GraphicsArray[{{mt[[1]], mt[[2]]},
                        {mt[[3]], mt[[4]]}}]]];
If[nt == 5,
 gr = Show[GraphicsArray[{
     {mt[[1]], mt[[2]], mt[[3]]},
     {mt[[4]], mt[[5]]}}]] ];
If[nt == 6,
 gr = Show[GraphicsArray[{
     {mt[[1]], mt[[2]], mt[[3]]},
     {mt[[4]], mt[[5]], mt[[6]]}}]] ];
If[nt == 7,
 gr = Show[GraphicsArray[{
     {mt[[1]], mt[[2]], mt[[3]], mt[[4]]},
     {mt[[5]], mt[[6]], mt[[7]]}}]] ];
If[nt >= 8,
 gr = Show[GraphicsArray[{
     {mt[[1]], mt[[2]], mt[[3]], mt[[4]]},
     {mt[[5]], mt[[6]], mt[[7]], mt[[8]]}}]] ];
    Display["fig2_5", gr];
(* ------------End of the block ----------------- *)
          ];
```

appr1.ma enables one to compute the symbolic form of the first differential approximation, or the modified equation, of a difference scheme approximating a scalar PDE of the form (2.5.1).

sch1 sch1 represents the left hand side of a difference equation $L_{h,\tau} u^n = 0$ approximating the PDE (2.5.1), where $L_{h,\tau}$ is a difference operator corresponding to a specific difference scheme.

The variables j and n in the body of the function sch1 represent the number of a grid node on the x-axis and the number of a time level, respectively.

eqn eqn gives the left-hand side of the PDE

$$\frac{\partial u}{\partial t} - P\left(u, \frac{\partial u}{\partial x}, \frac{\partial^2 u}{\partial x^2}, \ldots, \frac{\partial^{m_x} u}{\partial x^{m_x}}\right) = 0$$

which is approximated by the difference scheme specified with the aid of the function sch1.

u u[k_,m_] enables one to perform the Taylor expansion of a grid function u_k^m at point ($x =$ xc, $t =$ tc).

The meaning of the input data is explained in the following cell of the Notebook appr1.ma:

Parameter	Description
kt	order of expansion into the Taylor series, kt > 1
eq	the left-hand side of the differential equation eq=0
sc	the left-hand side of the difference equation sc=0

Below we present the listing of the program appr1.ma.

```
ClearAll[appr1];
appr1[kt_, eq_, sc_] := Module[
{eq1},

ClearAll[u, ss, eq1, sc1, sc2, dcs];
   u[k_, m_]:= Block[{x, t, ss},
                               ss = v[x, t];
  Do[
           Do[ i2=ktj - i1;
      dr1 = If[i1==0, 1, (k*h - xc)^i1];
      dr2 = If[i2==0, 1, (m*dt - tc)^i2];
      dr = dr1*dr2/(i1!*i2!);
     ss = ss + Derivative[i1, i2][v][x, t]*dr,
```

```
                        {i1, 0, ktj}],
              {ktj, kt}];
         ss ];

(* -- Specification of the x- and t-coordinates of
       the center of the Taylor series expansion - *)
       centexp = {xc -> j*h, tc -> n*dt};
eq1 = eq;
   pv = Derivative[0, 1][v][x, t] - eq1;
Print["The right-hand side of the differential"];
Print["equation (2.5.1)"];
         Print[pv];

(* -- Determining mx, the highest order of the
       x-derivative in the original partial
       differential equation ------------------ *)

 Do[k1 = 20- k + 1;
    mk = Exponent[pv, Derivative[k, 0][v][x, t]];
 If[mk==0, Goto[endloop], mx = k; Break[ ]];
 Label[endloop],
 {k, 20}];
Print["The highest order of x-derivative in PDE",
      " is ", mx];
 OutputForm[pv] >> ap.m;

 (* -- Initialization of the table dcs for the
        differential consequences of the
        differential equation ------------- *)
    dcs = {}; ktp = kt + 1;
    indcs = Table[99, {i, 1, ktp}, {k, 1, kt}];

 (* -- Symbolic computation of dcs[[i,j]]
         for 2<=i+j<=kt -------------------- *)
 ind = 1; indcs[[1, 1]] = ind; AppendTo[dcs, pv];
 ind = 2; indcs[[2, 1]] = ind;
 sc2 = D[pv, x]; AppendTo[dcs, sc2];
 Print["The differential consequences"];
         Print[dcs];
 Do[
    Do[ nxj = ktj - ntj;
               If[nxj==1&&ntj==1, Goto[endloop]];
 If[ntj==1, Goto[ntj1]];
    k1 = mx*ntj;
```

```
  k2 = indcs[[nxj+1,ntj-1]]; vtt = D[dcs[[k2]], t];
       Do[
 vtt = vtt /. Derivative[k, 1][v][x, t] ->
               D[dcs[[1]], {x, k}],
     {k, k1}];
Goto[ready];
Label[ntj1]; k= indcs[[nxj, ntj]];
               vtt = D[dcs[[k]], x];
Label[ready]; ind = ind + 1;
             indcs[[nxj+1, ntj]] = ind;
               AppendTo[dcs, vtt];
OutputForm[nxj] >>> ap.m;
OutputForm[ntj] >>> ap.m;
OutputForm[vtt] >>> ap.m;
Label[endloop],
               {ntj, ktj}],
   {ktj, 2, kt}];

(* -- Substitution of the differential
   consequences into the modified equation -- *)
     sc1 = sc /. centexp;
                       sc1 = Simplify[sc1];
 Print["The left-hand side of the differential"];
 Print["approximation equation fda = 0"];

OutputForm[sc1] >>> ap.m;
     Print[sc1]; sc2 = Expand[sc1];
     r1v = sc2 - eq1;
 Do[ Do[ nxj = ktj - ntj; k = indcs[[nxj+1, ntj]];
 r1v =
 r1v /. Derivative[nxj, ntj][v][x, t] -> dcs[[k]],
               {ntj, ktj}],
     {ktj, kt}];
 r1v = r1v /. Derivative[0, 1][v][x, t] -> pv;
   r1v = Expand[r1v]; r1v = Simplify[r1v];
   sc2 = eq1 + r1v;
Print["The left-hand side of the differential"];
 Print["approximation equation fda = 0 after "];
Print["the substitution of differential",
     " consequences"];
                   Print[sc2];
 OutputForm[sc2] >>> ap.m;
                   r1v = Expand[r1v];
Print["The expression for the leading terms of"];
```

```
Print["the truncation error of difference scheme"];
  Print[r1v];
  If[r1v===0,
Print["The modified equation of scheme coincides",
      " with the original PDE at kt = ",kt, "."];
  k1 = kt +1;
Print["Please take the value kt = ",k1," and",
      "repeat the computation with this new",
      " value of kt."]; Interrupt[ ]];

  (* -- Determination of the local approximation
        order of the difference scheme ------- *)
  k1 = mx*kt; ntx = 0;
  Do[ ak = Coefficient[r1v, h, k];
   If[ak===0, Goto[nextk]]; nxord = k;
  ntx= Exponent[ak, dt]; If[ntx > 0, Goto[nextk]];
      Break[ ];
Label[nextk],
      {k, k1}];
  ntx = 0;
  Do[ ak = Coefficient[r1v, dt, k];
   If[ak===0, Goto[nextk]]; ntord = k;
  ntx= Exponent[ak,h]; If[ntx > 0, Goto[nextk]];
  Break[ ];
          Label[nextk],
      {k, k1}];

Print["The approximation order of scheme is O(h^",
        nxord, ") + O(dt^", ntord, ")"];
  napmax = Max[nxord, ntord];
      If[napmax > 1,
  Do[ak = Coefficient[r1v, dt, k];
    If[ak=!=0, ntx1 = k; ntx2 = 0;
    Do[bj = Coefficient[ak, h, j];
     If[bj=!=0, ntx2 = j; Break[ ]],
    {j, napmax}]];
    If[ntx2!=0&&ntx1 + ntx2 == napmax,
  Print["+ O(dt^", ntx1, "*h^", ntx2, ")"]]],
 {k, napmax}]];

(* ----- End of the module ------ *)
   ClearAll[u, ss, eq1, sc1, sc2, dcs]; ];
```

appr2.ma enables one to compute the symbolic expression for the leading terms of the truncation error of the finite-volume approximation of the differential operator $\partial F(x,y)/\partial x$ or $\partial F(x,y)/\partial y$ on an irregular mesh of quadrilateral cells.

F The function F[x_, y_] enables one to perform the Taylor expansion of a function $F(x,y)$ at point (x =xc,y = yc).

The notebook appr2.ma needs no file of the input data.

appr3.ma enables one to compute the symbolic expression for the leading terms of the truncation error of the finite-volume approximation of the differential operator $\partial F(x,y)/\partial x$ or $\partial F(x,y)/\partial y$ on an irregular or regular mesh of triangular cells.

F The function F[x_, y_] has the same meaning as in the notebook appr2.ma.

The notebook appr3.ma needs no file of the input data.

appr31.ma enables one to compute the symbolic expression for the leading terms of the truncation error of the finite-volume approximation of the differential operator $\partial F(x,y)/\partial x$ or $\partial F(x,y)/\partial y$ on an irregular or regular mesh of triangular cells, see also subsection 4.6.2 of the book. The uniform mesh of triangular cells is made from the uniform rectangular grid by drawing one diagonal in each rectangular cell.

F The function F[x_, y_] has the same meaning as in the notebook appr2.ma.

The file appr31.ma needs the specification of only one input parameter, the order **kt** of the Taylor expansion.

appr4.ma enables one to compute the symbolic expression for the leading terms of the truncation error of the finite difference approximation of the differential operator $\partial F(x,y)/\partial x$ or $\partial F(x,y)/\partial y$ on a curvilinear logically rectangular grid. The difference approximation under consideration involves the central approximations of the derivatives x_ξ, x_η, y_ξ, y_η.

f The function f[x_, y_] has the same meaning as the function F[x_,y_] in the notebooks appr2.ma and appr3.ma.

The notebook appr4.ma needs no file of the input data.

col.ma The main program of this notebook

col[ix_, jy_, h0_, m1int_, m2int_]

creates two different color maps corresponding to the test function

$$v(x,y) = |\sin 2x \cos 2y|, \quad 0 \le x,y \le 2\pi$$

(see also Section 2.12). The meaning of the input data is explained in
the following cell of the notebook col.ma:

Parameter   Description

---

ix   the number of cells of an uniform mesh in the
          interval 0 <= x <= 2 Pi; ix > 1;
jy   the number of cells of an uniform mesh in the
          interval 0 <= y <= 2 Pi; jy > 1;
h0   the number in the interval 0 < h0 <= 1 to
          specify the color corresponding to the
          minimum value of the function u(x, y) in
          the spatial region under consideration;
m1int   the number of cells of a fine mesh in the
          interval 0 <= x <= 2 Pi; m1int >= ix;
m2int   the number of cells of a fine mesh in the
          interval 0 <= y <= 2 Pi; m2int >= jy. --- *)

---

disp1.ma The main program of this notebook

disp1[kt_, cpl_, cpr_, sc_]

enables one to perform the analysis of the phase errors, or dispersion,
of scalar explicit two-level difference schemes approximating the one-
dimensional advection equation (2.1.1). The meaning of the input data
is explained in the following cell of the notebook disp1.ma:

Parameter   Description

---

kt   the number of terms in the Tailor expansion of the
          phase error err; kt > 1;
cpl   the left end of the stability interval of the
          difference scheme under consideration;
cpr   the right end of the stability interval of the
          difference scheme under consideration; cpl < cpr.
sc   the left-hand side of the difference equation sc=0

---

disp1ad.ma The main program of this notebook

```
displad[kt_, cpl_, cpr_, cp20_, sc_]
```

enables one to perform the analysis of the phase errors, or dispersion, of scalar explicit two-level difference schemes approximating the one-dimensional advection-diffusion equation (3.3.2). The meaning of the input data is explained in the following cell of the notebook `displad.ma`:

Parameter	Description

---

kt	the number of terms in the Tailor expansion of the phase error err; kt > 1;
cpl	the left end of the stability interval of the difference scheme under consideration;
cpr	the right end of the stability interval of the difference scheme under consideration; cpl < cpr;
cp20	the user-specified value of the variable κ_2 for plotting the 3D section of the hypersurface $E = E(\kappa_1, \kappa_2, \xi)$.
sc	the left-hand side of the difference equation sc=0

---

displadc.ma differs from the foregoing notebook `displad.ma` in that it considers the dispersion of the Crank-Nicolson scheme (3.8.15) approximating the advection-diffusion equation (3.3.2). The meaning of the input data for the notebook `displadc.ma` is the same as in the case of the notebook `displad.ma`.

fem1.ma The main program of this notebook

```
fem1[M1_, M2_, sor_, TOL_, xview_, yview_, zview_]
```

solves numerically the Poisson equation (4.4.37) by the finite element in the square region $D = \{-\pi/2 \leq x, y \leq \pi/2\}$. The meaning of the input data is explained in the following cell of the notebook `fem1.ma`:

Parameter	Description

M1	the number of the grid nodes on the x-axis, M1 > 2;
M2	the number of the grid nodes on the y-axis, M2 > 2;
sor	the relaxation parameter entering the successive over-relaxation method, 0 < sor < 2;

TOL   the user-specified tolerance entering the criter-
        ion for the termination of the SOR iterations,
        TOL > 0;
xview   the coordinates of the viewpoint for the surface
yview   u = u(x, y). At least one of the coordinates
zview   xview, yview, zview should be greater than 1 in
        modulus.

---

uex uex[x_,y_] computes numerically the exact solution of the boundary-value problem (4.4.37), (4.4.2) at any point $(x, y) \in D$.
Below we present the listing of the program fem1.ma.

```
ClearAll[uex, fem1];
 uex[x_, y_] :=
(* --- The function to compute the exact solution
       of the Poisson equation ---------------- *)
( z1 = zpi + x*x; z0 = -1.0; z2 = 0.0;
  Do[ z0 = -z0; k1 = 2*k - 1;
      zk = N[Cosh[k1*y]*Cos[k1*x]/Cosh[k1*Pi/2]];
  zk = z0*zk/(k1^3); z2 = z2 + zk, {k, nt}];
  z1 = z1 + zk1*z2; z1 )


fem1[M1_, M2_, sor_, TOL_, xview_, yview_, zview_]:=
(
(* -- Computation of the mesh steps h1 and h2 -- *)
  m1m = M1 - 1; m2m = M2 - 1; n = 0;
  dr = N[Pi]; h1 = dr/m1m; h2 = dr/m2m;
(* -- Specification of the initial iteration
     u^{(0)}(x, y)-- *)
     v = Table[0.0, {i, M1}, {j, M2}];
                 vp = Table[0.0, {i, M1}, {j, M2}];
cel = h1/h2; c1 = 1/(2*(1 + cel^2)); cel2 = cel^2;
        h12 = 2*h1^2;  h1q = h1^2; h2q = h2^2;
  xl = -dr/2; mm3 = (M1 - 2)*(M2 - 2);
(* Computation of the necessary constants for
    the exact solution of the Poisson equation *)
 eps = 10^(-7); z = N[0.5*(1 + (1/eps)^(1/3))];
 nt = Floor[z]; zpi = -N[(0.5*Pi)^2]; zk1 = N[8/Pi];
(* ------------------------------------------- *)
Print["M1 = ", M1, ", M2 = ", M2, ", sor = ",
       sor, ", TOL = ", TOL, ", nt = ", nt];
Print["h1 = ", h1, ", h2 = ", h2];
a0 = uex[0.0, 0.0]; Print ["u(0,0) = ", a0];
```

```
(* ------- Begin of the SOR iterations ------- *)
 Label[iter];
(* ------ Do-loop over the (i,j) nodes ------- *)
 Do[
     Do[ vs = c1*(vp[[i-1, j]] + v[[i+1, j]]
        + cel2*(vp[[i, j-1]] + v[[i, j+1]]) - h12);
  vp[[i, j]] = sor*vs + (1-sor)*v[[i, j]],
               {i, 2, m1m}],
    {j, 2, m2m}];
(* --------- Exchange of the lists ----------- *)
v = vp; Rhnorm = 0;
(* ----- Computation of the RMS residual ----- *)
 Do[
 Do[ Rh = (v[[i+1,j]]-2*v[[i,j]]+v[[i-1,j]])/h1q
 + (v[[i,j+1]] - 2*v[[i,j]] + v[[i, j-1]])/h2q - 2;
  Rhnorm = Rhnorm + Rh^2, {i, 2, m1m}],
    {j, 2, m2m}];
               Rhnorm = Sqrt[Rhnorm/mm3];
 n = n + 1;
Print["n = ",n,", Norm of Rh = ", Rhnorm];
        If[n > 300, Goto[endt]];
   If[Rhnorm > TOL, Goto[iter]];
(* --------- End of the SOR iterations -------- *)
Label[endt];
(* --- Plotting the numerical solution surface
   v = v(x,y)--------------------------------- *)
 gr3 = ListPlot3D[v,
               ViewPoint -> {xview, yview, zview},
        AxesLabel -> {"y", "x", "v"}];
Display["fig4_40", gr3];
                   gr4 = ListContourPlot[v];
Display["fig4_41", gr4];
(* ----- The numerical computation of the exact
          solution in the grid nodes ----------- *)
Print["Computing the exact solution in the grid",
      " nodes"];
  Do[ yj = xl + (j - 1)*h2;
 Do[ xi = xl + (i - 1)*h1; vp[[i, j]] = uex[xi, yj];
  zij = vp[[i, j]];
 Print["i = ", i, ", j = ", j, ", u(xi, yj) = ",
      zij],
{i, 2, m1m}],
   {j, 2, m2m}];
(* - The computation of a mean-square error
```

```
      of the numerical solution ---------------- *)
 rmser = 0;
Do[
   Do[ du = (vp[[i, j]] - v[[i, j]])^2;
                      rmser = rmser + du,
        {i, 2, m1m}],
           {j, 2, m2m}];
 rmser = Sqrt[rmser/mm3];
Print["The mean-square error of the numerical",
      " solution = ", rmser];
 gr5 = ListPlot3D[vp,
                ViewPoint -> {xview, yview, zview},
           AxesLabel -> {"y", "x", "u"}];
)
```

fem2.ma The main program of this notebook

fem2[M1_,M2_,sor_,TOL_,exct_,xview_,yview_,zview_,h0_,mfine_]

differs from the foregoing notebook fem1.ma only in that it produces the color map of the obtained FEM numerical solution. The meaning of the input data is explained in the following cell of the notebook fem2.ma:

Parameter   Description

```
M1     the number of the grid nodes on the x-axis, M1 > 2;
M2     the number of the grid nodes on the y-axis, M2 > 2;
sor    the relaxation parameter entering the successive
       over-relaxation method, 0 < sor < 2;
TOL    the user-specified tolerance entering the criter-
       ion for the termination of the SOR iterations,TOL>0;
exct   if exct = 0, then the exact solution is not
       calculated; if exct >= 1, then it is calculated;
xview  the coordinates of the viewpoint for the surface
yview  u = u(x, y). At least one of the coordinates
zview  xview, yview, zview should be greater than 1 in
       modulus;
h0     the number in the interval 0 < h0 <= 1 to
       specify the color corresponding to the
       minimum value of the function u(x, y) in
       the spatial region under consideration;
mfine  the number of cells of a fine mesh in the
```

```
interval -Pi/2 <= x or y <= Pi/2;
m1int >= min(ix, jy).
```

---

grid1.ma The main program of this notebook

$$\text{grid1[hc\_, cy0\_, cR0\_, cR\_, cx0\_, cx4\_, thet1\_, thet2\_, ix\_,}$$
$$\text{jy]}$$

implements a simple algebraic procedure of curvilinear grid generation in the upper half of a given axisymmetric Laval nozzle. The meaning of the input data is explained in the following cell of the notebook **grid1.ma**:

Parameter   Description

---

hc	the nozzle wall height in the critical section of the nozzle, hc > 0;
cy0	the ordinate of the nozzle wall in the inlet section x = 0 related to hc;
cR0	the radius of the curvature of the nozzle wall in its subsonic part related to hc;
cR	the nozzle wall radius of the curvature in the critical section related to hc;
cx0	the abscissa of the critical nozzle section related to hc;
cx4	the nozzle length related to hc, cx4 > cx0;
thet1	the inclination angle of the nozzle wall in the subsonic part of the nozzle; the angle is specified in radians;
thet2	the inclination angle of the nozzle wall in the supersonic part of the nozzle; the angle is specified in radians;
ix	the number of grid points on the xi-axis, ix > 1;
jy	the number of grid points on the eta-axis, jy > 1

---

fw **fw[x_]** computes the ordinate y of the nozzle wall in accordance with Fig. 2.41.

gridttm.ma The main program of this file

$$\text{grid2[hc\_, cy0\_, cR0\_, cR\_, cx0\_, cx4\_, thet1\_, thet2\_, ix\_,}$$
$$\text{jy\_, Saf\_, eps]}$$

generates a curvilinear grid in the given axisymmetric Laval nozzle with
the aid of the elliptic grid generator due to Thompson, Thames and
Mastin. The corresponding PDEs are solved with the aid of the ADI
method.The meaning of the input data is explained in the following cell
of the notebook `gridttm.ma`:

Parameter	Description

---

hc	the nozzle wall height in the critical section of the nozzle, hc > 0;
cy0	the ordinate of the nozzle wall in the inlet section x = 0 related to hc;
cR0	the radius of the curvature of the nozzle wall in its subsonic part related to hc;
cR	the nozzle wall radius of the curvature in the critical section related to hc;
cx0	the abscissa of the critical nozzle section related to hc;
cx4	the nozzle length related to hc, cx4 > cx0;
thet1	the inclination angle of the nozzle wall in the subsonic part of the nozzle; the angle is specified in radians;
thet2	the inclination angle of the nozzle wall in the supersonic part of the nozzle; the angle is specified in radians;
ix	the number of grid points on the xi-axis, ix > 1;
jy	the number of grid points on the eta-axis, jy > 1;
Saf	the safety factor in the formula for the computation of the pseudo time step dt; Saf > 0;
eps	a small positive number entering the criterion for the termination of the iterations by the ADI method.

---

fw `fw[x_]` computes the ordinate *y* of the nozzle wall in accordance
with Fig. 2.41.

Below we present the listing of the program `gridttm.ma`.

```
ClearAll[fw, grid2];
fw[x_] := ( f1 = If[0 <= x && x <= x1,
        y0 - R0 + Sqrt[(R0 - x)*(R0 + x)], 0];
f2 = If[x1 <= x && x <= x2, y1 - ct1*(x - x1), f1];
```

```
f3 = If[x2 <= x && x <= x3,
    hc1 + R - Sqrt[(R - x + x0)*(R + x - x0)]], f2];
f4 = If[x3 <= x && x <= x4, y3 + ct2*(x - x3), f3]; f4 )

grid2[hc_, cy0_, cR0_, cR_, cx0_, cx4_, thet1_,
      thet2_, ix_, jy_, Saf_, eps_] :=
(
(* -- Specification of the nozzle geometry -- *)
R0 = cR0*hc; y0 = cy0*hc; x0 = cx0*hc; R = cR*hc;
    hc1 = hc;
x1 = R0*N[Sin[thet1]]; ct1 = N[Tan[thet1]];
                  ct2 = N[Tan[thet2]];
x2 = x0-R*N[Sin[thet1]]; x3 = x0+R*N[Sin[thet2]];
x4 = cx4*hc; y1 = y0 - R0*(1 - N[Cos[thet1]]);
    y3 = hc + R*(1 - N[Cos[thet2]]);
Print["The values of the nozzle geometry",
      " parameters"];
Print["hc = ",hc, ", R0 = ", R0, ", R = ", R,
      ", x0 = ",x0];
Print["thet1 = ", thet1, ", thet2 = ", thet2,
      ", nozzle length = ", x4];

(* -- yw is the list for the ordinates of the
   nozzle wall ---------------------------- *)
   yw = {};  xorg = 0.0;
  Do[ xi = (i - 1)*x4/(ix - 1); yi = fw[xi];
      AppendTo[yw, yi], {i, ix}];
(* -- x, y are the lists for the abscissas x[[i]]
      and the ordinates y[[i, j]] of the nodes of
      the curvilinear spatial grid ----------- *)
  x = Table[N[(i - 1)*x4/(ix - 1)], {i, ix}];
y = Table[N[yw[[i]]*(j-1)/(jy-1)], {i,ix}, {j,jy}];
(* - The graphical output of the initial grid - *)
Print["The initial grid"];
(* -------- The lines eta = const ----------- *)
   mt = {}; xtab = {};
 Do[xi = (i - 1)*x4/(ix - 1);
       AppendTo[xtab, xi], {i, ix}];
 Do[ytab = {};
  Do[yi = y[[i,j]]; AppendTo[ytab, yi], {i, ix}];
  xytab = Table[{xtab[[i]], ytab[[i]]}, {i, ix}];
  mt1 = ListPlot[xytab,
    AxesLabel  -> {"x", "y"},
            AxesOrigin -> {xorg, 0},
```

```
      PlotJoined -> True,
      PlotRange  -> {{0, x4}, {0, y0}},
            DisplayFunction -> Identity];
            AppendTo[mt, mt1],
                                    {j, jy}];
(* -- mt3 = Show[mt,
      DisplayFunction -> $DisplayFunction]; -- *)
(* --------- The lines xi = const ------------ *)
   Do[ xtab = {}; ytab = {};
    xi = (i - 1)*x4/(ix - 1); AppendTo[xtab, xi];
   AppendTo[ytab, 0.0];
 yi = yw[[i]]; AppendTo[xtab, xi];
            AppendTo[ytab, yi];
 xytab = Table[{xtab[[j]], ytab[[j]]}, {j, 2}];
      mt1 = ListPlot[xytab,
                     PlotJoined -> True,
                     DisplayFunction -> Identity];
        AppendTo[mt, mt1],
                          {i, ix}];
mt3=Show[mt,DisplayFunction -> $DisplayFunction];
         Display["fig4_6a", mt3];


  (*%%%%%%%%%%%%%%%%%%%%%%%%%%%%%%%%%%%%%%%%%%%%
  %%                                         %%
  %%          Elliptic Grid Generator        %%
  %%                                         %%
  %%%%%%%%%%%%%%%%%%%%%%%%%%%%%%%%%%%%%%%%%%%%%*)


(* - Specification of the initial guess xx, yy - *)
yy = y; xx = y; dt = Saf*x4/(ix - 1); dt2 = 0.5*dt;
ix1 = ix - 1;  jy1 = jy - 1; ifix = Floor[ix/2];
Do[ xi = x[[i]];
 Do[ xx[[i, j]] = xi, {j, jy}], {i, ix}];
(* - The ADI solver in the form of a function - *)
ADI[nxy_] :=
( xg = x; yg = x; xn1 = xx; xn2 = xn1;
(* - The first half-step
    passage from t = n*dt to t = (n+0.5)*dt) - *)
xg[[1]] = 0.0;
   Do[ yg[[1]] = 0.0;
     If[nxy != 1, yg[[1]] = yw[[1]]*(j-1)/(jy-1)];
     xn1[[1, j]] = yg[[1]]; xn2[[1, j]] = yg[[1]];
Print["nxy = ", nxy,
   ", yy[[",ifix,", ", j, "]] = ", yy[[ifix,j]] ];
```

```
(* ---- The scalar sweeps along the lines
      eta = const ----------------------------- *)
       Do[ xet = (xx[[i, j+1]] - xx[[i, j-1]])/2;
           yet = (yy[[i, j+1]] - yy[[i, j-1]])/2;
   xks = (xx[[i+1,j]] - xx[[i-1,j]])/2;
   yks = (yy[[i+1,j]] - yy[[i-1,j]])/2;
If[nxy == 1,
fkset = xx[[i+1, j+1]] - xx[[i+1, j-1]] -
xx[[i-1, j+1]] + xx[[i-1, j-1]],
        fkset = yy[[i+1, j+1]] - yy[[i+1, j-1]] -
        yy[[i-1, j+1]] + yy[[i-1, j-1]] ];
a1 = xet*xet + yet*yet;   b1 = xks*xks + yks*yks;
If[nxy == 1,
 fet = xx[[i, j+1]] - 2*xx[[i,j]] + xx[[i, j-1]],
 fet = yy[[i, j+1]] - 2*yy[[i,j]] + yy[[i, j-1]] ];
g1 = b1*fet - (1/2)*(xks*xet + yks*yet)*fkset;
a2 = dt2*a1;    b2 = -1 - dt*a1;
If[nxy == 1, fcij=xx[[i, j]], fcij = yy[[i, j]] ];
f2 = -fcij - dt2*g1;   den = b2 + a2*xg[[i - 1]];
xg[[i]]=-a2/den; yg[[i]]=(f2-a2*yg[[i - 1]])/den,
    {i, 2, ix1}];
If[nxy == 1, xn1[[ix, j]] = x4,
   xn1[[ix, j]] = yw[[ix]]*(j - 1)/jy1];
   xn2[[ix,j]] = xn1[[ix, j]];
  Do[ i1 = ix - i + 1;
xn1[[i1, j]] = xg[[i1]]*xn1[[i1+1, j]] + yg[[i1]],
 {i, 2, ix1}],
          {j, 2, jy1}];
(* - The second half-step(passage from
     t = (n + 0.5)*dt to t = (n + 1)*dt) ----- *)
If[nxy == 1, xn2[[1, jy]] = 0; xn2[[ix, jy]] = x4;
            xn2[[ix, 1]] = x4,
     xn2[[1, jy]] = yw[[1]]; xn2[[ix, 1]] = 0;
     xn2[[ix, jy]] = yw[[ix]] ];

Do[ xg[[1]] = 0;
   If[nxy == 1, yg[[1]] = x[[i]], yg[[1]] = 0];
     xn2[[i,1]] = yg[[1]];
(* The scalar sweeps along the lines xi = const *)
       Do[ xet = (xx[[i, j+1]] - xx[[i, j-1]])/2;
           yet = (yy[[i, j+1]] - yy[[i, j-1]])/2;
   xks = (xx[[i+1,j]] - xx[[i-1,j]])/2;
   yks = (yy[[i+1,j]] - yy[[i-1,j]])/2;
fkset = xn1[[i+1, j+1]] - xn1[[i+1, j-1]] -
```

```
xn1[[i-1, j+1]] + xn1[[i-1, j-1]];
a1 = xet*xet + yet*yet;  b1 = xks*xks + yks*yks;
g1=a1*(xn1[[i+1,j]] - 2*xn1[[i,j]] + xn1[[i-1,j]]);
g1 = g1 - (1/2)*(xks*xet + yks*yet)*fkset;
a2 = dt2*b1;   b2 = -1 - dt*b1;
f2 = -xn1[[i,j]] -dt2*g1; den = b2 + a2*xg[[j-1]];
xg[[j]] =-a2/den; yg[[j]]=(f2 - a2*yg[[j-1]])/den,
   {j, 2, jy1}];
xn2[[i, jy]] = x[[i]];
If[nxy != 1, xn2[[i, jy]] = yw[[i]] ];
  Do[ j1 = jy - j + 1;
xn2[[i, j1]] = xg[[j1]]*xn2[[i, j1+1]] + yg[[j1]],
     {j, 2, jy1}],
{i, 2, ix1}];
  Return[xn2] );
(* - The check-up of the convergence criterion - *)
nit = 0;
Print["The start of the ADI iterations, eps = " ,
      eps];
Label[iter];
nit = nit + 1;
xn2 = ADI[1];
dxmax = 0.0;  dymax = 0.0;
  Do[
      Do[ dxij = Abs[xn2[[i, j]] - xx[[i, j]] ];
dxmax=Max[dxmax, dxij], {i, 2, ix1}], {j, 2, jy1}];
(* ---- Exchange of the lists ---- *)
xx = xn2;   xn2 = ADI[2];
Do[
    Do[ dyij = Abs[xn2[[i,j]] - yy[[i, j]] ];
dymax=Max[dymax, dyij], {i, 2, ix1}], {j, 2, jy1}];
(* ---- Exchange of the lists ---- *)
yy = xn2;
Print["n = ", nit,",dxmax = ",dxmax,",dymax = ",
      dymax];
If[dxmax > eps || dymax > eps, Goto[iter] ];
(* --- End of the elliptic grid generator --- *)

(* ---- The graphical output of the grid ---- *)
(* --------- The lines eta = const ---------- *)
  mt = {}; xtab = {};
(* Do[xi = (i - 1)*x4/(ix - 1);
          AppendTo[xtab,xi], {i, ix}]; *)
  Do[xtab = {}; ytab = {};
```

```
    Do[ xi = xx[[i,j]]; AppendTo[xtab, xi];
 yi = yy[[i,j]]; AppendTo[ytab, yi], {i, ix}];
 xytab = Table[{xtab[[i]], ytab[[i]]}, {i, ix}];
 mt1 = ListPlot[xytab,
   AxesLabel  -> {"x", "y"},
           AxesOrigin -> {xorg, 0},
   PlotJoined -> True,
   PlotRange  -> {{0, x4}, {0, y0}},
           DisplayFunction -> Identity];
           AppendTo[mt, mt1],
                                  {j, jy}];
mt3 = Show[mt,DisplayFunction -> $DisplayFunction];
(* -------- The lines xi = const ------------- *)
   Do[xtab = {}; ytab = {};
     Do[ xj = xx[[i,j]]; AppendTo[xtab, xj];
     yj = yy[[i,j]]; AppendTo[ytab, yj], {j, jy}];
   xytab = Table[{xtab[[j]], ytab[[j]]}, {j, jy}];
     mt1 = ListPlot[xytab,
                    PlotJoined -> True,
                    DisplayFunction -> Identity];
     AppendTo[mt, mt1],
                      {i, ix}];
mt3 = Show[mt,DisplayFunction -> $DisplayFunction];
        Display["fig4_62", mt3];
(* ------- End of the procedure grid2 --------- *)
)
```

heat1.ma The main program of this notebook,

`heat1[nu_, beta_, A1_, A2_, a1_, b1_, M_, cap_, numbt_, Npic_]`

solves numerically the one-dimensional heat equation (3.1.2) under the initial condition (3.2.4) by the explicit scheme (3.2.1). The meaning of the input data is explained in the following cell of the notebook `heat1.ma`:

Parameter	Description
nu.	the positive coefficient ν in the heat equation (3.1.2);
beta	the constant entering the exact solution (3.2.3);
A1	the constant entering the exact solution (3.2.3);
A2	the constant entering the exact solution (3.2.3), |A1| + |A2| != 0;
a1	the abscissa of the left end of the interval on the x-axis;
b1	the abscissa of the right end of the above interval, a1 < b1;
M	the number of the grid node coinciding with the right boundary x = b1 of the spatial integration interval (see also (2.1.5));
cap	the Courant number, cap= nu*dt/(h*h), where dt is the time step of the difference scheme, and h is the step of the uniform grid in the interval a1 <=x <= b1;
numbt	the number of the time steps to be made by the difference scheme, numbt > 1;
Npic	the number of the pictures of the difference solution graphs for Npic moments of time, Npic > 1.

uinit uinit[x_, beta_, A1_, A2_] computes the numerical value of the initial function (3.2.4) at a given point x.

uexact The function uexact[x_, t_, nu_, beta_, A1_, A2_] computes the numerical value of the exact solution (3.2.3) at a given point (x, t).

Below we present the listing of the program heat1.ma.

```
ClearAll[uinit, uexact, heat1];

(* --- The function uinit[...] computes the initial
        condition (3.2.4) --- *)
uinit[x_, beta_, A1_, A2_] :=
(xb = beta*x; A1*Cos[xb] + A2*Sin[xb] )

(* --- The function uexact[...] computes the exact
        solution (3.2.3) ----- *)
 uexact[x_, t_, nu_, beta_, A1_, A2_] :=
```

```
(xb = beta*x;
E^(-nu*beta^2*t)*(A1*Cos[xb] + A2*Sin[xb]) )


heat1[nu_, beta_, A1_, A2_, a1_, b1_, M_, cap_,
      numbt_, Npic_]:=Block[
(* --- Definition of the local variables --- *)
{uj,up,ug,ut,tnt,y,j,k,n,M1,h,t,dt,xf0,Tmax,N},

(*  uj = the difference solution at time level n,
up = the difference solution at time level n+1  *)
    h = (b1 - a1)/M;   M1 = M + 1; M2 = M - 1;
        dt = cap*h*h/nu;   xf0 = a1 + jf0*h;
  Tmax = numbt*dt; N = numbt;
(* Computation of the initial values u(xj,0)
                                    at grid points *)
uj = Table[1.0, {j,0,M}];
Do[xj = a1 + j*h; uj[[j]] = uinit[xj, beta, A1, A2],
                                        {j, 0, M}];
up = uj; u0 =uj[[0]];
                ug = Table[u0,{j, 1, M1}];
        tnt = Table[Tmax, {j, Npic}];
        mt = Table[j, {j, Npic}];
(* - Tmax is the right end of the time interval - *)
    nplot = Max[1, Floor[N/Npic]];
(* - The solution graphs u(x, tk) will be given at
    tk = (k-1)*nplot*dt, k = 1,...,Npic ------ *)
    um0 = Abs[A1] + Abs[A2];
    Print["um0 = ",um0];
t= 0.0;  n2 = nplot;  nt = 1;  tnt[[1]] = 0.0;


(* - Generation of the two-dimensional table ut - *)
        ut = Table[u0, {k, 1, Npic}, {j, 1, M1}];
(* - Storing the initial profile u(x,0) in
    ut[[j, 1]] - *)
        Do[ ut[[1,j + 1]] = uj[[j]], {j, 0, M}];
(* --- Begin of the Do-loop for time steps --- *)
        Do[
            t = t + dt;
            Print["n = ", n, ", t = ", t];
(* -- Do-loop over grid nodes on the x-axis -- *)
    fxr = uj[[1]] - uj[[0]];
        Do[ fxl = fxr;


(* -- The explicit scheme (3.2.1) for the heat
```

```
      equation (3.1.2) ---------------------- *)
      fxr = uj[[j + 1]] - uj[[j]];
        up[[j]] = uj[[j]] + cap*(fxr - fxl),
          {j, 1, M2}];

(* -- The boundary condition on the left
                          boundary x = a1 -- *)
        up[[0]] = uexact[a1, t, nu, beta, A1, A2];
(* -- The boundary condition on the right
                          boundary x = b1 -- *)
        up[[M]] = uexact[b1, t, nu, beta, A1, A2];
(* --------- Exchange of the lists ----------- *)
                      uj = up;
(* ------ Find Max[Abs[u(j,n+1) - um] -------- *)
      um = E^(-beta^2*t)*um0;  umax = 0.0;
Do[ y = Abs[uj[[j]] -um]; umax = Max[y, umax],
      {j, 0, M, 1}];
            If[n == n2, Goto[store]];
  If[umax > 2.*um, Goto [endsteps], Goto [endn]];
Label[endsteps]; Tmax = t;
Label[store]; n2 = n2 + nplot;
      nt = nt + 1;
                    If[nt <= Npic,
      tnt[[nt]] = t;
(* -- Storing the solution u(x,t) in ut[[nt,j]] --*)
      Do[ ut[[nt, j + 1]] = uj[[j]], {j, 0, M}]];
      If[umax > 2.0*um||nt == Npic, Break[]];
 Label[endn],

(* -------- End of the Do-loop in n ------------ *)
            {n, 0, N, 1}];
(* -- Plotting the solution graphs for several
                        moments of time -- *)
(* -- Determining min(ut) and max(ut) ---------- *)
          umin = 10^6;  umax = -umin;
      nt = If[nt > Npic, Npic, nt];
      Do[
          y = ut[[k, j]];
          umin = If[y < umin, y, umin];
            umax = If[y > umax, y, umax],
        {k, nt}, {j, M1}];
        Print["umin = ", umin, ", umax = ", umax];

(* -- Storing the pictures in the elements mt[[j]],
```

```
    j = 1,...,Npic -- *)
    umin = If[umin > 0, 0.0, umin];
    umax = If[umax < 0, 0.0, umax];
    Do[
        ug = ut[[j]];
        t  = tnt[[j]];
    Print["The solution graph at t = ", t];
 mt[[j]] = ListPlot[ug,
                AxesLabel -> {"x", "u"},
                PlotJoined -> True,
                PlotRange -> {{0, M1}, {umin, umax}}],
        {j, nt}];
If[nt == 2,
gr = Show[GraphicsArray[{mt[[1]], mt[[2]]}]] ];
If[nt == 3,
gr = Show[GraphicsArray[{{mt[[1]], mt[[2]]},
                         {mt[[3]]}}]]];
If[nt == 4,
gr = Show[GraphicsArray[{{mt[[1]], mt[[2]]},
                           {mt[[3]], mt[[4]]}}]]];
If[nt == 5,
 gr = Show[GraphicsArray[{
    {mt[[1]], mt[[2]], mt[[3]]},
    {mt[[4]], mt[[5]]}}]] ];
If[nt == 6,
 gr = Show[GraphicsArray[{
    {mt[[1]], mt[[2]], mt[[3]]},
    {mt[[4]], mt[[5]], mt[[6]]}}]] ];
If[nt == 7,
 gr = Show[GraphicsArray[{
    {mt[[1]], mt[[2]], mt[[3]], mt[[4]]},
    {mt[[5]], mt[[6]], mt[[7]]}}]] ];
If[nt >= 8,
 gr = Show[GraphicsArray[{
    {mt[[1]], mt[[2]], mt[[3]], mt[[4]]},
    {mt[[5]], mt[[6]], mt[[7]], mt[[8]]}}]] ];
    Display["fig3_14", gr];
(* ----------- End of the block -------------- *)
        ];
```

lw1.ma The main program of this notebook

advbc[a_,a1_,b1_,jf0_,ul_,ur_,M_, cap_, Npic_]

solves numerically the advection equation (2.1.1) under the initial condition (2.4.7) representing a semi-ellipse pulse by the one-step Lax-Wendroff scheme (2.4.3). The meaning of the input data is explained in the following cell of the file `lw1.ma`:

Parameter	Description

---

a	the advection speed in eqn (2.1.1), a > 0.
a1	the abscissa of the left end of the interval on the x-axis;
b1	the abscissa of the right end of the above interval, a1 < b1;
jf0	the number of the grid node, 0 < jf0 < M, such that at x = jf0*h there takes place the maximum of the initial semi-ellipse pulse;
ul	and ur are the constants entering the initial condition, the value ul corresponds to the constant background, and ul + ur is the extremal value of the initial function u(x, 0);
M	the number of the cell whose right boundary coincides with the right boundary x = b1 of the spatial integration interval;
cap	the Courant number, cap = a*dt/h, where dt is the time step of the difference scheme, and h is the step of the uniform grid in the interval a1 <=x<= b1;
Npic	the number of the pictures of the difference solution graphs for Npic moments of time, Npic > 1."

---

uinit `uinit[x_,ul_,ur_,jf0_,h_]` computes the numerical value of the initial function (2.4.7) at a given point x.

lw2.ma differs from the foregoing notebook `lw1.ma` by a more detailed graphical display of the numerical solution in the subregion of a travelling semi-ellipse pulse. The meaning of the input data for the notebook `lw2.ma` is the same as in the case of the notebook `lw1.ma`.

mc2d.ma The main program of this notebook,

```
mc2d[U_, phi_, xf0_, ul_, ur_, xa_, yb_, ix_, jy_, Saf_,
          Npic_, xview_, yview_, zview_]
```

solves numerically the two-dimensional advection equation (2.12.2) under the initial condition (2.12.11) of a step form (the square wave test)

by the MacCormack difference scheme (2.12.3). The meaning of the input data is explained in the following cell of the notebook mc2d.ma:

Parameter   Description

U   the modulus of the convection speed,
        U = Sqrt[A^2 + B^2], A and B are the constant comp-
        onents of the convection velocity along the x- and
        y-axes, respectively; U > 0;
phi   the angle between the convection velocity vector
        and the positive direction of the x-axis,
        0 <= phi <= Pi/2;
xf0   the abscissa of a point at the disconti-
        nuity front at t = 0 on the upper boundary y = yb,
        xf0 >= 0;
ul   the value of u(x, y, 0) behind the discontinuity
        front;
ur   the value of u(x, y, 0) ahead of the discontinuity
        front;
xa   the size of the rectangular spatial computational
        domain along the x-axis, xa > 0;
yb   the size of the rectangular spatial domain along
        the y-axis, yb > 0;
ix   the number of cells of an uniform rectangular
        spatial grid along the x-axis, ix > 1;
jy   the number of cells of an uniform rectangular
        spatial grid along the y-axis, jy > 1;
Saf   the safety factor in the formula for the computat-
        ion of the time step dt; 0 < Saf <= 1 to ensure a
        stable computation;
Npic   the number of the pictures of the difference sol-
        ution surfaces u = u(x,y,t) for Npic moments of
        time, Npic > 1;
xview   the coordinates of the viewpoint for the surface
yview   u = u(x, y, t). At least one of the coordinates
zview   xview, yview, zview should be greater than 1 in
        modulus.

---

uex uex[x_, y_, t_] computes numerically the exact solution of the
initial-value problem (2.12.2), (2.12.11) at any (x, y, t) point.

Below we present the listing of the program mc2d.ma.

```
ClearAll[uex, mc2d];
uex[x_, y_, t_] :=

(* -- The function to compute the exact solution
        of the problem on a plane parallel convection
        of the substance property u(x, y, t) ----- *)
( xfn = xf1 - (y - yf1)*ckf + U1*t/cphi;
  z = If[x < xfn, ul1, ur1]; z )


mc2d[U_, phi_, xf0_, ul_, ur_, xa_, yb_, ix_, jy_,
         Saf_, Npic_, xview_, yview_, zview_] :=
(

(* -- Computation of the mesh steps hx,hy,dt -- *)
        cphi = N[Cos[phi]]; sphi = N[Sin[phi]];
A = U*cphi; B = U*sphi; ckf = sphi/cphi;
               yf0 = yb; hx = xa/ix; hy = yb/jy;
 dt = Saf/(Abs[A/hx] + Abs[B/hy]);
Tmax = ((xa - 4*hx)*cphi - xf0*cphi - yf0*sphi)/U;

(* Tmax is the right end of the time interval *)
 t1 = Tmax/2;    cp1 = A*dt/hx;    cp2 = B*dt/hy;
Nt = Floor[Tmax/dt]; Nt = Max[1, Nt];
n1 = Floor[Nt/Npic]; n1 = Max[1, n1];
 xf1 = xf0; yf1 = yf0; ul1 = ul; ur1 = ur; U1 = U;
Print["A = ", A, ", B = ", B, ", n1 = ", n1,
       ", Tmax = ", Tmax];
Print["Nt = ", Nt, ", kappa1 = ", cp1,
       ", kappa2 = ", cp2];

(* --- Specification of the initial function
     u(x, y, 0) ---------------------------- *)
     u = Table[ul, {i, ix}, {j, jy}];
               up = Table[ul, {i, ix}, {j, jy}];
 Do[ y1 = (j - 0.5)*hy;
       Do[ x1 = (i - 0.5)*hx;
             u[[i, j]] = uex[x1, y1, 0.0],
   {i, ix}],     {j, jy}];
 Print[u[[1,1]],",",u[[2,1]],",", u[[3,1]],",",
       u[[4,1]] ];

(* Plotting the initial surface u = u(x, y, 0) *)
Print["The initial surface u = u(x, y, 0)"];
 gr1 = ListPlot3D[u,
```

```
            ViewPoint -> {xview, yview, zview},
        AxesLabel -> {"y", "x", "u"}];
  Display["fig2_26", gr1];

(* gr2 = ListDensityPlot[u];  *)
gr3 = ListContourPlot[u];
  tn = 0; n3 = n1;
um = Max[Abs[ul], Abs[ur]];
  Print["umax = ", um];

(* --- Begin of the Do-loop for time steps --- *)
      Do[

(* --- Computation of the fluxes across the
       lower boundary y = 0 ------------------- *)
  fb = Table[ul, {i, ix}]; y1 = 0.0;
  Do[ fb[[i]] = u[[i,1]], {i, ix}];

(* -- The computation of the predictor values
     up[[i, j]] ------------------------------- *)
(* ------- Do-loop over the (i, j) nodes ------ *)
    Do[ fr = u[[1, j]];
        Do[ fl = fr; uz = u[[i, j]];
    fr = If[i == ix, uz, u[[i + 1, j]] ];
If[j == jy, x1 = (i - 0.5)*hx; y1 = (jy + 0.5)*hy;
    gt = uex[x1, y1, tn], gt = u[[i, j + 1]] ];
    gb = fb[[i]]; fb[[i]] = gt;
  up[[i, j]] = uz - cp1*(fr - fl) - cp2*(gt - gb),
                                      {i, ix}],
          {j, jy}];

(*  Computation of the corrector values u[[i, j]] *)
  tn = tn + dt; umax = 0.0; umax1 = 0.0; y1= -0.5*hy;
   Do[ x1 = (i - 0.5)*hx; fb[[i]] = uex[x1, y1, tn],
                                      {i, ix}];
   Do[ x1 = -0.5*hx; y1 = (j - 0.5)*hy;
    fr = uex[x1, y1, tn];
      Do[ fl = fr; uz = up[[i, j]];
        fr = uz; gt = uz; gb = fb[[i]]; fb[[i]] = gt;
  u[[i, j]] = 0.5*(u[[i, j]] + uz - cp1*(fr - fl)
            - cp2*(gt - gb));
  y1 = Abs[u[[i,j]] ]; umax1 = Max[y1, umax1];
  y = Abs[u[[i, j]] - um]; umax = Max[y, umax],
                {i, ix}], {j, jy}];
```

```
Print["n = ",n,", t = ", tn, ", max[u] = ", umax1];
  If[n == n3, Goto[graph]];
        If[umax > 2.*um, Goto[endsteps], Goto[endn]];
    If[n != n3, Goto[endn]];
     If[n == Nt, Goto[endsteps] ];
Label[endsteps]; Tmax = tn;
  Label[graph]; n3 = n3 + n1;
(* --- Plotting the numerical solution surface
                u = u(x,y,t) -------------------- *)
Print["The numerical solution surface",
      " u = u(x,y,t) at t =", tn];
  gr3 = ListPlot3D[u,
                ViewPoint -> {xview, yview, zview},
          AxesLabel -> {"y", "x", "u"}];
If[n == n1, Display["fig2_28a", gr3],
                         Display["fig2_28b", gr3] ];


 gr4 = ListContourPlot[u];
If[umax > 2.*um, Break[]];
Label[endn],
(* ---------- End of the Do-loop in n ----------- *)
                      {n, Nt}];
(* --------- End of the rogram mc2d.m ----------- *)
)
```

mc2dc.ma is a version of the above notebook mc2d.ma, which enables one to obtain the color map of the numerical solution in a region of the spatial variables x, y. This notebook can be used together with the same input data as the notebook mc2d.ma.

multlev.ma reduces a multi-level difference equation (2.8.12) to a system of two-level difference equations (2.8.21) and finds the symbolic expressions for the entries of the matrix difference operator C in Eq. (2.8.21).

sch The function sch[j_,n_] specifies the left-hand side of the scalar difference equation sch = 0.

The notebook multlev.ma does not have a cell of input data.

st1.ma The main program of this file

$$\text{stab1[eps\_, n1plot\_, ac1\_, bc1\_, sc\_]}$$

performs the Fourier stability analysis of a given explicit scalar difference scheme in the case of one spatial variable x and the time variable t.

The meaning of the input data is explained in the following cell of the notebook st1.ma:

Parameter	Description
eps	the given accuracy of the computation of the coordinates of points of the stability region boundary;
n1plot	the number of graphics pictures showing the curves (Re[lambda(kappa, xi)], Im[lambda(kappa, xi)], xi = 0, 2Pi, where lambda(kappa, xi) is the amplification factor; n1plot >=0;
ac1	the left end of the interval in which the stability region boundary is to be determined;
bc1	the right end of the above interval ac1 <= c1 <= bc1;
sc	the left-hand side of the difference equation sc=0

---

sch The function sch1 has exactly the same meaning as in the above described program appr1.ma.

fbis The function

```
fbis[x_, ss1_List, cc1_List, nspg_]
```

computes the value of the binary function (2.3.18) at a given point κ. The arguments ss1 and cc1 represent the one-dimensional lists of the numerical values of the functions $\cos \xi_1$ and $\sin \xi_1$ in the nspg nodes of an uniform mesh on the interval $0 \leq \xi_1 < 2\pi$.

bisec The function

```
bisec[eps_, ac1_, bc1_, ss1_List, cc1_List, nspg_]
```

implements the numerical solution of equation $f(\kappa) = 0$ with the aid of a bisection method. eps is the user-specified accuracy with which the root of equation $f(\kappa) = 0$ is to be determined. This equation is solved in the given interval [ac1, bc1].

Below we present the listing of the program st1.ma.

```
ClearAll[fbis, rel, iml, bisec];
fbis[x_, ss1_List, cc1_List, nspg_]:=
( fbs1 = 1.0; deps = N[1.0 + 1.0/(10^8)];
     j = 1; nspg1 = nspg;
```

```
Label[loop];
 rj = N[r /. {sa->ss1[[j]], ca->cc1[[j]], c1->x}];
   If[rj[[1]]> deps, nspg1 = j];
   j = j + 1; If[j <= nspg1, Goto[loop]];
   If[nspg1 < nspg, fbs1 = -1.0];Clear[j];fbs1 )
(* -- The function providing a list {Re(lambda(xi)),
     Im(lambda(xi))} for the Mathematica function
                       ParametricPlot  ------- *)


bisec[eps1_, ac1_, bc1_, ss1_List, cc1_List, nspg_]:=
(* -- This function determines the stability region
     boundary by the bisection method --------- *)
   ( fa = N[fbis[ac1, ss1, cc1, nspg]];
       fb = N[fbis[bc1, ss1, cc1, nspg]];
         isl = Sign[fa]; isr = Sign[fb];
       ac2 = ac1; bc2 = bc1;
   If[isl!= isr, Goto[bisproc]];
   If[isr== -1, Goto[unstable]];
Print["The scheme is stable in the interval ", ac1,
       "<= c1 <= ", bc1];
   xbs = bc1; Goto[endbis];
Label[unstable];
   Print["The scheme is unstable"];
   xbs = ac1; Goto[endbis];
Label[bisproc];
xbs = N[(ac2 + bc2)*0.5];
yy = N[fbis[xbs, ss1, cc1, nspg]];  ny = Sign[yy];
If[ny==isr, bc2=xbs, ac2=xbs];
If[N[Abs[ac2 - bc2]] >= eps1, Goto[bisproc]];
Label[endbis]; xbs )


ClearAll[stab1];
  stab1[eps_, n1plot_, ac1_, bc1_,sc_] := Module[
 {sc1, sc2, mt,rel,iml},
(* --- The representation of the difference solution
       in the form of a Fourier harmonic --------- *)

(* --- Substitution of a Fourier harmonic into the
       difference scheme --------------------------- *)
Print["The difference scheme under consideration"];
Print[sc," = 0"];
(* -- Divide the both sides of obtained equation by
                               1^n Exp[I*j*xi] -- *)
sc1=sc/.{u[j_,n_]->1^n Exp[I j xi]};
```

```
sc3 = sc1/(1^n*Exp[I*j*xi]) //Expand;
sc2 = Simplify[sc3];
sol = Solve[sc2==0, l];
l1 = Simplify[l /. sol];
Print["The amplification factor of scheme has the form"];
Print["lambda = ", l1];
rel=ComplexExpand[Re[l1]]/. c1rule //Simplify;
iml=ComplexExpand[Im[l1]]/. c1rule //Simplify;
  Print["Re(lambda) =",rel];
  Print["Im(lambda) =",iml];
      If[n1plot != 0,

(* --- Specification of the values of the
   nondimensional parameter c1 ----------- *)
      c1j = Table[N[2*j/n1plot], {j, n1plot}];
      Print["c1j=",c1j];

(* --- Storing the pictures in the elements mt[[j]],
                                 j = 1,...,n1plot - *)
      mt = {};
      Do[capp = c1j[[j]];
  zz1 = N[rel /. {c1 -> capp, xi -> t}];
  zz2 = N[iml /. {c1 -> capp, xi -> t}];
 mt1 = ParametricPlot[{zz1[[1]],zz2[[1]]},{t,0,2Pi},
                     DisplayFunction -> Identity];
mt2 = ParametricPlot[{Cos[t], Sin[t]}, {t, 0, 2Pi},
           PlotStyle ->{Dashing[{0.04,0.03}]},
           DisplayFunction -> Identity];
 obj= Polygon[Table[{Cos[n Pi/40],Sin[n Pi/40]},{n,80}]];
mt4 = Show[Graphics[{RGBColor[0.3,0.9,1], obj}],
           Axes -> True,
           Ticks -> Automatic,
           DisplayFunction -> Identity];
mt3 = Show[mt4, mt1, mt2,
           Axes -> True, Ticks -> Automatic,
           AspectRatio -> Automatic,
           DisplayFunction -> $DisplayFunction];
AppendTo[mt, mt3],
                     {j, n1plot}];
Clear[gr];
If[n1plot == 2,
gr = Show[GraphicsArray[{mt[[1]], mt[[2]]}]] ];
If[n1plot == 3,
gr = Show[GraphicsArray[{{mt[[1]], mt[[2]]},
```

```
       {mt[[3]]}}]] ];
If[n1plot == 4,
gr = Show[GraphicsArray[{{mt[[1]], mt[[2]]},
                         {mt[[3]], mt[[4]]}}]] ];
If[n1plot == 5,
gr = Show[GraphicsArray[{{mt[[1]], mt[[2]], mt[[3]]},
                         {mt[[4]], mt[[5]]}}]] ];
If[n1plot == 6,
gr = Show[GraphicsArray[{{mt[[1]], mt[[2]], mt[[3]]},
                         {mt[[4]], mt[[5]], mt[[6]]}}]] ];

(* -- End of If[n1plot != 0... -------------------- *)
     ];
<<Algebra'Trigonometry';
casa = {Cos[xi] -> ca, Sin[xi] -> sa};
iml = TrigReduce[iml]/.casa;
rel = TrigReduce[rel]/.casa;
Print["Re(lambda) = ", rel];
Print["Im(lambda) =", iml];
 r = rel^2 + iml^2 //Expand //Simplify;
    Print["|lambda|^2 =", r];


       (*%%%%%%%%%%%%%%%%%%%%%%%%%%%%%%%%%%%%%%%%
       %%                                      %%
       %%   The numerical stage of the method  %%
       %%                                      %%
       %%%%%%%%%%%%%%%%%%%%%%%%%%%%%%%%%%%%%%%%%%*)
(* ------ Determination of nspg ------------ *)
  nspg0 = 2; c1sb = 10000; eps1 = eps;
     Do[nspg = nspg0 + 2*j; j1 = j;
   ss1 = Table[N[Sin[(k-1)*2*Pi/nspg]], {k, nspg}];
   cc1 = Table[N[Cos[(k-1)*2*Pi/nspg]], {k, nspg}];
   c1s0 = bisec[eps1, ac1, bc1, ss1, cc1, nspg];
dabsc1 = 100;
If[j1 > 1, dabsc1 = N[Abs[c1s0-c1sb]]];
   Print["nspg, cpold, cpnew,dabsc1 = ",
   nspg, "  ",c1sb," ",c1s0," ", dabsc1];
         If[dabsc1<eps, Break[]];
   c1sb = c1s0,
                 {j, 48}];
   Print["j        Xj          sin(Xj)        cos(Xj)"];
Do[ss= ss1[[j]]; cc= cc1[[j]]; xj= N[(j - 1)*2*Pi/nspg];
    Print[j, "  ", xj, ," ", ss, , "  ", cc],
           {j, nspg}];
```

```
cap = Table[N[fbis[(k-1)*(bc1-ac1)/15,ss1,cc1,nspg]],
                {k, 15}];
Print["The values of the binary function fbis(kappa)"];
Print[cap[[1]], cap[[2]], cap[[3]], cap[[4]], cap[[5]],
cap[[6]], cap[[7]], cap[[8]], cap[[9]], cap[[10]],
cap[[11]], cap[[12]], cap[[13]], cap[[14]], cap[[15]]];
c1s = bisec[eps1, ac1, bc1, ss1, cc1, nspg];
Print["The stability region boundary is ", c1s];
(* --------- End of the block --------------- *)
  ];
```

st2.ma The main program of this notebook

```
stab2[xl_, xr_, yb_, yt_, ix_, jy_, nsptr_, eps_, sc_]
```

performs the Fourier stability analysis of a given scalar difference scheme in cases of one or two spatial variables and the time t. The form of a planar stability region is arbitrary, in particular, this region can be multiply connected. The meaning of the input data is explained in the following cell of the notebook **st2.ma**:

Parameter	Description
xl	the left end of the interval on the kappa1-axis (the horizontal axis), in which the stability region is to be determined;
xr	the right end of the above interval on the kappa1-axis; xl<=kappa1<= xr;
yb	the lower end of the interval on the kappa2-axis (the vertical axis), in which the stability region is to be determined;
yt	the upper end of the above interval on the kappa2-axis; yb<=kappa1<= yt; the stability region is thus to be determined in a rectangle in the (kappa1, kappa2) plane;
ix	the number of cells of an uniform mesh in the interval xl<= kappa1<=xr; ix >= 3;
jy	the number of cells of an uniform mesh in the interval yb<= kappa2<=yt; jy >= 3;
nsptr	the number of spectral variables, nsptr = 1 or nsptr = 2;

```
eps    the given accuracy of the computation of the coordi-
       nates of points of the stability region boundary;
sc     the left-hand side of the difference equation sc=0
```

---

sc The function sc has exactly the same meaning as the function sch1 in the above described notebooks appr1.ma and st1.ma.

inds The function inds[is_, i_, j_] determines the indices (i1, j1) of the next cell along the contour basing on the indices (i, j) of a given cell and the given search direction is, $1 \leq$ is≤ 8.

tracer The function

$$\text{tracer[nsg\_, i0\_, j0\_]}$$

traces a contour departing from a given contour point (i0, j0). It generates a sequence of the $(\kappa_{1i}, \kappa_{2i})$ points belonging to a specific contour.

atan3 The function atan3[zy_, zx_] computes $arctan(zy/zx)$ in such a way that the value of $arctan(\cdot)$ changes smoothly when the argument changes its sign.

bisec The function bisec[x_, y_, ik_, k_, mk_, adir_List] performs the refinement of the position (x, y) of a contour point with the aid of a bisection method.

Below we present the listing of the program st2.ma.

```
ClearAll[fbis1, fbis2, binfun, bingen, inds, atan3,
         tracer, bisec, stab2];

fbis1[x_, y_, ss1_List, cc1_List, nspg_] :=
( bin1 = 1; deps=N[1.0+1.0/(10^8)];
  jb1 = 1; msp = nspg;
Label[loop]; vals = {sa -> ss1[[jb1]],
ca -> cc1[[jb1]], cp1 -> x, cp2 -> y};
rlj = N[rel /. vals]; imj = N[iml /. vals];
rr = rlj*rlj + imj*imj;
 If[rr[[1]] > deps, msp = jb1];
  jb1 = jb1 + 1; If[jb1 <= msp, Goto[loop]];
  If[msp < nspg, bin1 = 0]; Return[bin1] );

fbis2[x_, y_, ss1_List, cc1_List, nspg_] :=
( bn2d = 1; deps = N[1.0 + 1.0/(10^8)];
  kb2 = 1; nspg2 = nspg;
 Label[lpk2]; jb2 = 1; msp = nspg;
Label[lpj]; vals = {sa -> ss1[[jb2]],
```

```
ca -> cc1[[jb2]], sb -> ss1[[kb2]],
    cb -> cc1[[kb2]], cp1 -> x, cp2 -> y};
reljk = N[rel /. vals]; imljk = N[iml /. vals];
    rjk = reljk*reljk + imljk*imljk;
  If[rjk[[1]] > deps, msp = jb2];
    jb2 = jb2 + 1; If[jb2 <= msp, Goto[lpj]];
  If[msp < nspg, Return[0]];
  kb2 = kb2 + 1; If[kb2 <= nspg2, Goto[lpk2]];
    If[nspg2 < nspg, bn2d = 0]; bn2d );


binfun[x_, y_, ss1_List, cc1_List, nspg_] :=
(
If[nsptr == 1, fbs3 = fbis1[x, y, ss1, cc1, nspg];
   Return[fbs3]];
fbs3 =fbis2[x, y, ss1, cc1, nspg]; Return[fbs3] );


bingen:=
(* --- Generation of the bilevel image fbin --- *)
( fbin = Table[0, {i, ixt}, {j, jyt}];
jsum = 0; hx = (xrt - xlt)/ixt; hy = (ytt-ybt)/jyt;
Print["Bilevel picture generation"];
 Do[x = xlt + (i - 0.5)*hx; binlist = {};
    Do[y = ybt + (j -0.5)*hy;
    fbin[[i, j]] = binfun[x, y, ss1, cc1, nspg];
    AppendTo[binlist, fbin[[i, j]] ];
       jsum = jsum + fbin[[i, j]], {j, jyt}];
       Print[binlist], {i, ixt}];
                {jsum, fbin} );


inds[is_, i_, j_]:=
(* - This procedure determines the indices (i1,j1)
     of the next cell along the contour basing on
     the indices (i,j) of a given cell and the
     given search direction "is", 1<= is<= 8 - *)
( jb = 0; is2 = is; If[is2 > 8, is2 = is - 8];
  If[is2 == 1, i1 = i + 1; j1 = j;
 If[i1 < 1||i1 > ixt||j1 < 1 || j1 > jyt, jb = 1];
 Return[{i1, j1, jb}]];
   If[is2 == 2, i1 = i + 1; j1 = j + 1;
   If[i1 < 1||i1 > ixt||j1 < 1||j1 > jyt, jb = 1];
   Return[{i1, j1, jb}]];
    If[is2 == 3, i1 = i; j1 = j + 1;
    If[i1 < 1||i1 > ixt||j1<1||j1 > jyt, jb = 1];
    Return[{i1, j1, jb}]];
```

```
      If[is2 == 4, i1 = i - 1; j1 = j + 1;
 If[i1 < 1||i1 > ixt||j1 < 1 || j1 > jyt, jb = 1];
      Return[{i1, j1, jb}]];
        If[is2 == 5, i1 = i - 1; j1 = j;
 If[i1 < 1||i1 > ixt||j1 < 1 || j1 > jyt, jb = 1];
      Return[{i1, j1, jb}]];
        If[is2 == 6, i1 = i - 1; j1 = j - 1;
 If[i1 < 1||i1 > ixt||j1 < 1 || j1 > jyt, jb = 1];
      Return[{i1, j1, jb}]];
      If[is2 == 7, i1 = i; j1 = j - 1;
 If[i1 < 1||i1 > ixt||j1 < 1 || j1 > jyt, jb = 1];
      Return[{i1, j1, jb}]];
     If[is2 == 8, i1 = i + 1; j1 = j - 1;
 If[i1 < 1||i1 > ixt||j1 < 1 || j1 > jyt, jb = 1];
      Return[{i1, j1, jb}]]);


tracer[nsg_, i0_, j0_]:=
(* ------------ Tracing a contour ------------ *)
( i = i0;   j = j0;   is = 0;
fbin[[i, j]] = fbin[[i, j]] + 1;
is = is + 1; jc[[is, nsg]] = 9; iser = 7; ifir = 1;
     xs[[is, nsg]] = xlt + (i - 0.5)*hx;
     ys[[is, nsg]] = ybt + (j - 0.5)*hy;
Label[m8]; icnt = 0; idect = 0;
Label[m13]; is1 = iser - 1; If[is1 == 0, is1 = 8];
   {i1, j1, jb} = inds[is1, i, j];
 If[jb == 1|| fbin[[i1, j1]] == 0, Goto[m9]];
   iser = iser - 2; If[iser <= 0, iser = iser + 8];
Label[m11]; is = is + 1; i=i1; j=j1;
           fbin[[i, j]] = fbin[[i, j]] + 1;
xs[[is, nsg]] = xlt + (i - 0.5)*hx;
ys[[is, nsg]] = ybt + (j - 0.5)*hy;
jc[[is, nsg]] = is1; idect = 1;
        If[i!= i0 || j != j0, Goto[m8], Goto[m7]];
Label[m9]; {i1, j1, jb} = inds[iser, i, j];
        If[jb == 1||fbin[[i1, j1]] == 0, Goto[m10]];
is1 = iser; Goto[m11];
Label[m10]; is1=iser+1; If[is1 > 8, is1 = is1 - 8];
{i1, j1, jb} = inds[is1, i, j];
If[jb==1||fbin[[i1,j1]]==0, Goto[m12]]; Goto[m11];
Label[m12]; iser=iser+2; If[iser>8,iser=iser - 8];
           ifir = 0;
If[idect == 1, Goto[m7]]; icnt = icnt + 1;
                    If[icnt < 3, Goto[m13]];
```

```
Label[m7]; kss[[nsg]] = is );


remv[istar_, ifin_, k_] :=
( istar2 = istar; If[istar<=ifin,ic=jc[[istar,k]]];
istar2 = istar2 + 1; Return[{ic, istar2}]];
ic = -1; {ic, istar2} );


 atan3[zy_, zx_] :=
( rof = 0.000000001;
If[Abs[zx]>rof,Goto[m2]]; If[Abs[zy]<rof,Goto[m3]];
a = N[Pi/2.0]; If[zy < 0.0, a = N[a + Pi]];
   Return[a];
Label[m3]; a = 0.0; Return[a];
Label[m2]; If[Abs[zy] < rof, a = 0.0;
            If[zx < 0.0, a = N[Pi]]; Return[a]];
a = ArcTan[zx,zy]; If[zx>0.0&&zy>0.0, Return[a]];
If[zx > 0.0&& zy < 0.0, a = N[a + 2*Pi]; Return[a]];
    If[zx < 0.0, a = N[a + Pi]]; Return[a]   );


bisec[x_,y_,ik_,k_,mk_,adir_List]:=
(* The refinement of the (x,y) coordinates of the
   ik th point of the k th contour; mk is the
   length of th kth contour --------------------- *)
( i = Floor[1.0 + (x - xlt)*hx1];
                       j = Floor[1.0 + (y -ybt)*hy1];
i1 = ik + 1; i2 = ik - 1;
   If[ik == mk, i1 = i1 - 1; i2 = i2 - 1];
      If[ik == 1, i2 = 1];
(* -- Determining the number k2 of the principal
      direction corresponding to the direction of
      a normal to the contour at (i, j) point. - *)
(* x1 = xs[[i1, k]]; y1 = ys[[i1, k]];
        x2 = xs[[i2, k]]; y2 = ys[[i2, k]];  *)
  x1 = xxxx[[i1]]; y1 = ytab[[i1]];
                  x2 = xxxx[[i2]]; y2 = ytab[[i2]];
     dx = x2 - x1; dy = y1 - y2;
anrm = atan3[dx, dy];    zm = 1000.0; il = -2;
Do[ zk = Abs[adir[[ja]] - anrm];
                     If[zk < zm, zm = zk;k2 = ja],
     {ja, 9}];
                If[k2 == 9, k2 = 1];
(* -- Determining the indices (il, jl) of a cell
      lying on a normal to the contour -------- *)
If[k2 == 1, il = i + 1; jl = j; ir = i - 1; jr = j];
```

```
  If[il > -2, Goto[refin]]];
If[k2 == 2, il=i+1; jl=j+1; ir = i - 1; jr= j - 1];
     If[il > -2, Goto[refin]]];
If[k2 == 3, il = i; jl = j+ 1; ir = i; jr = j - 1];
       If[il > -2, Goto[refin]]];
If[k2 == 4, il = i - 1; jl=j+1; ir=i+1; jr = j - 1];
          If[il > -2, Goto[refin]]];
If[k2 == 5, il = i - 1; jl = j; ir = i + 1; jr= j];
           If[il > -2, Goto[refin]]];
If[k2 == 6, il=i-1; jl=j-1; ir=i + 1; jr = j + 1];
        If[il > -2, Goto[refin]]];
If[k2 == 7, il = i; jl = j - 1; ir = i; jr = j + 1];
   If[il > -2, Goto[refin]]];
If[k2 == 8, il= i + 1; jl = j-1; ir=i-1; jr=j + 1];
Label[refin];
x1n = xlt + (il - 0.5)*hx; y1n = ybt+(jl - 0.5)*hy;
x2n = xlt + (ir - 0.5)*hx; y2n = ybt+(jr - 0.5)*hy;
tl = 0.0; tr = 1.0; a1 = x2n - x1n; a2 = y2n - y1n;
 isl = binfun[x1n, y1n, ss1, cc1, nspg];
isr = binfun[x2n,y2n,ss1,cc1,nspg]; x3 = x; y3 = y;
     If[isl != isr, Goto[bisproc]];
                         If[isl== 0, Goto[recalc]];
If[il == 0||ir == 0, x3 = xlt];
      If[il == ixt + 1||ir == ixt + 1, x3 = xrt];
If[jl == 0||jr == 0,y3 = ybt];
       If[jl == jyt + 1||jr == jyt + 1, y3 = ytt];
Return[{x3, y3}];
Label[recalc]; ir=i;jr=j; x2n=x; y2n= y; isr = 1;
               a1 = x2n - x1n; a2 = y2n - y1n;
Label[bisproc];
t = N[(tl+tr)*0.5]; xc=a1*t+x1n; yc = a2*t + y1n;
         nc = binfun[xc, yc, ss1, cc1, nspg];
If[nc == isr, tr = t, tl = t];
  If[N[Abs[a1*(tl - tr)]] > epst, Goto[bisproc]];
  If[N[Abs[a2*(tl - tr)]] > epst, Goto[bisproc]];
  x3 = a1*t + x1n; y3 = a2*t + y1n;
 Return[{x3, y3}]   );

 stab2[xl_,xr_,yb_,yt_,ix_,jy_,nsptr_,eps_,sc_] :=
 (
(*%%%%%%%%%%%%%%%%%%%%%%%%%%%%%%%%%%%%%%%%%%%%%%%%%%
%%%%%%%%%%%%%%%%%%%%%%%%%%%%%%%%%%%%%%%%%%%%%%%%%%%%
%%% Part 1. Symbolic computations with the pur- %%%%
%%% pose of obtaining the expressions for the   %%%%
```

```
%%% coefficients of the characteristic equation %%%
%%% of difference scheme                         %%%
%%%%%%%%%%%%%%%%%%%%%%%%%%%%%%%%%%%%%%%%%%%%%%%%%%%%
%%%%%%%%%%%%%%%%%%%%%%%%%%%%%%%%%%%%%%%%%%%%%%%%%%*)

<< Algebra'ReIm';
Print["The package Algebra'ReIm' is loaded"];

(* The variables xi, xi1, xi2, cp1, cp2 are declared
                                      to be real *)
xi/: Im[xi] = 0;
xi1/: Im[xi1] = 0;
xi2/: Im[xi2] = 0;
cp1/: Im[cp1] = 0;
cp2/: Im[cp2] = 0;

(* ---- Substitution of a Fourier harmonic into
        the difference scheme ----------------- *)
  If[nsptr == 1,
sc0 = sc/. c12rule ;
Print["The difference scheme under consideration"];
Print[sc0, " = 0"];
sc1 = sc0 /. {u[j_, n_] -> l^n*Exp[I*j*xi]};

(* -- Divide the both sides of obtained equation
          by l^n Exp[I*j*xi] ----------------- *)
     sc3 = sc1/(l^n*Exp[I*j*xi]) //Expand;
     sc2 = Simplify[sc3] ];
  If[nsptr == 2, sc0 = sc/. c12rule;

Print["The difference scheme under consideration"];
Print[sc0, " = 0"];
sc1 = sc0 /.
{u[j_, k_, n_] -> l^n*Exp[I*j*xi1]*Exp[I*k*xi2]};

(* -- Divide the both sides of obtained equation by
      l^n Exp[I*j*xi1]*Exp[I*j*xi2] ------------ *)
sc3 = sc1/(l^n*Exp[I*j*xi1]*Exp[I*k*xi2]) //Expand;
     sc2 = Simplify[sc3] ];

  sol = Solve[sc2==0, l];
l1 = Expand[l /. sol] ;
(* l1 = Simplify[l1]; *)
```

```
Print["The amplification factor of scheme has",
" the form"];
Print["lambda = ", l1];
l1 >> str2.m;
l4 = ComplexExpand[l1];
l = l4; lnum = Numerator[l];
OutputForm[lnum] >>> str2.m;
lden = Denominator[l]; imd = Im[lden];
OutputForm[lden] >>> str2.m;
OutputForm[imd] >>> str2.m;
If[lden == 1, iml = Im[l]; rel = l - I*iml];
If[lden =!= 1 && imd == 0,
imn = Im[lnum]; relnom = lnum - I*imn;
imn = Simplify[imn]; relnom = Simplify[relnom];
iml = imn/lden; rel = relnom/lden];

(* ---- The complex denominator ---- *)
If[lden =!= 1 && imd =!= 0, relden = lden - I*imd;
ld2 = relden*relden + imd*imd;
imn = Im[lnum]; relnom = lnum - I*imn;
imn = Simplify[imn]; relnom = Simplify[relnom];
rel = (relnom*relden + imn*imd)/ld2;
iml = (imn*relden - relnom*imd)/ld2];
rel = Simplify[rel]; iml = Simplify[iml];
OutputForm[rel] >>> str2.m;
OutputForm[iml] >>> str2.m;
          Print["Re(lambda) = ", rel];
          Print["Im(lambda) = ", iml];
   rel >>> str2.m;
   iml >>> str2.m;
      << Algebra`Trigonometry`;
casa = {Cos[xi] -> ca, Sin[xi] -> sa};
cacb = {Cos[xi1] -> ca, Sin[xi1] -> sa,
        Cos[xi2] -> cb, Sin[xi2] -> sb};
If[nsptr == 1,
rel = TrigReduce[rel] /. casa;
iml = TrigReduce[iml] /. casa,
 rel = TrigReduce[rel] /. cacb;
 iml = TrigReduce[iml] /. cacb];
If[nsptr == 1, rel = rel /. {ca^2 -> 1 - sa^2};
               iml = iml /. {ca^2 -> 1 - sa^2}];
rel = Simplify[rel];

Print["Re(lambda) = ", rel];
```

```
iml = Simplify[iml];
 Print["Im(lambda) =", iml];

(*%%%%%%%%%%%%%%%%%%%%%%%%%%%%%%%%%%%%%%%%%%%%%%%%%%%%%%%%%%%
%%%%%%%%%%%%%%%%%%%%%%%%%%%%%%%%%%%%%%%%%%%%%%%%%%%%%%%%%%%%
%%%  Part 2. Numerical computations with the purpose  %%%%
%%%  of determining the coordinates of points of all  %%%%
%%%  the contours of the stability region boundary    %%%%
%%%%%%%%%%%%%%%%%%%%%%%%%%%%%%%%%%%%%%%%%%%%%%%%%%%%%%%%%%%%
%%%%%%%%%%%%%%%%%%%%%%%%%%%%%%%%%%%%%%%%%%%%%%%%%%%%%%%%%%*)

xlt = xl; xrt = xr; ybt = yb; ytt = yt;
ixt = ix; jyt = jy; epst = eps;
(* - Determination of nspg, the number of grid
     points along the axis of each spectral
     variable xi1, xi2 ---------------------- *)
     fbin = Table[0, {i, 1, ix}, {j, 1, jy}];
         nspg0 = 2; npold = 0; dnabs = 1000;
   Print["Determination of nspg"];
     Do[nspg = nspg0 + 2*j;
   ss1 = Table[N[Sin[(k-1)*2*Pi/nspg]], {k, nspg}];
   cc1 = Table[N[Cos[(k-1)*2*Pi/nspg]], {k, nspg}];
  {npnew, fbin} = bingen;
     If[j == 1, npold = npnew,
 dnabs = N[Abs[npnew - npold]]];
Print["nspg, pointsnew, pointsold, dnabs = ",
nspg, "  ", npnew, " ", npold, " ",dnabs];
         If[dnabs == 0, Break[]];
  npold = npnew, {j, 48 }];
  nparea = npold; lenmax = ix*jy;
        Do[k = jy - j + 1; frow = {};
      Do[AppendTo[frow, fbin[[i, k]]], {i, ix}];
      Print[frow], {j,jy}];
If[npold == 0, Print["No stability points"];
Interrupt[ ]];
If[npold == ix*jy,
Print["The difference scheme is stable in the",
" given region"];
Print["{(kappa1,kappa2)|",xl,"<=kappa1<=",xr,
",", yb,"<=kappa2<=",yt,"}"];
Print["Computation is interrupted."];
Interrupt[] ];

   Print["j       Xj        sin(Xj)     cos(Xj)"];
```

```
Do[ss= ss1[[j]]; cc= cc1[[j]]; xj= N[(j - 1)*2*Pi/nspg];
Print[j," ", xj, " ", ss, " ", cc], {j, nspg}];


z1 = npold*hx*hy;
Print["Area of stability region = ", z1];
      nparea = npold; ifir = 0; nsg = 0;
nsgmax = 10; lenmax = ix*jy;
(*  nsg is the number of contours of the stability
    region boundary  *)
(* -- lenmax is the maximum length, that is the
       maximum number of pixels, for each detected
       contour --------------------------------- *)
(* --- Initialization of a number of lists ----- *)
   xs = Table[999.99999, {i, lenmax}, {j, nsgmax}];
   ys = Table[999.99999, {i, lenmax}, {j, nsgmax}];
(* -- (xs[[i, j]], ys[[i, j]]) are the (kappa1, kappa2)
    coordinates of the i th point of the j th contour --- *)
      jc = Table[99, {i, lenmax}, {j, nsgmax}];
(*  jc is a two-dimensional array of integers used
    to mark different points of different contours *)
      kss = Table[9999, {j, nsgmax}];
(*  kss[[j]] is the length of the j th contour,
    1 <= j <= nsgmax  *)


(* ------- Tracing the external contours ------- *)
      Do[
        Do[
           If[fbin[[ii, jj]] == 0, Goto[endloop]];
(* - Search for the starting point of the contour - *)
If[fbin[[ii, jj]] > 1, Goto[endloop]];
                         If[ii == 1,Goto[start]];
If[fbin[[ii - 1, jj]] != 0, Goto[endloop]];
Label[start]; i0 = ii; j0 = jj; nsg = nsg + 1;
 If[nsg > nsgmax,
    Print["The number of contours exceeds ", nsgmax];
Print["Detection of further contours is interrupted."];
Print["To detect the remaining contours please increase",
" the value of nsgmax."]; Break[ ]];
    tracer[nsg, i0, j0];
        Label[endloop], {ii, ix}];
If[nsg > nsgmax, nsg = nsg - 1; Break[ ]],
    {jj, jy}];
(* ------- Tracing the contours of holes ------- *)
 k1 = 1; k2 = nsg; nsg1 = nsg; hx = (xr - xl)/ix;
```

```
hy = (yt - yb)/jy; hx1 = 1/hx; hy1 = 1/hy; k = 1;
Label[120];
     ig = kss[[k]]; istar = 1; ifin = ig;
Label[117]; {id, istar2} = remv[istar, ifin, k];
        istar = istar2;
  If[id == -1, Goto[13]];  If[id != 9, Goto[115]];
id0 = id; {id,istar2} = remv[istar, ifin, k];
                 istar = istar2;
Label[115]; If[id0 < 5||id0 > 8, Goto [116]];
            If[id < 6||id > 8, Goto[116]];
x1 = xs[[istar, k]]; y1 = ys[[istar, k]];
  ilef = Floor[0.51 + (x1 - xl)*hx1];
                  j2 = Floor[0.51 + (y1 - yb)*hy1];
ilp = ilef + 1; irm = ix - 1;  i2 = ilp;
  Label[labilp];
   ia = fbin[[i2 - 1, j2]]; ib = fbin[[i2, j2]];
   ic = fbin[[i2 + 1, j2]]; If[ia != 0, Goto[137]];
   If[ib != 1||ib != 2||ic != 0, Goto[127]];
         i0 = i2; j0 = j2; nsg1 = nsg1 + 1;
 Print["The contour No. ", nsg1, " is detected."];
 tracer[nsg1, i0, j0]; Goto[116];
Label[127];
        If[ia == 0 && ib > 2 && id ==-1, Goto[116]];
Label[137];
   If[ia == 1 && ib == 2&&id == -1, Goto[116]];
 i2 = i2+1; If[i2 <= irm, Goto[labilp]]; Goto[13];
Label[116]; id0 = id; Goto[117];
Label[13];
          k = k + 1; If[k <= k2, Goto[120]];
If[nsg1 == nsg, Goto[121]];
nsg = k2; k1 = nsg + 1; k2 = nsg1; Goto[120];
Label[121]; nsg = nsg1;
Print["Number of contours = ",nsg];
  Do[ isg = kss[[k]];
Print["The length of the contour No. ", k, " is ",
      isg, " points"], {k, nsg}];

(* -- Storing the pictures in the elements mt[[k]],
        k = 1,...,nsg ---------------------- *)
              nsg1 = nsg + 1;  mt = {};
         xxxx = Table[99.9999, {j,lenmax}];
         ytab = Table[99.9999, {j,lenmax}];
    Do[ isg = kss[[k]];
Print["The x- and y-coordinates of points of the contour ",
```

```
         "No. ", k];
          Do[ xx = xs[[j, k]]; yy = ys[[j, k]];
              Print[j," ", xx, " ", yy], {j, isg}],
              {k, nsg}];
      Do[ isg = kss[[k]];
   Do[xxxx[[j]] = xs[[j, k]]; ytab[[j]] = ys[[j, k]]],
              {j, isg}];
       xytb = Table[{xxxx[[j]], ytab[[j]]}, {j, isg}];
       plj = ListPlot[xytb, PlotJoined -> True,
                        AxesLabel->{"kappa1","kappa2"},
                        AspectRatio -> Automatic,
                        DisplayFunction -> Identity];
           AppendTo[mt, plj],
              {k, nsg}];
   xxxx = Table[99.999,{j, nparea}];
   ytab = Table[99.999,{j, nparea}];
      k = 0;
      Do[
         Do[ k1 = fbin[[i, j]];
      If[k1 != 0, k=k+1; xxxx[[k]]=xl + (i - 0.5)*hx;
             ytab[[k]] = yb + (j - 0.5)*hy],
                          {j, jy}],
             {i, ix}];
   xytb=Table[{xxxx[[j]], ytab[[j]]}, {j, nparea}];
      pldot = ListPlot[xytb,
              AxesLabel->{"kappa1","kappa2"},
              AspectRatio -> Automatic,
              DisplayFunction -> Identity];
      AppendTo[mt, pldot];
   gr = Show[mt,DisplayFunction->$DisplayFunction];
    Display["fig2_38a", gr];


(*  Specification of the angles adir[[j]] corresponding to
    the eight principal directions; the direction
    adir[[9]] = 2 Pi is an auxiliary direction, which is
    identified with the direction adir[[1]] = 0 in subseq-
    uent computations. *)
   adir = {}; AppendTo[adir, 0.0]; z = ArcTan[hx, hy];
        AppendTo[adir, z]; pi2 = N[Pi/2.]; z2 = N[pi2 - z];
   z3 = N[pi2 + z2]; z4 = N[Pi + z]; z7 = N[Pi + pi2];
   z8 = N[z7 + z2];
        AppendTo[adir, pi2];
   AppendTo[adir, z3]; AppendTo[adir, N[Pi]];
        AppendTo[adir, z4];
```

```
      AppendTo[adir, z7]; AppendTo[adir, z8];
                z9 = Pi + Pi; AppendTo[adir, N[z9]];
Print["The values of angles corresponding to nine",
              " principal directions"];
  Print[adir];
  Print["The refinement of coordinates of contour points"];
(*  Refinement of coordinates of boundary points  *)
Do[  mk = kss[[k]]; xxxx = {}; ytab = {};
   Do[ xx6 = xs[[ii,k]]; yy6 = ys[[ii, k]];
   AppendTo[xxxx, xx6]; AppendTo[ytab, yy6],
       {ii, mk}];
   Do[x1 = xs[[ii, k]]; y1 = ys[[ii, k]];
   {x3, y3} = bisec[x1, y1, ii, k, mk, adir];
Print["The point no. ", ii, " has been refined"];
   xs[[ii, k]] = x3; ys[[ii, k]] = y3, {ii, mk}];
   Print["The contour No.", k," has been refined"],
   {k, nsg}];
(* ------- Closing the refined contours ------- *)
Do[ isg = kss[[k]]; z1=0.5*(xs[[1,k]]+xs[[isg,k]]);
                    z2 = 0.5*(ys[[1,k]] + ys[[isg, k]]);
   xs[[1,k]] = z1; ys[[1, k]] = z2;
 xs[[isg, k]] = z1; ys[[isg, k]] = z2, {k, nsg}];
(* - Graphical display of the refined contours - *)
                        mt = {};
   Do[ isg = kss[[k]]; xxxx = {}; ytab = {};
Do[ AppendTo[xxxx, xs[[j, k]]];
     AppendTo[ytab, ys[[j, k]]], {j, isg}];
   xytb = Table[{xxxx[[j]], ytab[[j]]}, {j, isg}];
     plj = ListPlot[xytb, PlotJoined -> True,
                    AxesLabel->{"kappa1","kappa2"},
                    AspectRatio -> Automatic,
                    DisplayFunction -> Identity];
   AppendTo[mt, plj], {k, nsg}];
                          AppendTo[mt, pldot];
gr1 = Show[mt,DisplayFunction->$DisplayFunction];

  Display["fig2_38b", gr1];

(* ------- End of the procedure stab2 --------- *)
  )
```

st2t.ma is a version of the above notebook st2.ma, which enables one to perform a test at the binary function (2.13.6). All the basic functions are in the program st2t.ma the same as in st2.ma.

tvd1.ma The main program of this notebook

tvd[a_, a1_, b1_, jf0_, ul_, ur_, M_, cap_, beta_, bfl_,
                                                Npic_]

solves numerically the advection equation (2.1.1) under the initial condition (2.4.7) representing a semi-ellipse pulse by the TVD-type scheme of Osher and Chakravarthy (2.7.8)-(2.7.13). The meaning of the input data is explained in the following cell of the notebook tvd1.ma:

Parameter	Description
a	the advection speed in eqn (2.1.1), a > 0.
a1	the abscissa of the left end of the interval on the x-axis;
b1	the abscissa of the right end of the above interval, a1 < b1;
jf0	the number of the grid node, 0 < jf0 < M, such that at x = jf0*h there takes place the maximum of the initial semi-ellipse pulse;
ul	and ur are the constants entering the initial condition, the value ul corresponds to the constant background, and ul + ur is the extremal value of the initial function u(x, 0);
M	the number of the cell whose right boundary coincides with the right boundary x = b1 of the spatial integration interval;
cap	the Courant number, cap = a*dt/h, where dt is the time step of the difference scheme, and h is the step of the uniform grid in the interval a1 <=x <= b1;
beta	the weight parameter in the TVD scheme, -1<=beta<= 1;
bfl	the coefficient entering the flux limiters; 1 <= bfl <= (3 - beta)/(1 - beta);
Npic	the number of the pictures of the difference solution graphs for Npic moments of time, Npic > 1.

---

uinit uinit[x_,ul_,ur_,jf0_,h_] computes the numerical value of the initial function (2.4.7) at a given point x, see also the files lw1.ma and lw2.ma.

upw1.ma The main program of this notebook

upw1[a_,a1_,b1_,jf0_,ul_,ur_,M_, cap_, Npic_]

solves numerically the advection equation (2.1.1) under the initial condition (2.4.7) representing a semi-ellipse pulse by the explicit one-sided difference scheme (2.1.9). The input parameters for the program file upw1.ma are the same as for the file lw1.ma.

Table 1a. The *Mathematica* source files and the examples of the input data needed for obtaining the Figures of Chapter 2

Section	Figure number	Notebook	Example of input data
2.2	Fig. 2.5	advbac.ma	Example 1
2.2	Fig. 2.6	advbac.ma	Example 2
2.3	Fig. 2.7	st1.ma	Example 1
2.4	Fig. 2.10	lw1.ma	Example 1
2.4	Fig. 2.11	lw2.ma	Example 1
2.4	Fig. 2.12	lw2.ma	Example 2
2.4	Fig. 2.13	upw1.ma	Example 2
2.7	Fig. 2.15	tvd1.ma	Example 1
2.7	Fig. 2.16	tvd1.ma	Example 2

Table 1b. The *Mathematica* source files and the examples of the input data needed for obtaining the Figures of Chapter 2

Section	Figure number	Notebook	Example of input data
2.12	Fig. 2.26	mc2d.ma	Example 1
2.27	Fig. 2.27	mc2d.ma	Example 2
2.12	Figs. 2.28 a,b	mc2d.ma	Example 2
2.12	Fig. 2.29	col.ma	Example 1
2.12	Fig. 2.30	col.ma	Example 1
2.12	Fig. 2.31	mc2dc.ma	Example 2
2.12	Figs. 2.32 a,b	mc2d.ma	Example 3
2.13	Fig. 2.36	st2t.ma	Example 1
2.13	Fig. 2.37	st2t.ma	Example 1
2.13	Fig. 2.38 a,b	st2.ma	Example 5
2.14	Fig. 2.42	grid1.ma	Example 1
2.14	Fig. 2.43	grid1.ma	Example 2

Many Figures presented in this book were obtained with the aid of the above listed programs. For the convenience of the reader we present in Tables 1a, 1b, 2 and 3 the names of the *Mathematica* source files and the input data files, which were used by us to obtain many Figures of the book. The left column of each Table shows the number of the book section, in which a specific Figure is located.

The interested reader can easily reproduce the Figures whose numbers are indicated in Tables 1,2, and 3.

Table 2. The *Mathematica* source files and the examples of the input data needed for obtaining the Figures of Chapter 3

Section	Figure number	Notebook	Example of input data
3.2	Fig. 3.1	heat1.ma	Example 1
3.2	Fig. 3.2	heat1.ma	Example 2
3.2	Fig. 3.3	heat1.ma	Example 3
3.2	Fig. 3.4	heat1.ma	Example 4
3.2	Fig. 3.5	heat1.ma	Example 5
3.3	Fig. 3.6	st2.ma	Example 1
3.3	Figs. 3.8 a,b	st2.ma	Example 2
3.4	Fig. 3.9	st1.ma	Example 2
3.8	Fig. 3.13 a	disp1.ma	Example 1
3.8	Fig. 3.13 b	disp1.ma	Example 2
3.8	Fig. 3.14 a	disp1.ma	Example 3
3.8	Fig. 3.14 b	disp1.ma	Example 4
3.8	Fig. 3.15 a	disp1ad.ma	Example 1
3.8	Fig. 3.15 b	disp1adc.ma	Example 1

Table 3. The *Mathematica* source files and the examples of the input data needed for obtaining the Figures of Chapter 4

Section	Figure number	Notebook	Example of input data
4.4	Fig. 4.3	fem1.ma	Example 1
4.4	Fig. 4.4	fem1.ma	Example 1
4.4	Fig. 4.5	fem2.ma	Example 1
4.5	Fig. 4.6 a,b	gridttm.ma	Example 1

Index

Milton Keynes UK
Ingram Content Group UK Ltd.
UKHW031127141024
449569UK00006B/395